Media Rurality

Media Rurality

PATRICK BRODIE AND DARIN BARNEY EDITORS

DUKE UNIVERSITY PRESS
Durham and London 2026

Project Editor: Lisa Lawley
Designed by Matthew Tauch
Typeset in Garamond Premier Pro and IBM Plex Sans
by Westchester Publishing Services

Library of Congress Cataloging-in-Publication Data
Names: Brodie, Patrick, [date] editor |
Barney, Darin David, [date] editor
Title: Media rurality / Patrick Brodie, Darin Barney.
Description: Durham : Duke University Press, 2026. |
Includes bibliographical references and index.
Identifiers: LCCN 2025037078 (print)
LCCN 2025037079 (ebook)
ISBN 9781478033257 paperback
ISBN 9781478029793 hardcover
ISBN 9781478062004 ebook
Subjects: LCSH: Rural development—Technological innovations |
Communication in rural development | Rural development—
Economic aspects | Mass media in economic development |
Infrastructure (Economics) | Globalization
Classification: LCC HN49.C6 M395 2026 (print) |
LCC HN49.C6 (ebook) | DDC 307.1/4—dc23/eng/20251231
LC record available at https://lccn.loc.gov/2025037078
LC ebook record available at https://lccn.loc.gov/2025037079

Cover art: Aerial view of cityscape and farmland in
Turkey, 2024. Photograph by Ali via Pexels.com.

Contents

Media Rurality

PATRICK BRODIE AND DARIN BARNEY

Wind farms next to hearth-heated houses. Cell phone towers on isolated mountain tops. Data centers in sight of grazing cattle. CCTV-monitored mining access roads cutting through Indigenous land. Drones buzzing over croplands embedded by sensor arrays. Dams drawing energy from remote rivers and fiber-optic cables landing on quiet beaches. Ruined foundations of experimental colonial telecommunications facilities on either side of the Atlantic. Rural communities as hubs of high-tech industry and high-tech work. Border zones that serve as logistical exchanges for migrating media equipment, products, and workers. Informal networks of communication circulating information and cultural engagement with state development projects in metropolitan hinterlands. Low-tech, DIY, and fugitive energy systems in energy-impoverished rural sacrifice zones. Right-wing news and content streaming into rural households and stoking smoldering, misplaced resentments. Contrary to persistent stereotypes of wilderness and countryside, rural locations are heavily mediated and media intensive. What forms of mediation emerge when we foreground rurality in the function of media systems and technologies? How do these forms of mediation affect how we think about, inhabit, and relate to rurality?

Media Rurality investigates the centrality of rural places and people within the media systems and technologies that shape daily life in and across rural and urban settings alike in diverse global locations. Drawing from a range of humanities and social sciences fields and discourses that foreground rural media technologies, experiences, and practices, this book resituates our inquiries and reorients our gaze to spaces often conceived of as spatially marginal and temporally lagging. *Media Rurality* challenges these biases by demonstrating that rural spaces are both media intensive and instrumental to media systems typically associated with modern and transnational urban economies and experience. Confronting the epistemological and geographical fixity of the rural with a dynamic and relational condition of rurality, this book documents and theorizes a productive, unstable, and materially vibrant *media rurality* undergoing continual development and political transformation.

From the Rural to Ruralities

Rurality, in our view, responds to a problem of classification. Communication scholar Christopher Ali notes that "the rural" is often defined, unhelpfully, as that which is "not urban," determined on the basis of population metrics rather than relations to other economies and structures that occur within rural spaces and communities, and between rural and urban locations.[1] As Erin Morton observes, this way of thinking about rural places is baked in, both historically and etymologically: "The postwar modernization project rendered 'the rural' as a space of technological backwardness oriented toward subsistence. Even the linguistic root of its concomitant category, 'the country' (in French as *contrée* and in Latin as *contrara*), points to the idea of rurality existing in opposition to something, which, more often than not, means modern progress."[2] Reified in these ways, the rural conjures stereotypical images of agrarian and pastoral living, isolation and wilderness, and communities that are either quaint and quiet or violent, vaguely criminal, politically regressive, abandoned, and failing. Deployed in this manner, the rural tends to erase varied rural experiences of resource extraction, diverse forms of work and labor, technological change, infrastructural mediation, and political, cultural, and social heterodoxy.[3] This volume seeks to trouble these reifying descriptions, which fix the rural as the objectified other to the urban; it also seeks to endorse rurality as more dynamic, relational, and emergent category. *Media Rurality* reflects the diversity and instability of rural experience and the forms and practices of mediation by which it is constituted, proposing ways to work against the logics and practices that continue to foreclose just futures and consign rural spaces to an imaginary past or unlivable present.

As decades of interdisciplinary scholarship in rural studies has shown, to speak, think, or write of *the rural* as if the category refers to a singular, static, homogeneous, universal, and finished object is to engage in an exercise that is more ideological than empirical, more prescriptive than descriptive. Even the commonplace that defines the rural as "not urban" dissolves when faced with the reality of urban and rural spaces and experiences that are not only materially connected but also nested and hybrid. A night out with cinephiles at the Palace Theatre in Daysland, Alberta (population 824), is a curiously urban experience, just as tending lettuce and beans on a condo rooftop in downtown Montreal (population 4.3 million) and trading these for a neighbor's honey feels like a bit of country living in the heart of the city.[4] As chapters in this volume by Burç Köstem and Isihita Tiwary suggest, hybrid experiences in places that are not quite urban but also not quite rural are often the norm in peripheralized regions and liminal, connective spaces such as border zones and trade routes.[5] This is why recent scholarly turns toward articulating rurality (a social and material condition) and ruralization (a process) make more sense than studying the rural (an object) to guide the study of rural places and what happens in and to them.[6] *Ruralities* are plural, dynamic, diverse, situated, and emergent, and *ruralization* is the set of multiple ongoing relational processes by which rural places, people, and experiences are continuously formed, deformed, and reformed. Ruralities and ruralization are also highly mediated; they vary significantly from one geographic and temporal context to another, even as they sometimes face similar challenges and common factors, including the socioeconomic drivers of rural depopulation (and repopulation) through migration, industrial and infrastructural histories that have coevolved with certain cultures and practices, and political policies designed to produce, manage, and sustain these geographies and processes. This variety does not mean that specific ruralities have no substantive qualities or character distinctive of a place or context. It means the opposite: that the qualities and character of rurality in a particular place are historical, in the sense that ruralities are made, not found, and can also be unmade and remade.

In the global northern context (especially in the United States and Canada, our respective points of origin), contemporary commentary on rural places tends to focus on the politics of conservative resentment, homophobia, and nativism, sometimes at the expense of more complex engagements with rurality as a political formation demanding attention and care.[7] As Jordan Kinder's chapter in this volume on the rise of trucker convoys confirms, it is certainly the case that rural experience is often mobilized in support of the regressive politics associated with resurgent forms of right-wing populism. Similarly, in Europe, populist (and in some cases nativist) political formations, including erstwhile farmers' parties, have emerged in response to growing tensions between agrarian livelihoods, global agribusiness, and environmental regulation to challenge the authority of established farmers' organizations and state agencies, and have sought to anchor themselves in rural communities.[8] Together, these and other examples confirm that rural places are as susceptible as any others to right-populist political appeals—though, as Phil Neel's insightful analysis of class in America's hinterlands demonstrates, the recent success of these appeals has more to do with the structural conditions of hinterland economies and the organizational skill of radical right partisans than it does with any essential regressiveness attributable to rural people and places.[9]

However, there is now, and has always been, more than just conservatism to the politics, culture, and economies of rural places. As Jenna Burrell's chapter in this volume on the shifting dynamics of rural masculinity in the Pacific Northwest region of the United States shows, there is considerable cultural complexity even within traditionally conservative rural settings. More broadly, research in peasant studies and critical agrarian studies has established that rural settings have also been the site of progressive and often radical political subjectivities and movements organized around contesting post- and neocolonial underdevelopment, extractivism, and land dispossession.[10] This is especially true in rural settings in the global south, where the brunt of pathologies associated with industrial agriculture and mineral extraction continues to be felt and contested.[11] For over forty years, the Movimento dos Trabalhadores Rurais, Brazil's landless workers movement, has contested dispossession by massive *latifundios*, a struggle that has involved over 2,500 occupations by hundreds of thousands of families asserting their rights to confiscated rural lands.[12] More recently, in an action that lasted over sixteen months in 2020–21, farmers from across rural India mobilized in massive numbers to protest a suite of farm bills proposed by the national government that would have left them vulnerable to the predatory practices of global agribusiness firms.[13] Even in core economies, where farmers and rural dwellers more broadly are frequently posed as environmental villains or hindrances to sustainability, farmers are contesting the increasingly digital incursions of agribusiness. Some have even become hackers by sabotaging subscription-based tractor systems, sharing tactics to disrupt increasing capture by these systems.[14] Yet political ruralities are not only based on site-specific contestations and grievances. While distributed and unpredictable in political alignments, they are often transnational, liberatory, and frequently articulated in sophisticated ways against the uneven power and geographies that sustain their peripheralization, from imperialism to extractivism to various forms of dispossession, land grabs, and technological dependency.[15]

These diverse examples inform the focus in *Media Rurality* on the ambivalence of rurality as a political formation. Recent pathbreaking work by J. T. Roane demonstrates that reductive portraits of North American rural life are confounded by the rurality of insurgent "Black agoras" and everyday "Black ecologies" that form in places like Tidewater, Virginia.[16] We could say the same of historical alignments of solidarity across migrant agricultural labor struggles and environmental movements in California,[17] urban–rural coalitions of antiprison activists in the same state,[18] the construction of white masculinities across shifting modes of production in the US Northwest,[19] and the "combustive" politics of right-wing populism in rural Canada.[20] As chapters in this volume by Köstem on politics and technicity in the peripheries of Istanbul, Tiwary on the circulation of media in Indian and Nepalese border towns, Parks on do-it-yourself media systems in Tanzania, and Vemuri on the contested mediation of climate change in rural India demonstrate, the same goes for plural, diverse, and complex ruralities the world over.

Resource Ruralities

Contemporary ruralities are deeply implicated in the raw materials, infrastructures, technologies, modes of production, and supply chains of capitalist globalization.[21] As a result, they are also entangled materially with global media systems and cultures. The materials and energies that comprise contemporary media technologies are largely extracted in rural locations. Infrastructures that deliver media products and experiences to cities traverse rural lands and waters, including borderlands, and it is often back to rural locations that the unwanted material residues of media industries and culture typically return, in forms that include toxic wastes, dangerous working conditions, and climate collapse.[22] As recent work revising Marx's account of primitive accumulation has shown, the dispossession, enclosure, and occupation of rural lands and waters in colonial and settler colonial contexts is not only about turning peasants into proletarians but also about territorializing rural spaces as zones of extraction and circulation—and as sinks.[23] These ongoing relationships, and the logic of supply they materialize, form the fault lines of colonial, racial capitalist and technological modernity, in which rural people and ecologies have been designated as spatially and temporally peripheral to systems in which their role is, in fact, materially central.[24]

Rural places have long been produced as zones of extraction, exploitation, and disposability. Rurality in these forms has rested on cultural and legal constructions of rural people, lands and waters, and their elemental and biological constituents as features of the land and proprietary resources available for exploitation, commodification, and exchange. Such ruralities have also been predicated on false accounts of rural lands, waters, and inhabitations as empty, their economic potential wasted and untapped for want of their improvement.[25] These accounts, whether expressed in secular philosophies, religious doctrines, or cultural tropes, have given license to various techniques of dispossession, enclosure, and clearance that have been and continue to be instrumental in creating and recreating ruralities suited to capitalist, colonialist, and imperialist accumulations of value and power.[26] This means that rural settings have been primary sites where the inequalities and violence associated with extractive resource economies, including the slow violence of environmental degradation, have been made manifest.[27]

Resource extraction has been and remains a core organizing feature of colonial, imperial, and capitalist economies. The relationship between coloniality, capitalism, and rurality is a central focus of *Media Rurality*, especially the manner in which media technologies and practices are implicated in the spatial distribution of settler colonial and postcolonial power and industry.[28] Colonialism has always relied on the projection and production of emptiness that is often characteristic of metropolitan imaginaries of rural hinterlands, whether via the construction of colonized lands as terra nullius, wilderness, or wasteland, already empty of inhabitation and value, or via the subsequent emptying of rural territories and bodies by extractive industries.[29] In settler colonial contexts, this has followed racist dispossession and murder of Indigenous populations, wherein their association with nature justified the material and moral mission of colonization.[30] During the establishment of colonies, imperial rule took shape through the establishment of infrastructure and agriculture—the physical manifestations of settler ways of life that came to be identified with spatial and discursive construction of the rural itself.[31] As Indigenous scholars of infrastructure such as Anne Spice and Andrew Curley make clear, the process of infrastructural development and operation in the settler colony has always been one of land theft and the destruction of Indigenous lifeways.[32] Métis geographer Max Liboiron makes a similar contention that the condition of possibility for the toxic processes of colonial capitalism, from resource extraction to pollution, frequently enacted in and through territories categorized as rural, are premised on the logics of colonial land expropriation and use at differing levels of intensity.[33] These "violent inheritances," as communications scholar e. cram describes them in the context of settler existence in the North American West, then become the basis of forms of rural life and communality that persist in complex ways that both reproduce and exceed the destructive, oppressive historical processes that made them possible.[34]

However, as Imre Szeman and Jennifer Wenzel have argued, the meanings associated with the terms *extraction* ("a concrete, physical practice") and *extractivism* ("the cultural and ideological rationale that either motivates extraction or is the consequence of it") have proliferated in recent years, a dynamic that risks undermining their descriptive efficiency and analytical value as a way to understand the durabilities of such processes.[35] Narrowly defined, *extraction* refers to the removal of "natural" materials and life-forms configured as resources, in volumes that exceed their use value for nearby communities.[36] The purpose of extraction at this scale is resource commodification such that the resources might be accumulated, traded, or speculated on in order to generate exchange value, along with the wealth and power this affords. The value proposition of resource extraction thus entails removal in a second sense: the circulation of commodities and the value they generate for accumulation, processing, and expenditure elsewhere, in places—whether regional, national, or transnational metropoles—other than the sites from which the resource has been extracted. What remains local to sites of extraction are conditions of resource depletion; the social, political, and environmental residues of extractive industry; and various forms of resistance that arise in response to these conditions. For these reasons, rurally located extractive industries are key sites where political and economic inequality, environmental injustice, and dispossession are produced, reproduced, and contested. As the recent experience of so-called neoextractivism in Latin America shows, extractive industry's role in colonial, imperial, and capitalist formations has not prevented it from also featuring in postcolonial projects of national economic self-determination, including some with social welfarist intentions.[37]

The concept of extractivism expands the stakes of resource extraction in multiple ways, two of which are especially relevant here. First, it extends beyond the geophysical removals described above as definitive of extraction to encompass a broader set of social relationships, cultural practices, and subjective orientations. Reflecting on its emergence as an "organizing concept" across multiple disciplines in a piece for the *Journal of Peasant Studies*, Christopher Chagnon and colleagues describe extractivism as "a way of organizing life . . . ; a complex of self-reinforcing practices, mentalities, and power differentials underwriting and rationalizing socio-ecologically destructive modes of organizing life through subjugation, depletion, and non-reciprocity."[38] Second, the category of extractivism is increasingly applied to a range of economic practices and relations that exceed those typically associated with the removal of raw physical materials (i.e., forestry, mining, fisheries, agriculture), or, as Sandro Mezzadra and Brett Neilson put it, beyond "literal extraction."[39] Notable among these has been extension of the category to include proliferating forms of information, knowledge, and data harvesting enabled via digital networks.[40] This expansive conceptualization and application, which risks flattening and totalizing otherwise distinct social and economic practices and relationships, has prompted some to question the analytical utility of extractivism as a critical category.[41]

This is an important question, although one we do not seek to resolve here. What is made clear by several chapters in this volume is that rural locations remain primary sites for resource extraction and thus are profoundly shaped by, and responsive to, extractive activity. Moreover, if extractivism is "a way of organizing life," then its varied dimensions continue to be experienced most directly in rural places and by the people and species who inhabit them, as well as increasingly permeated by complex telecommunications technologies to make them legible by and for resource extraction.[42] As Chagnon and colleagues observe, "Extractivism has long been conceptually linked to capitalist processes and has recently been characterized as a fundamental expression of global capitalism, particularly in its manifestations across the rural realities of the Global South."[43] This has been and remains the case across the multiple forms that comprise literal extraction—industrial agriculture and aquaculture, forestry, mining, fisheries—and is also characteristic of extractive activity and industry "beyond literal extraction."[44] All of these extractive practices require mediation by technologies of prospecting, mapping, measurement, analysis, transportation, storage, and logistical coordination.[45]

Interestingly, Mezzadra and Neilson observe that "the productive front of data mining is particularly amplified in urban environments."[46]

Cities and their smart infrastructures are definitely key sites for extracting data from everyday social activity, a systematic harvesting of raw material for diverse processes of capital valuation, circulation, and coordination (and surveillance).[47] However, characterizing data mining as a predominantly urban phenomenon elides the manner in which emerging forms of resource extraction, including data harvesting, are reformatting rural places as extractive zones. This becomes evident if we replace *mining* with *farming* as the optic for catching sight of contemporary extractive activities. In the context of climate change and energy transition, the farm has become a commonplace way to describe extractive operations located in rural settings whose primary function (beyond capital accumulation) is to contribute to solving the environmental problems of cities, or the planet. Data centers (server farms), turbine complexes (wind farms), photovoltaic panel arrays (solar farms), replanted forests (tree farms), aquaculture operations (fish farms), and even land stewardship practices configured for carbon sequestration in soils (carbon farms) together comprise the infrastructures of decarbonization and sustainability that make demands on the rural landscapes where they are typically located. The degrees and manner to which they return observable benefits are ambiguous, unstable, and unevenly distributed. As sites of automated harvesting, computing, energy production, and carbon sequestration, these are often farms without farmers, at least in the traditional sense—a dramatic development in contexts where rurality has long been associated with farming communities.

At the same time, farms and farmers have not gone away, and rurality remains a category inextricably tied to food production. Agricultural practice and agrarian social relations in rural places remain central to future ways of growing food and managing land sustainably. As geographer Xiaowei Wang articulates in their remarkable book *Blockchain Chicken Farm*, "It is impossible to disentangle the countryside from food—food is at the core of the dynamic between the rural and the global. As humans, we eat to survive, and our appetite for food has carved new geographies and technologies into the world."[48] These geographies and practices have always been mediated by technology and infrastructure (both epistemological and material), but the media intensiveness of rural life has increased in recent decades as a result of emerging regimes of industrialization, efficiency, and automation. Wang demonstrates that from the large scale to the smaller scale—from industrial-scale precision agriculture to the artisanal blockchain chicken farm—food production is an increasingly digitally mediated prospect, and one that is crystallizing an increasingly dependent relationship between digital mediation and agrarian livelihoods. Consider the reverse migration of people from cities to the countryside in India during the COVID-19 pandemic, or a similar movement that has been occurring in China over the last decade. In the latter, incentivized by state and platform policies, creators and entrepreneurs have been returning to rural areas to establish businesses premised on promoting and building supposedly rural ways of life, typically foregrounding (and often idealizing) agroecological practices and traditional gender roles. This emerging platform economy requires the building of new infrastructure and the making available of new technologies in rural settings, including broadband cables and access to social media like Douyin, as intrinsic to the material and ideological construction of new economic prospects.[49] In other places, these new connectivities also potentially expose rural residents and producers to the novel extractive technologies of digital farm tech, like subscription-based John Deere tractors, autonomous harvesters, drone surveillance platforms, or high-capacity sensors, all premised on high-speed digital connectivity and often justified in the name of efficiency for more sustainable production.[50] Farm facilities, including contemporary precision agriculture operations and technologies of sensing and modeling weather and carbon cycles, are increasingly media intensive as the data they generate become significant additional sources of logistical and commercial value, enacting extractive relationships of harvesting that are decreasingly reciprocal to the land and places they extract from. The harnessing and application of agricultural knowledge, intricately tied to place-based ecological understanding, becomes not only a site of overt capture and extraction but also instrumentalization at scale. As societies adapt to the shifts of a climate-changed planet and the imperatives of transforming food systems and supply chains, rural agricultural areas are becoming key sites in which the material relations of sustainability are being imposed, negotiated, and contested.[51]

Widespread focus on urban ecologies and ways of living as sites of response to climate crisis serves to elide the impact of green capitalism and its mediating technologies on structures and cultures of rural and agrarian life, as well as the crucial role that rurality is playing in the construction of environmental, economic, and political futures in the context of climate change—futures in which agriculture and its media are deeply implicated, as Megan Wiessner, Anne Pasek, Nicole Starosielski, and Hunter Vaughan's chapter in this volume unpacks in detail. While it is certainly true that, according to Benjamin Peters, "agriculture has driven the need for both the physical and psychical channels for connection (primarily transportation and communication) between rural and urban population centers," it is also the case that emerging media ruralities in farm country are being driven as much by the communication of data, carbon, and nonrenewable energy as they are by trade in grain and livestock.[52] For example, Emily Duncan and colleagues document the integration of data-driven systems into food production as part of a more general effort to train AI models on increasingly complex environmental systems, such that "digitalized agriculture that is not necessarily about or for those we normally think of as agri-food governance actors."[53] This is one example of what Natacha Bruna describes as "green extractivism," in which information harvested from agrarian operations exceeds the purpose of optimizing agricultural production and instrumentalizes rural regions in the global south under broader regimes of "sustainable" governance.[54]

In a 2023 special issue of *New Media and Society* on "Farm Media," Zenia Kish and Benjamin Peters raise a provocative set of questions about the intersections of media and farming: "What does it mean to study farms, and agriculture more broadly, as sites of multivalent mediation? How are the histories and conceptual ordering of agriculture and media braided together? What points of intersection between media theory and critical agrarian studies might open up new modes of farm media analysis?"[55] *Media Rurality* extends these questions beyond conventional farming to include the broad range of industrial and economic activities that form the material basis of diverse ruralities across and between multiple national and continental settings. From farm data to the data farm, *Media Rurality* complicates the tight coupling of data with cities. It shows that media economies are deeply implicated in rural spaces, and further that ruralities are centrally positioned within regimes of datafication, sustainability, and energy transition on which smarter urban formations and governance continue to depend. Rural spaces thus mediate urban experiences in ways that are typically unseen or illegible to those in cities. It is not just that the extraction and disposal of the raw materials and fuels required for manufacturing and operating digital devices and networks are concentrated in rural places; it is also that extractive media systems reorganize rural places materially, in the image of data-driven capitalism, in ways that are not adequately reflected in accounts of datafication as a primarily urban phenomenon. As chapters in this volume by Wiessner et al., Jenna Burrell, and Cindy Lin illustrate, these technologies establish new frontiers for governance and capital accumulation within existing ruralities, adding digital layers onto sedimented relations of production and rural subjectivities. If data harvesting, storage, and circulation are indeed extractive, then they are yet another example of why extraction remains a key concept for understanding resource ruralities in general and media ruralities in particular.

Beyond the Urban-Rural Binary: Peripheralities of Imperial Globalization

Media rurality names a condition in which, contrary to persistent stereotypes, mediating infrastructures are materially present, not absent, from rural places. They include information, transportation, energy, financial, and knowledge infrastructures that bind rural peripheries to urban centers in complex circuits of extraction and supply, even as they pass through, over, or under rural communities without provisioning them. Henri Lefebvre, in his early Marxist writings analyzing the conceptual and geographical problems raised by the rural in mid-twentieth-century Europe, ascribes widespread ignorance of rural ways of life to bourgeois urban realities and parochialism: "For as long as the 'urban' reality and its institutions and ideologies, for as long as the successive modes of production, together with their superstructures bathed in a rural milieu and stood on a vast agricultural foundation, the middle and ruling classes paid scant attention to the peasantry."[56] As Raymond Williams describes in his canonical text *The Country and the City*, in the English context, this attention deficit cleared the way for ideology: "On the country has gathered the idea of a natural way of life: of peace, innocence and simple virtue. On the city has gathered the idea of an achieved center: of learning, communication, light." He hastily adds, "Yet the real history, throughout, has been astonishingly varied."[57]

In the European imaginary, urban cities are where workers are profitably assembled and disciplined for productive wage labor, while rural town and villages are where the residual peasantry works the land. However, these spaces, and the processes that comprise them, have always been as connected as they are separated. For example, the industrialization observed and theorized by Marx and rendered spatially by Lefebvre were fueled both by the migration of the European peasantry into the exploding regional metropolises of the eighteenth and nineteenth centuries and by the extraction of people and resources from the purportedly remote corners of expanding European empires. Observing both the agricultural devastation imposed on Ireland's peasantry and reflecting on the resulting migratory clearances, Engels writes in a letter to Marx in 1856, "One can already observe that the so-called liberty of English citizens is based on the oppression of the colonies."[58] In her chapter in this volume, Assatu Wisseh describes the "dark ruralities" of African enslavement generated by the violent extraction and forced migration of bodies to supply and nourish plantation capitalism. The structural tendency of capital to produce ruralities corresponding to the spatial organization of exploitable labor continues, including in ways that cannot be reduced to familiar stories of rural out-migration and urbanization. The rise of urban manufacturing in the European metropole itself relied heavily on an energy regime that entrenched spatial inequalities and class control in rural places, just as it intensified them in urban factories. As historian Cara New Daggett points out, coal, for example, was a more "mobile" fuel for energy production than water, whose flow was geographically dependent and largely seasonal, especially in rural areas where labor forces were more "static."[59] Coal, extracted far from urban centers, was a rurally sourced precondition of capitalism's signature productive relation: the concentration of an uprooted, expendable labor force in European industrial cities. As Ryan Cecil Jobson has shown, fossil capitalism was also racial capitalism, dependent not only on coal but on racialized, enslaved Africans who were violently displaced to support the manufactured rurality of colonial plantation economies.[60] In these cases, nonexploitative rural ecologies oriented to seasonal variability of resources and self-determined labor were actively foreclosed, subsumed by an industrial modernity built on territorial, class, and racialized divisions and unevenly defined mobilities between city and countryside, metropole and colony, that persist and intensify today.[61]

It was in this context that the cultural identification of the rural as hinterland, the distant other against whose difference the urban metropolitan seats and beneficiaries of colonial power were defined, became a core trope of Western modernity. *Rural* eventually marked internal divisions within European and North American states, but its genesis was in the imperial production of external places for extraction and harvesting, where people were treated at best as a source of physical labor that could be maintained and reproduced by means other than wages, or at worst as subject to displacement, dispossession, and extermination.[62] Like other, similar categories for designating places that are not cities but still perceived as sites of real and potential value—for example, *wilderness* and *wastelands*[63]—*rural* posits a spatially defined elsewhere subject to exceptional modes of economic and political governance designed to facilitate resource exploitation in the circuits of colonial and capitalist economies. In this respect, our account of ruralities resonates strongly with the formulation of Esther Peeren and colleagues of *hinterland* as "crucial for understanding the global and planetary present as a time defined by the lasting legacies of colonialism, increasing labor precarity under late capitalist regimes, and looming climate disasters . . . a lens to attend to the times and spaces shaped and experienced across the received categories of the urban, rural, wilderness, or nature, and to foreground the human and more than human lively processes that go on even in sites defined by capitalist ruin and political abandonment."[64] The work collected in *Media Rurality* aims to draw attention to how the forms, technologies, and practices Peeren et al. characterize as *hinterland thinking* are materialized in diverse ruralities.

Let's take a more contemporary example, one that brings us more directly to media. Lin Zhang, in her book *The Labor of Reinvention*, documents the planned rise of e-commerce entrepreneurship in rural China, mediated by proliferating digital platforms and associated infrastructure. As she describes, "Rural entrepreneurs, together with tech platforms and different levels of state apparatus, reinvented centuries-old practices of family-based petty capitalist production to construct a new regime of 'platformized family production' . . . Integrating the Chinese countryside into global digital capitalism via e-commerce created new opportunities for peasants and marginalized urban youth to achieve social mobility, but it also shaped a new regime of value that rewards some and marginalizes others."[65] Here, technological mediation and urban-to-rural migration generate a complex media rurality adapted to capitalist regimes of labor, productivity, and accumulation, but one that is far from the tropes of emptiness, pastoralism, and immobility that have long framed accounts of the rural and the proliferating hinterlands of human spatial practices.

Just over a decade ago, Monika Krause questioned the urbancentricity of prevailing accounts of "current sociospatial transformations," provocatively suggesting that "it is worth at least trying out the opposite perspective and analysing them from the perspective of that which is supposedly acted upon or being transformed."[66] *Media Rurality* takes up this suggestion. In their discussion of the durability of binary urban–rural thinking in academic geography, Jamie Gillen, Tim Bunnell, and Jonathan Rigg draw attention to the "porous relationality" between diverse spatial practices and designations across what is often characterized as the urban–rural divide.[67] Like rurality, urban technologies and ways of life are not self-contained, discrete activities and enterprises. Cities have long been engines, aggregators, and condensers of the forms of accumulation associated with imperialism, settler colonialism, and racial capitalism.[68] However, the rural settings of the dispossessive and extractive practices that undergird these formations have been and continue to be primary, not secondary. Martín Arboleda has evocatively described cities as "inverted mines" where resources drawn inward from the countryside are processed and consumed, their value accumulated.[69] This story is a familiar one, and its patterns and modalities are accelerating in response to the affordances and the demands of emerging media technologies and infrastructures, as well as to environmental imperatives that seem to recommend intensified urbanization and further relegation of rural locations to extractive zones as the path to sustainability.[70] However, even critical formulations such as Arboleda's inverted mine risk naturalizing the city as primary and reducing the future of rurality to eventual amalgamation into a planetary metabolism of connected, efficient global circulations centered in cities. This points to a structuring tension that is definitive of media ruralities and that surfaces throughout the chapters in this volume: one between the enrollment of rural places and people in economic, environmental, and political formations that exceed local scales and the persistent relationships and modes of life that continue to distinguish diverse ruralities from urban geographies (and from each other).

Geographer and sociologist Neil Brenner has influentially dubbed this process "planetary urbanization,"[71] a dynamic wherein spaces are crunched into a global system of metropoles that absorb and accumulate wealth and resources extracted from scattered but increasingly connected hinterland milieus.[72] As Peeren and colleagues have remarked, such accounts "end up making the rural disappear altogether," and "ruralization remains largely invisible."[73] The places most affected by this dynamic—whether oil palm plantations in Brazil's cleared rain forests, coal mines in Appalachia, oil fields in Iraq, olive groves in occupied Palestine, copper mines in Central Africa, or lithium salt flats in the Atacama Plateau—are not urban; nor are they greenfields on which sites of extraction, production, and circulation have been established without friction. They are perhaps more akin to what geographer Michael Woods describes as the "global countryside: a rural realm constituted by multiple, shifting, tangled and dynamic networks, connecting rural to rural and rural to urban, but with greater intensities of globalization processes and of global interconnections in some rural localities than in others, and thus with a differential distribution of power, opportunity and wealth across rural space."[74] These "global" processes remain central to the construction and emergence of rural experiences and economies, as Tiwary's chapter in this volume makes clear in the context of peculiarly urban markets in the border hinterlands of India and Nepal, where the migration of people and media structure encounters inseparable from dynamics of simultaneously global and peripheralizing forces. Places in the global countryside have been, and remain, sites of existing life and practices simultaneously drawn into and in excess of resource extraction and exploitation—what we might call, following Williams, dominant, residual, and emergent ruralities—and these plural and divergent dynamics demand sustained attention, including to the forms of mediation that enable them and the situation of media technologies within them.[75]

In many respects, the urban–rural divide is an imperial residue or, to borrow anthropologist Ann Laura Stoler's term, a form of "imperial debris," a register of the decimation that colonialism and imperialism have rendered on landscapes and ways of life.[76] Especially when considering the destruction of landscapes and the subduing of Indigenous cultural practice under colonial rule, we also see that rurality has been a crucial space and experience of refuge, endurance, and reclamation, especially in the aftermath of resource extraction. Connecting with discourses in environmental studies, anthropologist Anna Lowenhaupt Tsing, among others, has conceived of the "possibility of life in capitalist ruins,"[77] trying to understand how environmental degradation and supply-chain capitalism have not only rendered rural landscapes such as industrial forests inhospitable and unlivable but also presented fertile territory for emergent forms of practice and community in the aftermath of extraction. Ruralities are "sites for brutalist experimentation and active laying waste," but they are "also characterized by survival, endurance, refusal, fugitivity, and even resistance."[78] Capitalism and imperialism, however wide-ranging the effects and tactics of their systems, always face friction, tension, and resistance as well as refusal, withdrawal, and exit. As the chapter by Patrick Bresnihan and Patrick Brodie suggests, the structure and durability of rurality in Ireland can be understood through the example of the contested history of Guglielmo Marconi's early twentieth-century radio transmission infrastructure in Connemara, Ireland—a private imperial infrastructure in stark conflict with contemporaneous anticolonial mobilization and later postcolonial experiments in public infrastructural provision through the rural peat industry. In this and similar contexts, rurality is sometimes the space in which to enact different ways of life in the apparently all-encompassing shadow of capitalist exploitation and imperialist expansion. Here, too, infrastructural mediation becomes a key process in shaping diverse media ruralities and potential futures that have the potential to exceed the otherwise constrained possibilities of rurality in the expanding ruins of empire.

As several contributions to *Media Rurality* demonstrate, connectivity and quality of life do not necessarily go hand in hand, especially when connectivity—and the value derived from it—are structurally biased in favor of domestic and transnational capitalist organizations. As media scholars Nick Couldry and Ulises A. Mejia describe in their theory of "data colonialism," these "costs of connection" are typically externalized to rural locations by infrastructures that support regimes of logistical control and capital accumulation across distances.[79] This manifests most visibly and materially in the physical structures of extraction, production, and circulation that sustain contemporary supply chains. The experience of being abandoned to certain kinds of disconnection and compelled to other kinds of hyperconnection is a common double bind facing contemporary media subjects. When it comes down to it, the same processes that deploy an array of media to gather more and more data from rural people and places are also those that contribute to emptying out rural places. In light of this, sites of resistance against these practices—Indigenous communities fighting for data sovereignty or energy autonomy amid historical injustices,[80] dissection of climate and soils for precision and autonomous agriculture,[81] and expansion of land-intensive renewable energy infrastructures[82]—become more urgent than what is allowed by prejudicial dismissals of rural political subjects as "irrational," antiprogressive, and even criminal.[83] These are struggles for ways of life across highly varied rural settings. Furthermore, understanding the dynamics by which dispossession and displacement continue to define life in rural places can help us understand the roots of antiurban and even antistate biases and resentment, and especially how those feelings are capitalized on and instrumentalized by nefarious political actors.[84]

Places that have been systematically excluded from connection, or that were connected on terms experienced as nonnegotiable, are also sites of intense media activism and disruption.[85] Whether agrarian or nomadic, rural political subjects across global settings have seized on connectivity and the tools it provides for publicizing harms, celebrating and sustaining rural cultures, and accessing commodities, technologies, and forms of development often unavailable to rural and disconnected places. Similarly, in many contexts in the rural and agrarian regions of the global south, media technologies enable forms and temporalities of communication that assist in maintaining diasporic connections or even conducting everyday business through regional markets that sustain their communities.[86] Yet these are also scenes of compromise and disappointment. For example, small rural communities with few options accept extractive multinational developments as a path toward connectivity and chimeric "prosperity,"[87] while many rural communities in otherwise overdeveloped North America are unable to access reliable broadband, even in cases where it is essential to conduct agricultural business.[88] Such sites, addressed across this volume, point to the ambivalence and complexity of rural life in an era of continuously expanding media intensiveness and call for attention to how people and places react to and challenge the powerful, historically durable directionality of these systems.

Mediating Rurality: Infrastructure and Its Subjects

The modes of expropriation and displacement, and the possible futures offered to rural inhabitants as compensation, are often expressed in imaginaries that are bundled with the "promise of infrastructure"—of modernity, connection, and prosperity.[89] In his account of the postcolonial politics of large-scale infrastructure projects in northern Nigeria, Brian Larkin observes that these projects, however much they acted to maintain or facilitate particular kinds of exclusion or exploitation, also bore a "sublime" quality.[90] Residents may recognize that such infrastructures will not solve all their problems, and in fact may create new ones, but the chimera of connection and prosperity can alter how people subjectively process what otherwise looks like abandonment and exclusion.[91] As several chapters in this volume illustrate, the tension between development and abandonment mediated by infrastructure is a structuring feature of many contemporary ruralities.[92] Larkin's pathbreaking work has established that the politics of infrastructure manifests not just in the physical arrangements, mobilities, and relationships mediated by infrastructures but also in the subjective orientations conditioned by their poetic and aesthetic operations.[93] As detailed in chapters in this volume by Emily Ng and Esther Peeren, Jenna Burrell, and Jordan Kinder, the subjective dimensions of infrastructural mediation are crucial to understanding the political and cultural dimensions of contemporary ruralities.

In the past two decades, media scholars have turned to the concept of infrastructure to trace the material expressions and circulatory routes of media economies.[94] Infrastructures mediate and materialize the political, economic, and environmental processes of extraction, development, production, circulation, disposal, and abandonment that constitute the uneven relations between urban and rural settings regionally, nationally, and globally.[95] Anthropologist Akhil Gupta has argued that in spite of the perceived "invisibility" of functioning infrastructure in the overdeveloped global north, "urban dwellers are more dependent than their rural counterparts on infrastructures that deliver to them food, water, electricity, natural gas, gasoline, and sanitation."[96] In a sense, this rings true. Many urban dwellers, especially in the global north, form discourses and politics around disruptions to the smooth operation of provisioning infrastructures they have come to take for granted. Such complaints about missed and lapsed connections are often taken for granted in rural infrastructural politics. However, this characterization of some sort of urban infrastructural glut against rural infrastructural lack also reiterates urban–rural and north–south binaries that obscure the qualities of media rurality that the chapters in this book aim to clarify. Rural places are as infrastructurally mediated and dependent as urban places, even if what matters as infrastructure differs across these categories. Infrastructure, for example, has also been a crucial means by which ruralities configured to serve colonial jurisdiction and capitalist accumulation are installed and reproduced.[97] Thus, media rurality often takes invasive forms that disrupt existing infrastructures of life and impose infrastructures designed and operated to extract resources, energy, and data from rural settings and deliver them for expenditure or accumulation in urban settings.[98] In these cases, highly mediated rural spaces become part of the infrastructure of urban spaces and the networks of circulation that connect them. As this introduction makes clear, the suggestion that rural places are those in which infrastructural mediation and dependency are absent risks erasure of both the specific and spatially diverse character of rural infrastructures and the ways in which rural mediation is implicated in urban configurations of service provision, security, and wealth.

Nevertheless, rural places are also routinely sites of infrastructural abandonment, ruin, and decay, as capital and the state respond to economic, technological, and environmental instability with misplaced and misguided interventions in rural economies.[99] As Ayesha Vemuri's chapter on Kaziranga National Park in northeastern India details, often state-led environmental programs are sites of mediating risk in ways that maintain rather than correct ruralized imbalances, considering rhinos to be more worthy of state care (for tourism dollars) than local residents. Cindy Lin's chapter similarly captures these regimes of mediation as remote risk management through close study of digitalized state fire monitoring systems in Indonesia's peaty rain forests. These infrastructural interventions by the state, even under the guise of care and development, frequently leave rural people with little control and few options.

Rural places are also more vulnerable to, and frequently conditioned by, the temporalities of delay, suspension, deferral, and abandonment that are characteristic of infrastructural experience everywhere. The "promise of infrastructure"—and its spectacle—is as much about anticipation, duration, and absence as it is about connection and fulfillment.[100] *Media Rurality* offers a lens through which to register and understand the slowness and disconnection, the experience of waiting for access and experiencing the frustration of "half-built assemblages" that is typical of infrastructure time more generally.[101] As geographers Ashley Carse and David Kneas argue about the temporalities of stalled infrastructure, "The infrastructure projects that concern scholars are often considered complete, or, if not, their materialization is assumed to be imminent. And yet, many—if not most—of the dams, roads, railroads, ports, airports, and pipelines generally classified as infrastructure exist in states aptly characterized as unbuilt or unfinished. Planned, blocked, delayed, or abandoned, such projects are ubiquitous—the norm, rather than the exception."[102] In the rural areas of the world, these experiences of belated, uneven, and suspended connection to urban, national, or planetary systems, to which rural areas relate in ambiguous and contradictory ways, are constitutive of infrastructural life and politics. As Lisa Parks's chapter on homemade media systems in Tanzania documents, this means that rural places are also sites of ongoing infrastructural adaptation, resilience, and innovation—qualities that spill out from the material practices of media and infrastructure into social, political, and economic forms more broadly. In the context of hardening market logics, the individuation of economic risk, and the instability imposed by environmental degradation, the experience of rural communities in diverse geographies as zones of both sacrifice and refusal over the past several decades places them in the position of being ahead of urban centers, not behind them. Ranging across accounts of video bazaars, development projects, high-tech rural workplaces, agricultural media, energy infrastructures and systems, telecommunications, and conservation zones, *Media Rurality* aims to provide an interdisciplinary and international encounter with the complex relationships, subjectivities, and experiences across media, infrastructure, environments, and rural life at a time when the rural tends to be a mark of abjection, disavowal, and erasure.

Rurality and Media Studies

As we have made clear, *Media Rurality* responds to a persistent bias within media studies toward urban formations and the pasts, presents, and futures they represent—what Jack Halberstam refers to as *metronormativity*.[103] This response recalls pathbreaking work in media ethnography by Lila Abu-Lughod, whose fine-grained account of television viewing practices among women in a rural Egyptian village troubled *culture* as an abstract descriptive category in and across rural settings, instead pointing to "the kinds of cosmopolitanisms one finds in many rural areas around the post-colonial world and that confound the concept of 'cultures.'"[104] It is grimly fitting that Abu-Lughod closes her essay about the cultural politics of this agricultural village by describing the threat of its bulldozing in service of projects animated by modernist and national developmental ideals of progress, justified by the promise of cultural tourism and heritage dollars brought in by European visitors. Following this poignant prompt, *Media Rurality* asks what it would mean to recognize and truly, deeply engage with the town, village, and countryside as sites and thresholds of intensive and meaningful media activity and operations. Accordingly, *Media Rurality* is simultaneously broad and grounded in its geographical reach, presenting accounts of the material, social, and environmental conditions and relationships that characterize actually existing rural places in their great diversity. In its complexity and ambiguity, media rurality is a plural, situated, relational, dynamic, and intensively mediated condition whose social and political character cannot be decided in advance or from afar.

This is reflected in the deliberately sited, emplaced, and often field-based methods used by many authors in this volume. It goes without saying that hermeneutic analyses of mediated representations of rural people and places remain indispensable to critical study of media ruralities.[105] However, as the chapters by Ng and Peeren, Wisseh, Burrell, and Kinder demonstrate, the richness of such work is increased considerably when it is grounded in specific historical and material places and when it resists reification of the rural as an abstract category. Beyond this, *Media Rurality* also seeks to build on recent work in media studies—much of it addressed to media infrastructures, including in rural settings—that extends the field methodologically into *the field*, engaging with the atmospherics and ecologies of rural media practices.[106] In this, *Media Rurality* also takes a cue from the "dirt research" associated with earlier traditions in communication studies,[107] an inclination toward the field (in both senses) that prompts many of our authors' engagements. Such disjuncture between the field and the ideal singularization of the rural helps to defamiliarize ideologies that rest on the negation and denial of complex geographies of rural mediation, enabling in-depth engagements with the confounding spatial contradictions of capitalism, colonialism, and imperialism as well as the many enduring processes and afterlives of these structures across the rural world.

The studies in this book are immersed in the relationality and ambivalence of rurality, exposed by inquiring into sites of media and practices of mediation that make and remake rurality on the ground in the places where it is lived. This orientation implies neither a lament for a lost pastoralism nor panic about the alleged backwardness of rural places and people. Rather than reproducing binary distinctions and reductions that flatten the coexistence of different kinds of experience within and across rural areas—or, worse, shying away from the messy geographical and historical problems they raise—this book interrogates them and resituates rurality at the center of critical attention, a positioning that, as we have argued throughout this introduction, is uniquely afforded to those of us who study media. Here, *media rurality* refers to a complex set of contemporary practices, technologies, infrastructures, relationships, and processes that are always underway, incomplete, and evolving, often contradictory, and intensively mediated. How are rural places, people, and resources implicated in the mediation of global circuits of culture, technology, and capitalist circulation and exchange? How is urban life thereby enabled by particular material and cultural configurations of rural places? How do the infrastructures and technologies that mediate these relationships emerge as sites of politics and struggle? Ali's chapter on rural broadband in the United States, for example, draws attention to the ways in which connectivity (and its lack) to major media infrastructures, including electrical grids, telephone and broadcast networks, and internet service, have long structured rural experience in that country, as elsewhere.[108] *Media*

Rurality expands this focus on rural media, reaching beyond the frame of connectivity lack and exclusion to examine forms of mediation by which rural communities and economies are (often involuntarily) connected to global flows and circuits, typically serving priorities located elsewhere. Following geographers Jesse Heley and Laura Jones' call for attention to "relational rurals" in order to understand the co-constituency and mobility of rural life across material and discursive processes,[109] we thus see media, and particularly media practices, infrastructures, and technologies, as materials through which to unravel the differential and uneven modes and relationalities of communication and connection that make up rural life and experience.

As Geoff Hobbis, Marc Esteve Del Valle, and Rashid Gabdulhakov remark in a 2023 article in *Media, Culture and Society*, "It is time for a rural turn in media studies."[110] *Media Rurality* is both a provocation and a corrective, a way for scholars to approach rurality as a set of materially and geographically specific relations through the lens of media, its cultures, and its technologies. The volume speaks to pressing political issues facing the humanities and social sciences more broadly. Questions of environmental and climate justice and concerns about global economic and political disparity, as well as the mapping of these along historical and contemporary lines of racial capitalism, intersectional precarity, colonialism, and decolonization, are central to the book's contributions. This is because rural spaces shoulder many of the burdens of growth and environmental turbulence driven by these systems. Together, by bringing media technologies, infrastructures, practices, and politics to the fore as structuring and productive forces in these geographies of rurality, *Media Rurality*'s contributors reattune the conceptual and methodological tools of media studies toward these diverse implications, experiences, and politics of rurality. We hope that others will soon join us.

Organization of the Book

The first four chapters of the book, "Extractive Mediations," are united by the concept and experience of *extraction* across different periods and geographies.[111] How can media rurality be glimpsed and defined through extractive processes and industries, and what are the sometimes frictional and turbulent relationships that form when capitalism and colonialism encounter and are tasked with managing people and ecologies in places designated remote, unproductive, or otherwise requiring the material and discursive processes that ripen territories and peoples for extraction? What forms of mediation enable this, and how are media used to represent this? Megan Wiessner, Anne Pasek, Nicole Starosielski, and Hunter Vaughan comment on the rise of green digital technologies in the contemporary landscape of production, with a specific focus on rural geographies. From precision agriculture to soil carbon measurement, media technologies are increasingly being used to more efficiently generate and withdraw value from the countryside—with profound implications for emerging green digital economies. The next chapter, by Emily Ng and Esther Peeren, turns our gaze to rural areas in the Netherlands and China, specifically looking at documentary approaches to work, ecology, and community experience in and of the extractive industries, using Raymond Williams and Gayatri Chakravorty Spivak to articulate the hinterland as a central geography of extraction. Patrick Bresnihan and Patrick Brodie provide an extended version of the history of the Guglielmo Marconi radio station in colonial Connemara as informing contemporary geographies of tech capital in rural Ireland, theorizing "energetic mediation" as a process of imperial geography that continues to manufacture relationships in and across rural areas today. Assatu Wisseh traces a similarly historical account of the American Colonization Society and its mediated construction of "dark ruralities" in western Africa, which acted to racially justify the construction of the colony of Liberia through the ur-logic of extraction of people and things from the African continent by European colonialism.

In the next section, "Practicing Rurality," the chapters focus on life and practices of media and mediation in rural and ruralized places. Burç Köstem's chapter, for example, takes an ethnographic view of the politics of development in the hinterlands of contemporary Istanbul, arguing that these developmental and construction politics depend on processes of "mediating the periphery" experienced in and through communities and ecologies. Lisa Parks takes us to rural Tanzania, where small-scale homemade energy systems perform vital functions for lives and livelihoods, reflecting on how the politics of off-grid life can tell us about the promises of development and sustainability in and through systems of mediation. Ishita Tiwary's contribution focuses specifically on the movement of media content and technologies across the borders of China and India, demonstrating that the informal practices and experiences of the global in such peripheralized places of exchange shifts our conceptual frameworks of the politics and pathways of media systems. Jenna Burrell then reflects on the politics of masculinity in rural Oregon at the site of a Facebook data center, where its development and operations have mapped onto durable—and problematic—histories of masculine labor and "cowboy-ism" as embodied and experienced by local workers.

The final section, "Political Ruralities," complicates and disrupts taken-for-granted ideas of media politics by interrogating how governance, power, and subjectivity are assembled and negotiated through distinctive forms of mediation across diverse rural settings. Jordan Kinder's chapter is a case in point. Kinder, directly addressing the elephant in the room surrounding North American right-wing rural politics, takes up the case of the Canadian Freedom Convoy in 2022, an ostensibly antilockdown protest that, as Kinder finds, shared ideologies and personnel with pro-oil movements and other petroturfed right-wing movements in Canada over the past couple of decades. He argues that by appropriating indigeneity, these protestors used an ultimately fascist imaginary of Canadian rurality within which the hegemonic norms of white Canadian settler society sought to mobilize rurality toward supremacist ends. Christopher Ali, in the following chapter, explains how mapping mediates power in and through the development of rural broadband systems in the United States. Cindy Lin provides an account of state technological monitoring and detection of underground peatland fires in Indonesia, examining the politics of research labor and environmental governance enacted in these mediating practices. Finally, we conclude the collection with a chapter by Ayesha Vemuri on the politics of conservation in rural Kaziranga, India, in which several issues prevalent throughout the collection converge—rural territorialization, politics of indigeneity, and colonial politics of conservation and sustainability. Vemuri's contention is that financial and insurance infrastructures comprise forms of environmental that mediate between powerful global economic interests and rural dwellers in ways that reproduce historically entrenched relationships of environmental injustice.

Each of these chapters, together and individually, shows us something about how rurality and forms of mediation co-constitute contemporary relations and systems within and across multiple rural geographies and locations. Across the chapters, we see media play a formative, if sometimes elided, role in the construction and maintenance of uneven systems that have long foreclosed just futures and consigned rural spaces to an imaginary past or unlivable present. Rurality is ambivalent, subjective, relational, plural, emergent, systemic, mediated, and above all political. In presenting such a range of critical analyses of media rurality, we hope to suggest pathways toward engaging with rurality as a dynamic and lively political geography of media.

Notes

1 Ali, "Thoughts," 2.

2 Morton, "Rural."

3 See Epp and Whitson, *Writing Off the Rural West*. We are grateful to Roger Epp for presenting his work on the implications of artificial meat for livestock-based ruralities at our initial 2022 event at McGill University, "Media Rurality."

4 For an excellent account of the potentially liberatory politics of urban farming, see White, *Freedom Farmers*.

5 On this theme, see Cowan, *Subaltern Frontiers*.

6 See Krause, "Ruralization"; Parsons and Lawreniuk, "Geographies of Ruralisation or Ruralities?"; Gillen et al., "Geographies of Ruralization." See also Edensor, "Performing Rurality"; Woods, "Performing Rurality."

7 For more nuanced scholarly treatments of political identity in rural America, see Wuthnow, *Left Behind*; Hochshild, *Strangers*; Cramer, *Politics of Resentment*.

8 Matthews, "Farmer Protests."

9 Neel, *Hinterland*.

10 See, e.g., Edelman and Borras, *Political Dynamics*; Cauoette and Turner, *Agrarian Angst*; Padilla, *Rural Resistance*. See also the long-standing reservoir of radical academic work across disciplines on critical agrarian issues in the *Journal of Peasant Studies* as well as *Agrarian South: Journal of Political Economy*.

11 See, for example, Hetherington, *Government of Beans*; Macarena Gómez-Barris, *Extractive Zone*; Gordillo, "Metropolis"; Riofrancos, *Resource Radicals*.

12 Carter, "Landless Rural Workers."

13 Baviskar and Levien, "Farmers' Protests."

14 See Newman, "New Jailbreak." For a wide-ranging articulation of farmers' politics against agribusiness, including digital and antimonopoly struggles, see Gibson and Alexander, *In Defense of Farmers*.

15 See, for example, Ajl, "Does the Arab Region"; Dunlap, *This System*; Edelman, *Peasant Politics*.

16 See Roane, *Dark Agoras*; Roane, "Black Ecologies." We are grateful for having had the opportunity to engage with J. T. Roane at Media Rurality in 2022, where he presented his research on Black ecologies and rural life in the northeastern United States.

17 Bresnihan and Milner, *All We Want.*

18 Gilmore, *Golden Gulag.*

19 Burrell, chap. 8, this volume.

20 Kinder, chap. 9, this volume.

21 Gilmore, *Golden Gulag.*

22 See Iheka, *African Ecomedia*; Devine, *Decomposed*; Gabrys, *Digital Rubbish.*

23 Coulthard, *Red Skin*, 7–15; Nichols, *Theft Is Property!*, 52–84; Day, "Eco-Criticism"; Li, *Land's End*; Liboiron, *Pollution Is Colonialism.*

24 We should include in these conceptions, of course, John Bellamy Foster's famous theorization of the "metabolic rift"—a core eco-Marxist idea describing capital's constitutive severing of society from nature via intensive agriculture—as also one that crystallized an urban and rural divide in early capitalist society: "Marx provided a powerful analysis of the main ecological crisis of his day—the problem of soil fertility within capitalist agriculture—as well as commenting on the other major ecological crises of his time (the loss of forests, the pollution of the cities, and the Malthusian specter of overpopulation). In doing so, he raised fundamental issues about the antagonism of town and country, the necessity of ecological sustainability, and what he called the 'metabolic' relation between human beings and nature." Foster, "Marx's Theory," 373.

25 Bhandar, *Colonial Lives.*

26 Harvey, *New Imperialism.*

27 Adunbi, *Enclaves.* We were fortunate to engage with Omolade Adunbi at Media Rurality in 2022, where he presented his research on energy practices, climate crisis, and the social death of the environment in Nigeria. See also Nixon, *Slow Violence.*

28 Davies, "Coloniality of Infrastructure."

29 Voyles, *Wastelanding*; Bhandar, *Colonial Lives.*

30 See Daschuck, *Clearing the Plains*; Nichols, *Theft Is Property!*

31 Knoblach, *Culture of Wilderness.*

32 Spice, "Fighting Invasive Infrastructures"; Curley, "Infrastructures." We learned a great deal from our engagement with Andrew Curley at Media Rurality in 2022, where he presented his research on decolonizing the Colorado Compact, which regulates water distribution across Indigenous territories in the southwestern United States.

33 Liboiron, *Pollution Is Colonialism.*

34 cram, *Violent Inheritance.* See also Curley, *Carbon Sovereignty.*

35 Szeman and Wenzel, "What Do We Talk About."

36 For an excellent account of multispecies entanglements and extraction in rural contexts, see Chao, *In the Shadow.*

37 On neoextractivism in Latin America, see Gudynas, *Extractivisms*; Arboleda, *Planetary Mine*; Riofrancos, *Resource Radicals*; Svampa, *Neo-Extractivism in Latin America.*

38 Chagnon et al., "From Extractivism," 763.

39 See Mezzadra and Neilson, "On the Multiple Frontiers."

40 See Couldry and Mejias, *Costs of Connection*; Sadowski, "When Data Is Capital."

41 See Szeman, "On the Politics of Extraction," 444. Szeman argues that under the framework of extractivism, "extraction becomes the name for any process through which value is generated for capitalism" and, referring to equivalence between "coal mining and data mining," he suggests this is "a case of conceptual metaphor and allegory run (a little) wild." See also Ajl, "Theories of Political Ecology." Ajl similarly is skeptical of the ways that critiques of extractivism often seem to describe and stand in for capitalism across very different geographies, and thus are limited as political analyses.

42 Arboleda, *Planetary Mine*; Han, *Deepwater Alchemy.*

43 Chagnon et al., "From Extractivism," 761.

44 Mezzadra and Neilson, "On the Multiple Frontiers," 193.

45 See Wang, *Blockchain Chicken Farm*; Tollefson, "Staking a Claim."

46 Mezzadra and Neilson, "On the Multiple Frontiers," 195.

47 Halpern and Mitchell, *Smartness Mandate.*

48 Wang, *Blockchain Chicken Farm*, 6.

49 Zhang, *Labor of Reinvention.* See also the ongoing thesis research of Hanxiao Zhang, whose investigation into rural content creators on Douyin considers the platform-mediated (and often state-supported) prospects of this reverse migration and the construction of new rural livelihoods. Zhang, "Platformizing the Rural."

50 Barney, "Autonomous Agriculture?"; Hendrickson et al., "Power, Food, and Agriculture."

51 Hetherington, *Government of Beans.*

52 Peters, "Afterword," 1966.

53 Duncan et al., "New but for Whom?," 1195; see also Bronson, *Immaculate Conception.*

54 Bruna, "Going Beyond."

55 Kish and Peters, "Farm Media."

56 Lefebvre, *On the Rural*, 59.

57 Williams, *Country*, 1.

58 Engels quoted in Cleary, "Irish Studies," 42–43.

59 Daggett, *Birth of Energy*, 29–30; see also Malm, *Fossil Capital.*

60 Jobson, "Dead Labor."

61 See West, *Dispossession.*

62 Yusoff, *Billion Black Anthropocenes.*

63 Rasmussen and Lund, "Reconfiguring Frontier Spaces"; Voyles, *Wastelanding.*

64 Peeren et al., "Introduction," 4.

65 Zhang, *Labor of Reinvention*, 14.

66 Krause, "Ruralization," 233.

67 Gillen et al., "Geographies of Ruralization," 189.

68 Graham and Marvin, *Splintering Urbanism.*

69 Arboleda, *Planetary Mine.*

70 See Riofrancos, "What Green Costs"; Bresnihan and Brodie, "From Toxic Industries."

71 Brenner, *Implosions/Explosions.*

72 For an agrarian take on planetary urbanization, see Ajl, "Hypertrophic City."

73 Peeren et al., "Introduction," 11.

74 Woods, "Engaging the Global Countryside," 492.

75 Williams, *Marxism and Literature.*

76 Stoler, "Imperial Debris."

77 Tsing, *Mushroom.*

78 Peeren et al., "Introduction," 13.

79 Couldry and Mejias, *Costs of Connection.* See also Starosielski, *Undersea Network*; Cowen, *Deadly Life.*

80 Kukutai and Taylor, *Indigenous Data Sovereignty*; Kinder, "Solar Infrastructure."

81 Barney, "Autonomous Agriculture?"

82 Bresnihan and Brodie, "New Extractive Frontiers."

83 Lewis, *Scammer's Yard.* Rahul Mukherjee's discussion of mobile phone scams operating in rural India at Media Rurality in 2022 was particularly fruitful in this regard, mediating gestures of both connectivity and criminality in relation to uneven systems of finance, and we thank him for that early contribution.

84 See Kinder, chap. 9, this volume.

85 A compelling example of these dynamics was presented at Media Rurality in 2022, in a paper by Lisa Parks, Assatu Wisseh, Gaylene DuCharme, and Sarah DesRosier on "The Nuances of Network Sovereignty: A Collaborative Study of Internet and Digital Technologies in the Blackfeet Nation." See also Duarte, *Network Sovereignty.*

86 Hahn, *Media Culture*; Tsing, *Mushroom.*

87 Brodie, "Stuck in Mud"; see also Burrell, this volume.

88 See Ali, chap. 10, this volume.

89 See Anand et al., *Promise of Infrastructure*; Axel et al., *Coloniality of Infrastructure.*

90 Larkin, *Signal and Noise.*

91 Brodie, "Stuck in Mud."

92 See Gordillo, *Rubble.*

93 Larkin, "Politics and Poetics"; Larkin, "Promising Forms."

94 Barney, "Infrastructure"; Larkin, *Signal and Noise*; Parks and Starosielski, *Signal Traffic.*

95 Gillespie et al., *Media Technologies*; Hockenberry et al., *Assembly Codes*; Packer and Wiley, *Communication Matters*; Parks and Starosielski, *Signal Traffic*; Sharma and Singh, *Re-Understanding Media*; Towns, *On Black Media Philosophy.*

96 Gupta, "Future in Ruins," 66.

97 LaDuke and Cowen, "Beyond Wiindigo Infrastructure"; Ruiz, *Slow Disturbance.*

98 Spice, "Fighting Invasive Infrastructures."

99 Brodie and Velkova, "Cloud Ruins"; Dawney, "Decommissioned Places"; Storm, *Post-Industrial Landscape Scars*; Voyles, *Wastelanding.*

100 Anand et al., *Promise of Infrastructure.*

101 Burrell, "On Half-Built Assemblages."

102 Carse and Kneas, "Unbuilt and Unfinished," 9. See also Günel, *Spaceship.*

103 Halberstam, *In a Queer Time.*

104 Abu-Lughod, "Interpretation of Culture(s)," 123, emphasis in original.

105 See Fowler and Helfield, *Representing the Rural*; Strand and Barney, "Telling Their Stories."

106 See, for example, Larkin, *Signal and Noise*; Starosielski, *Undersea Network*; Ruiz, *Slow Disturbance*; Mukherjee, *Radiant Infrastructures*; Parks et al., "Media Fieldwork."

107 Young, "Innis's Infrastructure."

108 Ali, *Farm Fresh Broadband*; see also Ali, this volume.

109 Heley and Jones, "Relational Rurals."

110 Hobbis et al., "Rural Media Studies," 1489. We note that the case was arguably already made in 2010; see Andersson and Jansson, "Rural Media Spaces."

111 We share the terminology *extractive mediation* with Lisa Yin Han's exceptional 2024 book, *Deepwater Alchemy: Extractive Mediation and the Taming of the Seafloor*, surrounding the technologies and practices that condition deep-sea spaces as resource frontiers.

Bibliography

Abu-Lughod, Lila. "The Interpretation of Culture(s) After Television." *Representations* 59 (1997): 109–34.

Adunbi, Omolade. *Enclaves of Exception: Special Economic Zones and Extractive Practices in Nigeria.* Indiana University Press, 2022.

Ajl, Max. “Does the Arab Region Have an Agrarian Question?” *Journal of Peasant Studies* 48, no. 5 (2020): 955–83.

Ajl, Max. “The Hypertrophic City Versus the Planet of Fields.” In *Implosions/Explosions: Towards a Study of Planetary Urbanization*, edited by Neil Brenner. Jovis, 2014.

Ajl, Max. “Theories of Political Ecology: Monopoly Capital Against People and the Planet.” *Agrarian South: Journal of Political Economy* 12, no. 1: 12–50.

Ali, Christopher. *Farm Fresh Broadband: The Politics of Rural Connectivity*. MIT Press, 2021.

Ali, Christopher. “Thoughts on a Critical Theory of Rural Communication.” *CARGC Papers* 7 (2018). https://repository.upenn.edu/handle/20.500.14332/5679.

Anand, Nikhil, Akhil Gupta, and Hannah Appel, eds. *The Promise of Infrastructure*. Duke University Press, 2018.

Andersson, Magnus, and André Jansson. “Rural Media Spaces: Communication Geography on New Terrain.” *Culture Unbound* 2 (2010): 121–30.

Arboleda, Martín. *Planetary Mine: Territories of Extraction under Late Capitalism*. Verso, 2019.

Axel, Nick, Kenny Cupers, and Nikolaus Hirsch, eds. *The Coloniality of Infrastructure*. e-flux Architecture, 2021. https://www.e-flux.com/architecture/coloniality-infrastructure/.

Barney, Darin. “Autonomous Agriculture?” *Heliotrope*, February 8, 2023. https://www.heliotropejournal.net/helio/autonomous-agriculture.

Barney, Darin. “Infrastructure and the Form of Politics.” In “Materials and Media of Infrastructure,” edited by Aleksandra Kaminska and Rafico Ruiz, special issue, *Canadian Journal of Communication* 46, no. 2 (2021): 225–46.

Baviskar, Amita, and Michael Levien. “Farmers’ Protests in India: Introduction to the JPS Forum.” *Journal of Peasant Studies* 48, no. 7 (2021): 1341–55.

Bhandar, Brenna. *Colonial Lives of Property: Law, Land, and Racial Regimes of Ownership*. Duke University Press, 2018.

Brenner, Neil. *Implosions/Explosions: Towards a Study of Planetary Urbanization*. Jovis, 2014.

Bresnihan, Patrick, and Patrick Brodie. “From Toxic Industries to Green Extractivism: Rural Environmental Struggles, Multinational Corporations and Ireland’s Postcolonial Ecological Regime.” *Irish Studies Review* 32, no. 1 (2024): 93–122.

Bresnihan, Patrick, and Patrick Brodie. “New Extractive Frontiers in Ireland and the Moebius Strip of Wind/Data.” *Environment and Planning E: Nature and Space* 4, no. 4 (2021): 1645–64.

Bresnihan, Patrick, and Naomi Milner. *All We Want Is the Earth: Land, Labor, and Movements Beyond Environmentalism*. Bristol University Press, 2023.

Brodie, Patrick. "'Stuck in Mud in the Fields of Athenry': Apple, Territory, and Popular Politics." In "Media Populism," edited by Giuseppe Fidotta, Joshua Neves, and Joaquin Serpe, special issue, *Culture Machine* 19 (2020): 1–34. https://culturemachine.net/wp-content/uploads/2021/04/8.-Patrick-Brodie-revised-1.pdf.

Brodie, Patrick, and Julia Velkova. "Cloud Ruins: Ericsson's Vaudreuil-Dorin Data Center and Infrastructural Abandonment." *Information, Communication, and Society* 24, no. 6 (2021): 869–85.

Bronson, Kelly. *Immaculate Conception of Data: Agribusiness, Activists, and Their Shared Politics of the Future*. McGill-Queen's University Press, 2023.

Bruna, Natcha. "Going Beyond Efficiency-Driven Extractivism: A Climate-Smart World and the Rise of Green Extractivism." *Journal of Peasant Studies* 49, no. 4 (2022): 839–64.

Burrell, Jenna. "On Half-Built Assemblages: Waiting for a Data Center in Prineville, Oregon." *Engaging Science, Technology, and Society* 6 (2020): 283–305.

Carse, Ashley, and David Kneas. "Unbuilt and Unfinished: The Temporalities of Infrastructure." *Environment and Society: Advances in Research* 10 (2019): 9–28.

Carter, M. "The Landless Rural Workers Movement and Democracy in Brazil." *Latin American Research Review* 45, no. 1 (2010): 186–217.

Cauoette, Dominique, and Sarah Turner, eds. *Agrarian Angst and Rural Resistance in Contemporary Southeast Asia*. Routledge, 2009.

Chagnon, Christopher W., Francesco Durante, Barry K. Gills, et al. "From Extractivism to Global Extractivism: The Evolution of an Organizing Concept." *Journal of Peasant Studies* 49, no. 4 (2022): 760–92.

Chao, Sophie. *In the Shadow of the Palms: More-than-Human Becomings in West Papua*. Duke University Press, 2022.

Cleary, Joe. "Irish Studies, Colonial Questions: Locating Ireland in the Colonial World." *Outrageous Fortune: Capital and Culture in Modern Ireland*. Field Day Publications, 2007.

Couldry, Nick, and Ulises A. Mejias. *The Costs of Connection: How Data Is Colonizing Human Life and Appropriating It for Capitalism*. Stanford University Press, 2019.

Coulthard, Glen Sean. *Red Skin, White Masks: Rejecting the Colonial Politics of Recognition*. University of Minnesota Press, 2014.

Cowan, Thomas. *Subaltern Frontiers: Agrarian City-Making in Gurgaon*. Cambridge University Press, 2022.

Cowen, Deborah. *The Deadly Life of Logistics: Mapping Violence in Global Trade*. University of Minnesota Press, 2014.

cram, e. *Violent Inheritance: Sexuality, Land, and Energy in Making the North American West*. University of California Press, 2021.

Cramer, Katherine J. *The Politics of Resentment: Rural Consciousness in Wisconsin and the Rise of Scott Walker*. University of Chicago Press, 2016.

Curley, Andrew. *Carbon Sovereignty: Coal, Development, and Energy Transition in the Navajo Nation*. University of Arizona Press, 2023.

Curley, Andrew. "Infrastructures as Colonial Beachheads: The Central Arizona Project and the Taking of Navajo Resources." *Environment and Planning D: Society and Space* 39, no. 3 (2021): 387–404.

Daggett, Cara New. *The Birth of Energy: Fossil Fuels, Thermodynamics, and the Politics of Work*. Duke University Press, 2019.

Daschuck, James. *Clearing the Plains: Disease, Politics of Starvation, and the Loss of Aboriginal Life*. University of Regina Press, 2013.

Davies, Archie. "The Coloniality of Infrastructure: Engineering, Landscape, and Modernity in Recife." *Environment and Planning D: Society and Space* 39, no. 4 (2021): 740–57.

Dawney, Leila. "Decommissioned Places: Ruins, Endurance, and Care at the End of the First Nuclear Age." *Transactions of the Institute of British Geographers* 45 (2020): 33–49.

Day, Iyko. "Eco-Criticism and Primitive Accumulation in Indigenous Studies." In *After Marx: Literature, Theory, and Value in the Twenty-First Century*, edited by C. Lye and C. Nealon. Cambridge University Press, 2022.

Devine, Kyle. *Decomposed: The Political Ecology of Music*. MIT Press, 2019.

Duarte, Marisa Elena. *Network Sovereignty: Building the Internet Across Indian Country*. University of Washington Press, 2017.

Duncan, Emily, Alesandros Glaros, Denis Z. Ross, and Eric Nost. "New but for Whom? Discourses of Innovation in Precision Agriculture." *Agriculture and Human Values* 38 (2021): 1181–99.

Dunlap, Alexander. *This System Is Killing Us: Land Grabbing, the Green Economy, and Ecological Conflict*. Pluto Press, 2024.

Edelman, Marc. *Peasant Politics of the Twenty-First Century: Transnational Social Movements and Agrarian Change*. Cornell University Press, 2024.

Edelman, Marc, and Saturnino M. Borras Jr. *Political Dynamics of Transnational Agrarian Movements*. Fernwood, 2016.

Edensor, Tim. "Performing Rurality." In *The Rural Studies Reader*, edited by Paul Cloke. Sage, 2005.

Epp, Roger, and Dave Whitson. *Writing Off the Rural West: Globalization, Governments, and the Transformation of Rural Communities*. University of Alberta Press, 2001.

Foster, John Bellamy. "Marx's Theory of Metabolic Rift: Classical Foundations for Environmental Sociology." *American Journal of Sociology* 105, no. 2 (1999): 366–405.

Fowler, Catherine, and Gillian Helfield. *Representing the Rural: Space, Place, and Identity in Films about the Land.* Wayne State University Press, 2005.

Gabrys, Jennifer. *Digital Rubbish: A Natural History of Electronics.* University of Minnesota Press, 2011.

Gibson, Jane W., and Sara E. Alexander, eds. *In Defense of Farmers: The Future of Agriculture in the Shadow of Corporate Power.* University of Nebraska Press.

Gillen, Jamie, Tim Bunnell, and Jonathan Rigg. "Geographies of Ruralization." *Dialogues in Human Geography* 12, no. 2 (2022): 186–203.

Gillespie, Tarleton, P. Boczkowski, and K. A. Foot, eds. *Media Technologies: Essays on Communication, Materiality, and Society.* MIT Press, 2014.

Gilmore, Ruth Wilson. *Golden Gulag: Prisons, Surplus, Crisis, and Opposition in Globalizing California.* University of California Press, 2007.

Gómez-Barris, Macarena. *The Extractive Zone: Social Ecologies and Decolonial Perspectives.* Duke University Press, 2017.

Gordillo, Gaston. "The Metropolis: The Infrastructure of the Anthropocene." In *Infrastructure, Environment and Life in the Anthropocene,* edited by Kregg Hetherington. Duke University Press, 2019.

Gordillo, Gaston. *Rubble: The Afterlife of Destruction.* Duke University Press, 2014.

Graham, Steven, and Simon Marvin. *Splintering Urbanism: Networked Infrastructures, Technological Mobilities, and the Urban Condition.* Routledge, 2001.

Gudynas, Eduardo. *Extractivisms: Politics, Economy and Ecology.* Columbia University Press, 2021.

Günel, Gökçe. *Spaceship in the Desert: Energy, Climate Change and Urban Design in Abu Dhabi.* Duke University Press, 2019.

Gupta, Akhil. "The Future in Ruins: Thoughts on the Temporality of Infrastructure." In *The Promise of Infrastructure,* edited by Nikhil Anand, Akhil Gupta, and Hannah Appel. Duke University Press, 2018.

Hahn, Allison. *Media Culture in Nomadic Communities.* Amsterdam University Press, 2021.

Halberstam, Jack. *In a Queer Time and Place: Transgender Bodies, Subcultural Lives.* NYU Press, 2005.

Halpern, Orit, and Robert Mitchell. *The Smartness Mandate.* MIT Press, 2023.

Han, Lisa Yin. *Deepwater Alchemy: Extractive Mediation and the Taming of the Seafloor.* University of Minnesota Press, 2024.

Harvey, David. *The New Imperialism.* Oxford: Oxford University Press, 2003.

Heley, Jesse, and Laura Jones. "Relational Rurals: Some Thoughts on Relating Things and Theory in Rural Studies." *Journal of Rural Studies* 28 (2012): 208–17.

Hendrickson, Mary K., Philip H. Howard, and Douglas H. Constance. "Power, Food, and Agriculture: Implications for Farmers, Consumers, and Communities." In *In Defense of Farmers: The Future of Agriculture in the Shadow of Corporate Power*, edited by Jane W. Gibson and Sara E. Alexander. University of Nebraska Press, 2019.

Hetherington, Kregg. *The Government of Beans: Regulating Life in the Age of Monocrops*. Duke University Press, 2020.

Hobbis, Geoffrey, Marc Esteve Del Valle, and Rashid Gabdulhakov. "Rural Media Studies: Making the Case for a New Subfield." *Media, Culture, and Society* 45, no. 7 (2023): 1489–500.

Hochshild, Arlie Russell. *Strangers in Their Own Land: Anger and Mourning on the American Right*. New Press, 2018.

Hockenberry, Matthew, Nicole Starosielski, and Susan Zieger, eds. *Assembly Codes: The Logistics of Media*. Duke University Press, 2020.

Iheka, Cajetan. *African Ecomedia: Network Forms, Planetary Politics*. Duke University Press, 2021.

Jobson, Ryan Cecil. "Dead Labor: On Racial Capital and Fossil Capital." In *Histories of Racial Capitalism*, edited by Justin Leroy and Destin Jenkins. Columbia University Press, 2021.

Kinder, Jordan. "Solar Infrastructure as Media of Resistance, or Indigenous Solarities Against Settler Colonialism." *South Atlantic Quarterly* 120, no. 1 (2021): 63–76.

Kish, Zenia, and Benjamin Peters. "Farm Media: An Introduction." *New Media and Society* 25, no. 8 (2023): 1827–41.

Knoblach, Frieda. *The Culture of Wilderness: Agriculture as Colonization in the American West*. University of North Carolina Press, 1996.

Krause, Monika. "The Ruralization of the World." *Public Culture* 25, no. 2 (2013): 233–48.

Kukutai, Tahu, and John Taylor, eds. *Indigenous Data Sovereignty: Toward an Agenda*. Australian National University Press, 2016.

LaDuke, Winona, and Deborah Cowen. "Beyond Wiindigo Infrastructure." *South Atlantic Quarterly* 119, no. 2 (2020): 119 (2): 243–68 https://doi.org/10.1215/00382876-8177747.

Larkin, Brian. "The Politics and Poetics of Infrastructure." *Annual Review of Anthropology* 42 (2013): 327–43.

Larkin, Brian, "Promising Forms: The Political Aesthetics of Infrastructure." In *The Promise of Infrastructure*, edited by Nikhil Anand, Akhil Gupta, and Hannah Appel. Duke University Press, 2018.

Larkin, Brian. *Signal and Noise: Media, Infrastructure, and Urban Culture in Northern Nigeria*. Duke University Press, 2008.

Lefebvre, Henri. *On the Rural: Economy, Sociology, Geography*. University of Minnesota Press, 2022.

Lewis, Jovan Scott. *Scammer's Yard: The Crime of Black Repair in Jamaica*. Duke University Press, 2020.

Li, Tania Murray. *Land's End: Capitalist Relations on an Indigenous Frontier*. Duke University Press, 2014.

Liboiron, Max. *Pollution Is Colonialism*. Duke University Press, 2021.

Malm, Andreas. *Fossil Capital: The Rise of Steam Power and the Roots of Global Warming*. Verso, 2016.

Matthews, Alan. "Farmer Protests and the 2024 European Parliament Elections." *Intereconomics* 59, no. 2 (2024): 83–87.

Mezzadra, Sandro, and Brett Neilson. "On the Multiple Frontiers of Extraction: Excavating Contemporary Capitalism." *Cultural Studies* 31, no. 2–3 (2017): 185–204.

Morton, Erin. "Rural." In *Fueling Culture: 101 Words for Energy and Environment*, edited by Imre Szeman, Jennifer Wenzel, and Patricia Yaeger. Fordham University Press, 2017.

Mukherjee, Rahul. *Radiant Infrastructures: Media, Environment, and Cultures of Uncertainty*. Duke University Press, 2020.

Neel, Phil A. *Hinterland: America's New Landscape of Class and Conflict*. Reaktion Books, 2018.

Newman, Lily Hay. "A New Jailbreak for John Deere Tractors Rides the Right-to-Repair Wave." *Wired*, August 13, 2022. https://www.wired.com/story/john-deere-tractor-jailbreak-defcon-2022/.

Nichols, Robert. *Theft Is Property! Dispossession and Critical Theory*. Duke University Press, 2020.

Nixon, Rob. *Slow Violence and the Environmentalism of the Poor*. Harvard University Press, 2013.

Packer, Jeremy, and S. B. Crofts Wiley. *Communication Matters: Materialist Approaches to Media, Mobility and Networks*. Routledge, 2012.

Padilla, Tanalís. *Rural Resistance in the Land of Zapata: The Jaramillista Movement and the Myth of the Pax Priísta, 1940–1962*. Duke University Press, 2008.

Parks, Lisa, Lindsay Palmer, and Daniel Grinberg. "Media Fieldwork: Critical Reflections on Collaborative ICT Research in Rural Zambia." In *Applied Media Studies*, edited by Kirsten Ostherr. Routledge, 2017.

Parks, Lisa, and Nicole Starosielski, eds. *Signal Traffic: Critical Studies of Media Infrastructures*. University of Illinois Press, 2015.

Parsons, Laurie, and Sabina Lawreniuk. "Geographies of Ruralisation or Ruralities? The Death and Life of a Category." *Dialogues in Human Geography* 12, no. 2 (2022): 204–7.

Peeren, Esther, Hanneke Stuit, Sarah Nuttall, and Pamila Gupta. "Introduction: Conceptualizing Hinterlands." In *Planetary Hinterlands:*

Extraction, Abandonment and Care, edited by Pamila Gupta, Sarah Nuttall, Esther Peeren, and Hanneke Stuit. Palgrave Macmillan, 2024.

Peters, Benjamin. "Afterword: Medium America and the Grounds of a Transnational History of Farm Media." *New Media and Society*, 25, no. 8 (2023): 1960–70.

Rasmussen, Mattias Borg, and Christian Lund. "Reconfiguring Frontier Spaces: The Territorialization of Resource Control." *World Development* 101 (2018): 388–99.

Riofrancos, Thea. *Resource Radicals: From Petro-Nationalism to Post-Extractivism Ecuador*. Duke University Press, 2020.

Riofrancos, Thea. "What Green Costs." *Logic* 9 (2019), special issue, "Nature." https://logicmag.io/nature/what-green-costs/.

Roane, J. T. "Black Ecologies, Subaquatic Life, and the Jim Crow Enclosure of the Tidewater." *Journal of Rural Studies* 94 (2022): 227–38.

Roane, J. T. *Dark Agoras: Insurgent Black Social Life and the Politics of Place*. New York University Press, 2023.

Ruiz, Rafico. *Slow Disturbance: Infrastructural Mediation on the Settler Colonial Resource Frontier*. Duke University Press, 2021.

Sadowski, Jathan. "When Data Is Capital: Datafication, Accumulation, and Extraction." *Journal of Big Data* 6, no. 1 (2019): 1–12.

Sharma, Sarah, and Rianka Singh, eds. *Re-Understanding Media: Feminist Extensions of Marshall McLuhan*. Duke University Press, 2022.

Spice, Anne. "Fighting Invasive Infrastructures: Indigenous Relations against Pipelines." *Environment and Society* 9 (2018): 40–56.

Starosielski, Nicole. *The Undersea Network*. Duke University Press, 2015.

Stoler, Ann Laura. "Imperial Debris: Reflections on Ruin and Ruination." *Cultural Anthropology* 23, no. 2 (2008): 191–219.

Storm, Anna. *Post-Industrial Landscape Scars*. Palgrave, 2016.

Strand, Katherine, and Darin Barney. "Telling Their Stories: Ideology and the Subject of Prairie Agriculture." *Political Ideology in Parties, Policy, and Civil Society*, edited by David Laycock. UBC Press, 2019.

Svampa, Maristella. *Neo-Extractivism in Latin America: Socioenvironmental Conflicts, the Territorial Turn, and New Political Narratives*. Cambridge University Press, 2019.

Szeman, Imre. "On the Politics of Extraction." *Cultural Studies* 31, no. 203 (2017): 440–47.

Szeman, Imre, and Jennifer Wenzel. "What Do We Talk About When We Talk About Extractivism?" *Textual Practice* 35, no. 3 (2021): 505–23.

Tollefson, Hannah. "Staking a Claim: Mineral Mining, Prospecting Logics, and Settler Infrastructures." In "Materials and Media of Infrastructure," edited by Aleksandra Kaminska and Rafico Ruiz, special issue, *Canadian Journal of Communication* 46, no. 2 (2021): 177–99.

Towns, Armond. *On Black Media Philosophy*. University of California Press, 2022.

Tsing, Anna Lowenhaupt. *The Mushroom at the End of the World: On the Possibility of Life in Capitalist Ruins*. Princeton University Press, 2015.

Voyles, Traci Brynn. *Wastelanding: Legacies of Uranium Mining in Navajo Country*. University of Minnesota Press, 2015.

Wang, Xiaowei. *Blockchain Chicken Farm, and Other Stories of Tech in China's Countryside*. Farrar, Straus & Giroux, 2020.

West, Paige. *Dispossession and the Environment: Rhetoric and inequality in Papua New Guinea*. Columbia University Press, 2016.

White, Monica M. *Freedom Farmers: Agricultural Resistance and the Black Freedom Movement*. University of North Carolina Press, 2019.

Williams, Raymond. *The Country and the City*. Oxford University Press, 1973.

Williams, Raymond. *Marxism and Literature*. Oxford University Press, 1978.

Woods, Michael. "Engaging the Global Countryside: Globalization, Hybridity, and the Reconstitution of Rural Place." *Progress in Human Geography* 31, no. 4 (2007): 485–507.

Woods, Michael. "Performing Rurality and Practising Rural Geography." *Progress in Human Geography* 34, no. 6 (2010): 835–46.

Wuthnow, Robert. *The Left Behind: Decline and Rage in Small-Town America*. Princeton University Press, 2019.

Young, Liam Cole. "Innis's Infrastructure: Dirt, Beavers, and Documents in Material Media Theory." *Cultural Politics* 13, no. 2 (2017): 227–49.

Yusoff, Kathryn. *A Billion Black Anthropocenes or None*. University of Minnesota Press, 2018.

Zhang, Hanxiao. "Platformizing the Rural: the New Farmer Project on Douyin." *New Media and Society* (forthcoming).

Zhang, Lin. *The Labor of Reinvention: Entrepreneurship in the New Chinese Digital Economy*. Columbia University Press, 2023.

PART I EXTRACTIVE MEDIATIONS

1

Green Data Capitalism and Its Rural Extractions

MEGAN WIESSNER, ANNE PASEK, NICOLE STAROSIELSKI, AND HUNTER VAUGHAN

Rural landscapes have long been dense and complex sites of technological mediation. Telegraph networks and radio waves reshaped rural agricultural practice. Satellite television has had dramatic effects on rural communities. Today, following in the tradition of rurality conceptualized as a terra nullius for telecommunications expansion, rural spaces are targeted for data center placement and internet cables transit largely through rural landscapes. We argue in this chapter that beyond this historical continuity with other media infrastructures, the expanding infrastructures of digital media are now having a distinct political economic and ecological effect on rural locations. In response to climate change, turns toward sustainability are motivating new forms of digital practice and new forms of extraction and valuation in rural environments, what we term *green data capitalism*.

Challenging conventional framings of rural spaces as homogenous, this chapter unearths green data capitalism's extractive logics at work in two rural data projects in two distinct sectors—agriculture and forestry—where digital incursions and transformations of the rural are being driven by investments in sustainability. In agricultural practice, green data capitalism structures the uneven development of remote, digital, and on-field sensing capacities that enable precision agriculture and verify soil carbon removal. As previous scholars have revealed, in "data-driven agriculture" (from 1990s GPS "precision agriculture" to recent developments in big data sourcing and informatics), external stakeholders and digital platforms often define the needs and monetize the activity of rural communities and farmers.[1] "Climate farming," a similarly evolving set of agricultural practices designed to remove atmospheric carbon dioxide, also harnesses digital platforms in efforts to reshape farming methods. These two developments both rely on the tricky work of data verification. Advocates call for a potentially vast "Internet of Carbon" to quantitatively assess the qualitative benefits of regenerative farming techniques, rendering them legible to state and corporate carbon markets.[2] The infrastructural requirements to extract these data from the field, however, often fracture both scientific methodologies and business models.

Similarly, forest landowners, the forest products industry, and external investors across the United States are looking to digital tools to assess whether timber can be transformed from a quasi-extractive commodity to a marketable store of carbon, or even fulfill both of these roles at the same time. Machine learning and remotely sensed imagery are each becoming an integral part of decisions about how to optimize land use in rural places for carbon *and* for profit, underwriting a new class of productive carbon offsets, though not without their share of contradictions. Across these varied landscapes, we show how digital technology sets out to harness these environments' natural resources to power new and emerging forms of green data capitalism.

In the pages that follow, we first trace the emergence of data capitalism and green capitalism. *Green data capitalism*, a process that brings these two paradigms together, leverages networked systems to commoditize the traces of humans and nonhumans as valuable data for sustainable development. This concept helps explain connections between contemporary initiatives that see digital technologies and data as the necessary foundation of sustainable economic development, those that see the environment as a source of potentially lucrative data, and those that see sustainability as a use case for offering data services. While green data capitalism has spread around the world, harnessing the planetary reach of networked systems, we show that this framing has particular salience in rural environments. Drawing from literature on rurality and political ecology, we conceptualize green data capitalism as a ruralizing force: both a process that helps to produce the rural as a site of potential extraction and a vector through which rurality comes to shape other environments and knowledge systems.

We then turn to our two case studies, in agriculture and forestry, to examine the projects of green data capitalism and their reimagining of rural sites for the sustainable future. Contrary to framings that position these sites as lacking connectivity or as excluded from digital networks, we focus on the ways that they are also becoming new extractive zones for green data capitalism.

Our analysis of data platforms, their ruralizations, and the emergent paradigm of green data capitalism offer three broader interventions into our understanding of media systems and infrastructures. First, while theorizations of data capitalism tend to focus on consumer activity, user behavior, and urban and suburban subjects, our studies reveal how human practices and nonhuman data sources in rural areas are becoming—and arguably have long been—critical sites for the expansion of data capitalism. Second, as green data capitalist projects are pushed forward, they serve as vectors of ruralization, porting rural imaginaries, territories, and epistemologies into other spatial formations. And third, by foregrounding rural media systems and infrastructures, we show how the rural is a key locus of the imagination of a digital future, thus further building on this collection's aim to upend the underlying metronormativity of humanities-based critical scholarship.

The Emergence of Green Data Capitalism

As networked and digital media have become imbricated in everyday life, they have served as both products and vectors of capitalism. Scholars and critics have advanced different conceptual paradigms for this manifestation of contemporary capitalism, including informational capitalism, digital capitalism, and platform capitalism.[3] Across paradigms, data are instrumental; they are an important means of operation and accumulation. As Nick Srnicek writes, "In the twenty-first century advanced capitalism came to be centered upon extracting and using a particular kind of raw material: data."[4] While data are the material extracted, Srnicek describes, "the *activities* of users [are] the natural source of this raw material."[5] Christian Fuchs, in an analysis of corporate social media, offers a similar description of data capitalism, arguing that it is "based on the exploitation of the unpaid labour of Internet users and on the commodification of user-generated data and data about user behavior that is sold as commodity to advertisers."[6]

To historicize this paradigm, Sarah Myers West identifies data capitalism as a system and commercial logic that arose in the mid-1990s to early 2000s in which the commodification of data enabled the asymmetrical redistribution of power toward actors with the capacity to parse, analyze, and otherwise make sense of those data.[7] This logic of data capitalism "appeals to community and consumer power to mask the digital labor it relies on."[8] West's focus, among others studying similar dynamics, has largely been on the extraction of data from users' online activities. However, data capitalism not only extracts data from the digital traces of consumer behavior but also extracts data from and about a multitude of industrial operations and from the proliferating interfaces between technological systems and the nonhuman world.

A second paradigm of recent capitalist development, emerging in relation to environmentalist movements and environmental risks, is *green capitalism*. This paradigm, also branded as *ecocapitalism*, positions the consumption of green products and the development of green technologies as solutions to ecological problems. In the influential text *Natural Capitalism*, Paul Hawken, Amory Lovins, and Hunter Lovins argue that the industrial revolution and modern capitalism have led to the decline of "natural capital."[9] *Natural capitalism* is their name for an industrial system that regenerates this natural capital, in the process restoring human prosperity. In this vein, proponents of green capitalism insist that pricing and commodifying environmental functions is necessary "to factor the value of nature into the way markets operate to encourage producers to become more efficient and innovative in the way they use natural resources."[10]

The critiques of this paradigm are many. Some scholars have characterized green capitalist efforts as diversionary, acting as a kind of social siphon that channels innovators' creativity and engagement into profit, only to end up reproducing the status quo.[11] Others are doubtful that "capitalogenic" environmental problems like climate change will be meaningfully reversed through the profit-seeking economic model that caused them.[12] In this vein, Adrienne Buller, in *The Value of a Whale*, argues that green capitalism offers climate solutions and a decarbonized future that minimally disrupt existing distributions of wealth and power. Others in and beyond academia question the contradictory character of infinite economic growth on a finite planet, as in the expanding literature on degrowth.[13] Naomi Klein, in *This Changes Everything*, famously asks if it is possible to address climate change "without challenging the fundamental logic of deregulated capitalism." "Not a chance" is her reply.[14]

Neither data capitalism nor green capitalism is entirely new, and scholars have pointed out the long histories of both. For example, West points out the use of "political arithmetic" in England in the seventeenth century, the use of censuses by the Dutch East India Company, and the development of surveillance networks by commercial credit-reporting agencies in the nineteenth century. Hartmut Berghoff similarly demonstrates how "islands of green capitalism" developed in select industrial firms throughout the nineteenth and twentieth centuries.[15] What is distinctive and new, and what we focus on here, is the convergence of data capitalism and green capitalism as an increasingly powerful form governing life.

Green data capitalism, then, describes the ways that the traces left by humans and nonhumans in interactions with networked systems become valued data for green enterprise. It describes the converging tendencies of both data capitalism and green capitalism to create new markets and new kinds of accumulation. As digital technology advances and the effects of climate change become more dire, new spheres of human and nonhuman activity become the subject of data capitalist extraction in the name of sustainability.

In some cases, the extraction of user data from interactions with platforms meets the parallel goals of increasing profits and sustainability. As one example of this, take smart thermostats, which claim to learn from user input, generate an optimal indoor thermal environment, and increase energy efficiency. Environmental issues are addressed here through a new technology. At the same time, however, smart thermostat companies draw on user data to maximize profits by optimizing energy use, developing partnerships with utility companies, and providing data to third parties. The surplus value generated from user data is legitimated through green ideology, in which users are encouraged to adopt smart thermostats to save the environment.

As Maria I. Espinoza and Melissa Aronczyk show in their analysis of "Big Data for Climate Action"—an initiative in which companies share proprietary data with governmental and nongovernmental organizations invested in sustainability—green data side projects can often be "a strategy to legitimate extractive, profit-oriented data practices by companies."[16] Data philanthropy projects, they argue, often rely on a framing of social or ecological problems as problems of an information deficit, which companies can solve. This serves to legitimate infrastructures for the collection and processing of data, even as the direct climate benefits of these data are uncertain.

Nor is it clear if there will ever be enough data. The earth's system is unfathomably complex, consisting of biogeochemical cycles at multiple scales and rhythms and encompassing the metabolism and activity of every living being on earth. Data adequate to the task of modeling these systems could be sought forever. But in the meantime, what data do exist facilitate pricing, packaging, and trading *ecosystem services*—another version of green data capitalism. This term names those benefits like clean air and water, carbon sequestration, and nutrient cycling, which environments furnish for collective life, and which, in a green capitalist framework, can be better protected by being incorporated into markets. In framing environmental crises as problems of market valuation and price signals, the proponents of green capitalist approaches orient toward an endlessly receding horizon of perfect information. As ecosystems are turned into financial assets, more complex models and new datasets in turn become lucrative goods and services.

Finally, green data capitalism also takes form in the development of data infrastructures. As recent scholarship on the data center industry has shown, environments are often leveraged as a selling point to attract internet and data storage infrastructure to various regions.[17] In their study of server farms in the United States' Pacific Northwest, Anthony Levenda and Dillon Mahmoudi argue that "ecological metaphors paint data centers as a natural evolution in resource economies while maintaining a veneer of environmentally sustainable development."[18] And in their critical study of the postextractive futures of Irish peat boglands, Patrick Bresnihan and Patrick Brodie demonstrate that the "environmental strategies of big tech companies in Ireland are materially entwined within an expanding global supply chain of green capitalism."[19] In other words, the expansion of the infrastructures supporting data capitalism are justified in some parts of the world as a green capitalist development.

While the *what* of green data capitalism is constituted by intersecting promises of profit and precision in environmental management, the *where* of green data capitalism is less obvious. As we argue below, these dynamics are significantly shaped by urban–rural relations. This includes the pursuit of a spatial fix for the overaccumulation of capital, especially among tech giants struggling to maintain extreme rates of growth in saturated urban markets; these giants are now turning to rural use cases in pursuit of new forms of accumulation and capital investment.[20] In the process, green data capitalism also proposes a spatial fix for carbon, seeking to resolve the geophysical crisis of overaccumulated greenhouse gases through the transformation of rural spaces into carbon sinks that license further urban and industrial emissions elsewhere. Uneven geographies also facilitate the process of what Anna Tsing calls *salvage accumulation*, which goes hand in hand with the ruralization of data sensing and inference tools.[21] Green data capitalism, we demonstrate, tends to be built from the infrastructures and labors of noncapitalist commons, whether these are public datasets or more elemental configurations of air, energy, and soil. Rurality constitutes both the reason for these goods' persistence outside of dominant capitalist relations and the existing power relations by which they can be incorporated into systems of profit accumulation for an urban core.

Green Data Capitalism as Ruralizing Force

Green data capitalism is increasingly dependent on rural landscapes and processes as objects of value and sites of extraction. In turn, this has significant effects on the constitution of rural landscapes. While many have tracked digital transitions in relation to urbanization, few have looked at how digital and data capitalism are enmeshed in processes of ruralization and a new valuation of rural landscapes.

Here we follow Gillen et al. in moving beyond a static conception of the rural as inert countryside, considering "ruralization" as a process and "the rural–urban binary not as a set of oppositional categories but a relationality."[22] As they point out, this is not to define what is rural or urban "but rather to recognize that both drive transformations that are conventionally cast merely as urbanization."[23] Thinking about the production of ruralities creates space for thinking through the diversity of rural places and changes to rural life not subsumed under the rubric of urbanization while still thinking through rurality as a product of spatially uneven power relations.

Green data capitalist processes are also dependent on rural imaginaries. As Gillen et al. note, "Rural imaginings feed idealized futures that drive action."[24] The imagination of nature, of rural communities in need of employment, and of resource-rich rural zones fuels not only infrastructure development but also big data projects targeting rural zones that rely on legislative benefits and zoning permits afforded uniquely to these spaces. As Jenna Burrell observes, "digital infrastructures are built into and upon what is physically left behind of previous regimes of industrial capitalism," often in rural communities after previous investors have moved on.[25] Rural landscapes have long been mined, logged, and industrially modified for plant and animal agriculture. Green data capitalism continues to extract financial value from these landscapes, transforming them in new ways even as it relies on imaginaries of what they once were or might someday become again. What Raymond Williams notably investigates as a key cultural process of coproduction, whereby representations of the rural (or country) acted to define and ultimately reify city world views and modes of living, now operates in updated form across the emerging narratives offered by green data capitalism.[26]

At the same time, the capitalization of rural ecosystem services justifies the continuous expansion of data infrastructures into new environments, thus extending forms of surveillance that we might associate with the urban or with online spaces into remote rural areas. The apparent inevitability of expanding infrastructures for data collection and processing reduces the autonomy of rural communities and reproduces them as infrastructural elsewheres. In their genealogy of resource extraction and energy politics in the Columbia River Basin, for instance, Levenda and Mahmoudhi reveal how the operational impacts of data-driven societies are borne by rural places, whether as source of grid power, raw mined materials, or land for infrastructural development.[27] Green data capitalism's manifestations are not unique to rural places, but ruralized environments are most often home to the "stocks" of natural capital which have been newly discovered as financial assets, the source of legitimating narratives, and the places called on to subtend the infrastructural costs of a more generalized datafication of life and sociality. As we argue, green data capitalism offers a thin veil of environmental solutionism and a computational reproduction of the urban–rural duality.

FarmBeats, Spectral Surplus, and the Datafication of Soil

> A big reason agtech is not taking off is a lack of data. . . . If you don't have data, how can you drive insights? . . . If anyone can really solve for that, that could be a big breakthrough and drive disruption in the sector.
>
> —Ranveer Chandra, "Empowering Farmers with Affordable Digital Agriculture Solutions" (2021)

Ranveer Chandra's statement is not idle speculation; it's business strategy, shared with a bit of coy self-referentiality at an ag investment conference.[28] As the managing director of research for industry at Microsoft, Chandra's job is to find and develop such breakthroughs, either by acquiring young start-ups or by leading internal research and development efforts. He aims to solve agtech's data deficit and position Microsoft as an essential link in an increasingly digitized sector. Yet a paucity of data (and, more importantly, data infrastructures) characterizes the very landscapes agtech seeks to manage. US farms struggle with connectivity; roughly a third of farmers have poor or no internet service.[29] The marginal character of rural network developments thus stands in the way of a green data gold rush.

This poses an uncharacteristic challenge for the tech sector. While much of its essential early R&D and network buildout were heavily subsidized by public actors, new frontiers of digital business require the renovation or reinvention of enduringly marginalized rural spaces within network deployments.[30] This is expensive work; connecting a single farm to a fiber backbone can cost tens of thousands to hundreds of thousands of dollars. As such, this scale of infrastructural investment remains out of reach for most farmers, unattractive to private ISPs, and poorly considered in a patchwork of public policy efforts.[31] When it comes to telecommunications, there's not much of a free lunch to be found on the farm.

Yet to many policy and agtech actors, the future of farming remains stubbornly data intensive. Through expanded digital mapping and sensing technologies, precision agriculture paints a vision of croplands that can be virtually disaggregated into a plurality of microclimates, each with uniquely optimized management parameters. This promises green dividends: pesticides and fertilizers could be reduced wherever they are not needed or targeted to problem areas captured by optical sensors on a tractor. Advanced farm equipment could be automated to deliver this personalized care quickly and responsively, optimizing yield and minimizing costs. Although the significant capital investments such systems would require points to large agribusiness players as the most likely adopters (and thus to a future of increased consolidation), it is small holder farmers, often in the Global South, that are repeatedly featured in public presentations.[32] The potential efficiencies of precision agriculture are in turn interpreted as a means to feed the world's anticipated population boom without worsening ecological land use conflicts.[33] The language of "democratizing" sustainable technology abounds.[34]

The network deficiencies of rural spaces thus act as a dam to such disruptions. Chandra and his team are diligently trying to wrench this dam open. Their efforts, underway since 2014, go by the name FarmBeats. Now out in beta, it provides a "full stack" solution to digital agriculture. This includes the usual trappings of agtech systems, but with an infrastructural twist. Beneath the surveying drones, Internet of Things sensors, and machine learning is a humbler and more essential component that aims to solve the network deficits of rural spaces. It is a different kind of Wi-Fi: by tapping into UHF and VHF radio frequencies—the same used for TV broadcasts—extra bandwidth can be conjured out of unused spaces in the rural electromagnetic spectrum. This allows for local communications between smart equipment and on-site digital storage. Microsoft, in essence, had to invent a new kind of network to make data collection and processing possible on the farm.

Rurality, in this regard, is both the challenge and the means by which it might be overcome. In more populated spaces, TV spectrum rights can sell for millions of dollars at auction; they are often used by mobile providers to eke out additional bandwidth for subscribers in crowded urban markets. On farmlands, however, this space is unlicensed, unused, and essentially free for the taking.[35] All that is required is a regular TV antenna and a cheap transceiver. When attached to a farm shed where Microsoft places a base station and cloud edge node, the setup can extend signal connectivity for up to eight kilometers. Data from aerial and terrestrial sensors can then be computed, compiled, and packaged into smaller modular components ready to make their comparatively slow (or "leisurely," per Chandra) transit back and forth between the farm and Microsoft's distant data centers. While the farm may still lack a high-speed connection to digital elsewheres, thanks to salvaged spectrum, it might become a rich island of network capacity. Green data capitalism, in this agtech formation, thus creates the infrastructural preconditions for future value by organizing and incorporating the "latent commons" of air and energy.[36]

Linking local data to Microsoft's distant data centers allows for analytic products to be prepared and for third-party agtech companies to render and sell services and insights. The FarmBeats system has been open sourced to allow for such developments, though the company's cloud platform remains the mandatory infrastructural base of any related experiments. The role of off-site computational work also reduces the need for quite so many Internet of Things sensors on the farm itself; machine learning inferences happening in the cloud can help compensate for missing or scant data in the field.[37]

The FarmBeats system data and capital are thus articulated through multiple conjunctures. First, for the farmer, data are imagined to save or make money, reducing crop damages and expensive inputs while optimizing harvest times and yields. Second, to Microsoft, data brings in rent, attracting new enterprise tenants to its data centers, where the company acts as a platform monopoly and standard setter that connects a potentially vast ecosystem of agtech partners. A third, more speculative but potentially lucrative possibility lies in a more explicit turn to green data capitalism: using FarmBeats to monitor and commodify carbon offsets in agricultural fields.

Carbon offsets are a particularly exemplary value form under green data capitalism. They are not so much created by market producers as they are discursively performed by systems of technical measurement and governance. The result is "a complicated, contradictory, and only provisionally stabilized commodity form" that exists more as a data registry than as a concrete asset.[38] While early systems of producing offsets have been dogged by fraud, new efforts have arisen to define and establish a more rigorous trade in "carbon removal credits"—verified sequestrations of atmospheric carbon dioxide into durable sinks.[39] The market for removal credits is booming, with supply outstripped by the demand created by net-zero corporate commitments.

In this context, the potential for agricultural carbon sequestration has been both highly promoted and warily assessed. Practices like cover cropping, no-till, and adaptive multipaddock grazing can increase the amount of organic carbon stored in soils within a few years. The fastest way to scale the carbon removal market, therefore, may lie in the recruitment of farmers into new land management regimes and the expansion of sensors and network capacities into rural spaces.[40] However, farm soils are also dynamic and highly heterogenous sets of environments with sequestration capacities that are unevenly distributed across regional climates and land use histories, and that can always be undone by future decisions (such as by plowing a field or selling farmland for suburban development). To be credible, soil carbon sequestration thus requires rigorous and decades-long verification. Current players in carbon removal markets cannot provide this certainty. (Nori, one notable agricultural carbon removal platform, only guarantees that carbon removal work is maintained for ten years.) As a result, large market players are hesitant to put all their investments into a single soil carbon basket. Microsoft, for instance, has purchased soil carbon removal certificates since 2021. However, understanding the uncertainties that currently characterize this market, it limits the share of these kinds of investments in its portfolio and also internally discounts the impact of these removals by 10 percent.[41]

In this respect, Microsoft approaches the articulation of data and capital from multiple market perspectives, seeking to consume and produce environmental commodities made from digitized soil. Early work has begun on predictive algorithms to help farmers assess the potential stocks of carbon removal they might generate on their FarmBeats fields if they adopted regenerative practices. It is eventually hoped that these predictions might one day become pledged commodities, with carbon removal marketplaces joining the company's cloud ecosystem of FarmBeats partners, and with FarmBeats providing ongoing verification data. As such, in addition to extracting rent from both farmers and the companies administering carbon removal trading, Microsoft also stands to secure a wider and more secure supply of carbon removals for itself, solving the problem of its internal discounting and fast-arriving climate targets. In this way, the ruralization of green data capitalism finds new expressions and directions on both sides of the ledger.

Natural Capital Exchange, Automated Forest Inventory, and New Carbon Commodities

Microsoft's interests in both buying carbon and managing the infrastructure behind green data also explain its investment in Natural Capital Exchange (NCX), a US precision forestry start-up. NCX exemplifies the green data capitalist promise that the climate and the market can be reconciled through the mediations of data science. It also provides a recent and prominent example of the failure of this promise.

NCX began as SilviaTerra in 2010, a project of two graduates of the Yale School of Forestry. In their first decade, the cofounders developed several forestry tech and data products, such as an app where foresters could use smartphones to enter georeferenced data on species composition, age, and tree size to help automate inventory analysis. (Like other data capitalist projects, SilviaTerra at first gave away their tool for free in exchange for the user-generated data.[42]) In 2018, SilviaTerra was selected by Microsoft as an "AI for Earth" grantee, receiving Azure cloud computing support to explore the possibility of using manually collected inventory data to calibrate statistical predictions from remote sensing imagery.[43] Using decades of records from the US Department of Agriculture's Forest Inventory Analysis program, they created a proprietary system for using public satellite imagery to generate a simulated inventory of trees for any thirty-by-thirty-meter plot in the contiguous United States.[44] This synthetic inventory would be combined with other data like prices, regional harvest trends, road access, and distance to sawmills, establishing baseline harvest levels and estimating the carbon at risk of harvest on any given property. If payments could be made to landowners to avoid or defer these harvests, offsets could be sold. By connecting small rural landowners across the United States with large corporate entities looking to purchase carbon credits, the company aimed to be the main broker for an enormous but otherwise inaccessible supply of carbon credits on private lands.

SilviaTerra rebranded as Natural Capital Exchange and raised $50 million in Series B funding to pursue this business model. To generate the credits for their first few cycles, they held an open auction, in which landowners who averred that they were considering harvest could bid for the lowest they would be willing to accept in order to defer harvest for one year. Accepted landowners were paid on the basis of the estimated volume of avoided harvest, and NCX bundled these sales into carbon credits that could then be sold to corporate buyers. Although they initially focused on industrial lands where assessing baselines is fairly straightforward, they soon shifted to small rural landowners, a vast and diverse pool of people who might otherwise find the legal hurdles and time commitments involved in entering the traditional offset market too steep. The automation of inventory analysis and monitoring, NCX hoped, would make them the first to unlock this new supply of carbon credits.

Its leadership prominently and publicly bet that data and machine learning tools could offer a corrective to the well-documented flaws of other forest carbon projects.[45] Forest carbon offsets are notoriously open to abuse and difficult to quantitatively evaluate or measure, and this has made them frequent targets of critics.[46] NCX's focus on short-term harvest deferral and its ton-year accounting method have not eliminated controversy, however.[47] The calculated value of carbon depends on choices of discount rates and assumptions about how climate change will unfold in the coming decades. Even among proponents of offsets, there are substantial differences of opinion having to do with normative but highly abstract decisions about how to properly incentivize climate action through markets. In 2022, carbon offset verifier Verra—itself plagued by criticism and sensitive to new reputational risks—unexpectedly decided not to verify NCX's methodology, stranding it with obligations to landowners and unverified, worthless credits.[48]

While there is insufficient space here to summarize the debates around short-term harvest deferrals or ton-year accounting, both depend on data to facilitate perpetual monitoring and iterative adjustment. The collapse of longer temporal horizons to focus on short-term, data-driven reactivity can be seen everywhere from governance, finance, and urban planning to personal health and consumer devices.[49] Increasingly, NCX, Microsoft and others hope to bring this mentality to ecological management.[50] In place of structural changes, systems optimize between shifting parameters, even as changes of methodology from cycle to cycle make rigorous evaluation of impacts difficult.[51] It is not difficult to see why green capitalism embraces the logic of data optimization as it tries to endlessly square the circle of ecological health with profit. The elusive quest to simultaneously maximize these two radically disconnected goals feeds the search for more and better data, implicitly justifying the creation of systems and platforms to process and generate those data.

Revealingly, the more precisely complex social and ecological factors are quantified and tracked, the less secure the accord between ecological integrity and profitability appears. This can be seen in the quandary NCX created for itself over additionality. Additionality is the guarantee—crucial to the legitimacy of all carbon offsets—that payment corresponds to a scenario that would not have happened otherwise. Even with terabytes of satellite data, accurately assessing additionality is difficult on small family lands with informal management plans. Compared with predictably managed industrial plantations, in which rotation lengths are standardized and harvest decisions are tightly correlated with markets, smaller owners often make decisions about when to harvest and sell logs for far more arbitrary reasons: family deaths, medical bills, property taxes, college tuition. This kind of uncertainty makes it difficult to compute trustworthy baselines; compensating for uncertainty can reduce payments to negligible amounts. Furthermore, economically vulnerable sellers considering harvest because they need cash—those most likely to offer real guarantees of additionality—are likely to bid too high to participate or not be interested at all. In a market logic, a lower bid means harvest income is less urgent to the landowner and that the carbon is less "at risk" by definition. Low bids are more likely to be accepted but less likely to be truly additional, meaning that NCX's bundled credits risk representing nonadditional carbon.

The ability to process a wealth of data has not demonstrated a way to avoid these problems, helped NCX obtain third-party verification, improved the wider credibility of carbon markets, or resolved the conceptual slipperiness of selling a probabilistic scenario. But that does not stop this all from having an impact on rural places. Thanks in part to their outsize visibility in the press—from *Scientific American* to John Oliver's *Last Week Tonight*—thousands upon thousands of landowners are now touched by these projects, even if their only engagement with them is to consider carbon offsets alongside other revenue streams. Though NCX itself is scrambling for a new business model after Verra's decision, their approach has laid the groundwork for a new layer of accumulation in rural areas of the United States. As an exemplary case of green data capitalism, NCX uses forest and carbon data to work toward a future in which value can be accumulated from forests regardless of whether forests are logged for timber, managed for carbon sequestration, or both.

Green Data Capitalism and Rural Futures

Green data capitalist projects are full of such environmental inconsistencies, although the extent to which these contradictions can be deferred remains to be seen. The cases of NCX and FarmBeats suggest that in lieu of working these contradictions out, companies turn to the language of democratization, situating themselves as beneficent purveyors of new infrastructures of rural empowerment. In both instances, the promise of extending participation to rural businesspeople through data and connectivity is a central drive of company marketing. FarmBeats is engineered as a solution to the supposed network deficits of rural spaces as well as the supply constraints of net-zero corporate commitments. Yet contrary to the project's public rhetoric, prospects for democratizing public engagement with climate data remain subordinate to the digital rent- and ecological asset-seeking imperatives of the cloud company. Meanwhile, the promise of offering small forest landowners access to carbon markets nicely inverts NCX's actual business case, which is offering corporate and institutional purchasers access to a new supply of carbon credits. Language such as "democratizing access to forest carbon markets" and "empowering every landowner to participate" reframe exposure to volatile new markets as a public good.[52] Small payments for single-year harvest deferrals are unlikely to provide financial autonomy for rural landowners, but new sources of data about forest carbon are useful for managers of large portfolios of timberland. NCX's website prominently features case studies of rural families who received payments, but with that model now in question, there is little to stop the company from pivoting back to working with industrial forestry or working with investment management organizations owned by distant shareholders, allowing them to extract new profits from decisions to incrementally extend their harvest rotations.

This image of the newly empowered rural decision-maker obscures the appropriation, financialization, and reproduction of marginality at the heart of green data capitalism. Processes of salvage accumulation characterize key points in the production process of green data. Whether in Microsoft's appropriation of TV whitespaces or NCX's attempt to find value in nonindustrial lands and its use of old public datasets, both companies gain access to assets without needing to meaningfully create or sustain them. As Tsing points out, this dynamic is germane to all capitalist systems, which rely on the varied work of middlemen, translators, and other forms of mediation that bring salvaged resources—often from rural hinterlands where intensification of production is less extreme—into the circuits of capitalist production.[53] Within green data capitalism, computational platforms and processes act as automated versions of these middlemen, assembling free data and latent elemental infrastructures into rent-extracting platforms and digital commodities.

Financialization is also central to green data capitalism, with digital commodities fluctuating in value and circulating in exchanges far detached from the conditions on the ground. In the pursuit of financialized solutions to ecological and capital overaccumulation, data are tenuously harnessed to articulate new commodity forms out of rural spaces and temporalities. Through green data production and manipulation, Microsoft seeks to open new markets. In turn, it has had to reconfigure the logics and technologies of cloud computing to produce workable configurations in rural spaces. NCX, similarly, has sought to increase the value of existing rural data through new algorithmic means, generating predictions about tree species and human behaviors and rendering these into green data commodities.

In so doing, both companies work to make rural spaces newly available as a cheap carbon resource for largely urban actors, a resource that is in turn used to cancel out accumulations and emissions beyond the farm and forest. These cases, then, reveal the enduring and constitutive role of urban–rural relations, demonstrating how financial and infrastructural pressures in rural environments are not intrinsic to them but rather are formed through mounting urban demands on rural spaces. This follows the pattern of David Harvey's spatial fix for capital accumulation, but with a green twist.[54] Green data capitalism proposes to find both new sources of data for storage, analysis, and financialization (fueling the relentless growth of the cloud sector in the process[55]) and new frontiers in environmental goods and compensatory commodities to offset economic actions in varied elsewheres. In this way, rural spaces are approached as a site of urban claims-making, necessarily held apart from and imagined in service to the capital-intensive spaces that finance green data capitalist developments.

Rural realities and expertise shape the development of these systems, however, and to challenge the metronormativity of academic scholarship, we need to shift our critical focus accordingly. Although it has set out to automate inventory analysis rather than rely on embodied forestry labor and knowledge, for example, NCX is only trustworthy as an offset broker to the extent that its methodology respects foresters, landowners, and project developers' understanding of the salient factors in forest economics. As trained foresters themselves, NCX's cofounders know that their model must reflect everything from merchantable timber recovery volumes for different species to log prices and mill locations. The datasets, rules of thumb, and social practices of quantifying these factors in their models are built on decades and centuries of life and work in forest economies.

Microsoft's efforts to build new standards for soil carbon sequestration face a similar and potentially endless set of debates about their applicability to varied growing conditions and compatibility with equally varied farmer goals. The use of machine learning to compensate for a lack of in-field sensor data or qualitative assessments by the farmers themselves is likely to deepen debates rather than resolve them. With every new source of data comes a new potential site for disagreement about methodology and generalizability, underscoring rather than eliminating the importance of situated judgment. In automating decisions and articulating landscape's value in terms of its potential for carbon sequestration, these companies both depend on and devalue types of work that have often defined North American rural spaces since European settlement, creating an instability in who is authorized to define the value of the rural.[56]

These complexities trouble the company narratives of offering rural communities access to green data tools and markets. This is not to uncritically celebrate local and embodied practices or to insist that they will somehow inevitably thwart attempts to capitalize on activities in rural environments. But it is to insist that the increasingly hegemonic logic of green data capitalism sits uneasily in relation to many existing rural practices and values. Rural problems like an aging workforce, cash flow, network deficits, or uncertainty about forests and crops in a rapidly changing climate can be imagined as problems that green data can fix and as opportunities for green data to expand to new markets. However, this understanding of rural places as sites of lack or of rural communities as universally desiring integration into emerging markets ignores the forms of diversity and conflict that have always characterized the rural. In the face of increasingly autonomous workflows, people in these places may look to new stories and vocabularies to articulate the value of their experiences, knowledge, and labor.[57] For some landowners or workers, this might look like embracing green data capitalism and attempting to ride the waves of climate finance on their own terms. For others, it might look like rejecting these sorts of techniques and logics—a development that itself could take many different forms. Green data capitalism, therefore, is entangled with ongoing struggles to mediate the practice and meaning of rurality.

On this note, looking ahead to these terrains of struggle while also looking back on the scope of this chapter, one may be tempted to ask: what about the farmers and foresters? In this chapter, we have largely not detailed how the individuals, communities, or business entities construed as the imagined users of these systems have adopted, rejected, benefited from, or been harmed by them. Our goal instead has been to explore the logics of accumulation and mediation that drive these financial and technical experiments—logics that may be differently legible on the ground. As early experiments and evolving platforms, FarmBeats and NCX's engagement with rural landholders and workers have yet to scale and solidify into an archive of researchable data. Community-based research would be an excellent ongoing and future trajectory to expand understanding of these platforms and markets in rural—and ruralized—areas.

However, we leave the reader with a final cautionary note: Looking to the category of the farmer as a likely and legible site of resistance can do analytic harms even as it seeks productive political ground. Untangling the variety of relationships that different players might have to these rural green data initiatives is a valuable project; as these and allied initiatives take off, it is an increasingly necessary one. Nevertheless, we hold in mind that the persona of the farmer collapses different groups of people and workers: affluent landowners, independent or struggling smallholders, diverse employees of corporate agribusiness and timber companies. What FarmBeats affords varies significantly between a boutique organic farm owned by a former tech worker and a commodity sharecropping arrangement staffed by migrant farmworkers. Whether NCX's carbon markets are worth engaging with might be answered differently for a New England banker who has recently purchased timberland as a retirement project and is interested in green finance than for a fourth-generation landowner in the Deep South. So far, these green data capitalist systems have most often found their ready test subjects in the former groups rather than the latter. The encroachment of these new forms of capitalism, therefore, should not be exclusively understood as entrenching fixed inequalities between the urban and the rural. Green data capitalism will also likely lead to intrarural transitions and new cultural conflicts over lifeways and land ethics as well as new forms of finance and debt.

Notes

1 See Fairbairn and Kish, "Poverty of Data"; Ali, *Farm Fresh Broadband.*
2 Buck, "Best-Case Scenario."
3 For informational, digital, and platform capitalism, respectively, see Morris-Suzuki, "Capitalism"; Schiller, *Digital Capitalism*; Srnicek, *Platform Capitalism.*
4 Srnicek, *Platform Capitalism*, 23.
5 Srnicek, *Platform Capitalism*, 23.
6 Fuchs, *Social Media*, 122.
7 West, "Data Capitalism," 20.
8 West, "Data Capitalism," 21.
9 Hawken et al., *Natural Capitalism.*
10 Scales, "Green Capitalism," 1.
11 Goldstein, *Planetary Improvement*, 3.
12 Moore, "Capitalocene," 4.
13 Schmelzer et al., *Future.*
14 Klein, *This Changes Everything*, 31.
15 Berghoff, "Shades of Green," 19.
16 Espinoza and Aronczyk, "Big Data for Climate Action," 1.
17 Johnson, "Emplacing Data."
18 Levenda and Mahmoudi, "Silicon Forest," 1.
19 Bresnihan and Brodie, "Data Sinks," 378.
20 Harvey, "Globalization."
21 Tsing, *Mushroom.*
22 Gillen et al., "Geographies," 198.
23 Gillen et al., "Geographies," 194.
24 Gillen et al., "Geographies," 195.
25 Burrell, "On Half-Built Assemblages," 302.
26 Williams, *Country.*
27 Levenda and Mahmoudi, "Silicon Forest."
28 Burwood-Taylor, "Where Will Agtech Exits Come From Next?"
29 Mintert and Langemeier, "Ag Economy Barometer."

30 Greene, "Landlords."
31 Ali, *Farm Fresh Broadband*.
32 Chandra, "Empowering Farmers."
33 Gates, "Can the Wi-Fi Chip."
34 Chandra et al., "Democratizing Data-Driven Agriculture."
35 *Economist*, "TV Dinners."
36 Tsing, *Mushroom*, 135.
37 Chandra et al., "Democratizing Data-Driven Agriculture."
38 Huff, "Frictitious Commodities," 20.
39 Lovell and MacKenzie, "Accounting for Carbon."
40 Microsoft Corporation, "Microsoft Carbon Removal: Lessons" and "Microsoft Carbon Removal: Update."
41 Microsoft Corporation, "AI for Earth."
42 Myhrvold, "App Craze."
43 Reuben, "Foresters Now Monitoring"; see also Lin, "How to Make a Forest."
44 See Microsoft Corporation, "AI for Earth"; Parisa and Nova, "This AI Can See."
45 See Badgely et al., "Systematic Over-Crediting"; Murray et al., "Estimating Leakage."
46 See Cavanagh and Benjaminsen, "Virtual Nature"; Frewer, "What Exactly Do REDD+ Projects Produce?"; Gifford, "You Can't Value"; van Kooten et al., "Forest Carbon Offsets."
47 Cullenward et al., "Critique."
48 Jenkins, "Response."
49 Schüll, "Data for Life"; Ash, *Interface Envelope*; Halpern and Günel, "Demoing Unto Death."
50 Lukacz, "Dashboard Approach."
51 Halpern et al., "Smartness Mandate."
52 NCX, "NCX's Natural Capital Exchange."
53 Tsing, *Mushroom*, 66.
54 Harvey, "Globalization."
55 Hogan, "Data Center Industrial Complex."
56 Barney, "Autonomous Agriculture?"
57 Murray, "Agriculture Wars."

Bibliography

Ali, Christopher. *Farm Fresh Broadband: The Politics of Rural Connectivity*. MIT Press, 2021.

Ash, James. *The Interface Envelope: Gaming, Technology, Power*. Bloomsbury Academic, 2015.

Badgley, Grayson, Jeremy Freeman, Joseph J. Hamman, et al. "Systematic Over-Crediting in California's Forest Carbon Offsets Program." *Global Change Biology* 28, no. 4 (2022): 1433–45.

Barney, Darin. "Autonomous Agriculture?" *Heliotrope*, February 8, 2023. https://www.heliotropejournal.net/helio/autonomous-agriculture.

Berghoff, Hartmut. "Shades of Green: A Business-History Perspective on Eco-Capitalism." In *Green Capitalism? Business and the Environment in the Twentieth Century*, edited by Harmut Berghoff and Adam Rome. University of Pennsylvania Press, 2017.

Bresnihan, Patrick, and Patrick Brodie. "Data Sinks, Carbon Services: Waste, Storage, and Energy Cultures on Ireland's Peat Bogs." *New Media and Society* 25, no. 2 (2023): 361–83.

Buck, Holly Jean. "A Best-Case Scenario for Putting Carbon Back Underground." In "Geoengineering," special issue, *Science for the People*, Summer 2018. https://magazine.scienceforthepeople.org/geoengineering/best-case-scenario-carbon-underground/.

Buller, Adrienne. *The Value of a Whale*. Manchester University Press, 2022.

Burrell, Jenna. "On Half-Built Assemblages: Waiting for a Data Center in Prineville, Oregon." *Engaging Science, Technology, and Society*, vol. 6 (2020): 283–305. https://doi.org/10.17351/ests2020.447.

Burwood-Taylor, Louisa. "Where Will Agtech Exits Come From Next? 3 Corporates Weigh In." *AgFunderNews*, November 8, 2022. https://agfundernews.com/where-will-agtech-exits-come-from-next-3-corporates-weigh-in.

Cavanagh, Connor, and Tor A. Benjaminsen. "Virtual Nature, Violent Accumulation: The 'Spectacular Failure' of Carbon Offsetting at a Ugandan National Park." *Geoforum* 56 (2014): 55–65. https://doi.org/10.1016/j.geoforum.2014.06.013.

Chandra, Ranveer. "Empowering Farmers with Affordable Digital Agriculture Solutions." Paper presented at the Just Infrastructures Speaker Series, University of Illinois at Urbana–Champaign, April 2, 2021. https://www.youtube.com/watch?v=Ro7T3xkeSqo.

Chandra, Ranveer, Manohar Swaminathan, Tusher Chakraborty, et al. "Democratizing Data-Driven Agriculture Using Affordable Hardware." *IEEE Micro* 42, no. 1 (2022): 69–77. https://www.microsoft.com/en-us/research/publication/democratizing-data-driven-agriculture-using-affordable-hardware/.

Cullenward, Danny, Freya Chay, and Grayson Badgley. "A Critique of NCX's Carbon Accounting Methods." CarbonPlan (blog), January 30, 2022. https://carbonplan.org.

Economist. "TV Dinners: Unused TV Spectrum and Drones Could Help Make Smart Farms a Reality." September 17, 2016. https://www.economist.com/science-and-technology/2016/09/17/tv-dinners.

Espinoza, Maria I., and Melissa Aronczyk. "Big Data for Climate Action or Climate Action for Big Data?" *Big Data and Society* 8, no. 1 (2021): 1–15.

Fairbairn, Madeleine, and Zenia Kish. "'A Poverty of Data'? Exporting the Digital Revolution to Farmers in the Global South." In *The Nature of Data: Infrastructures, Environments, Politics*, edited by Jenny Goldstein and Eric Nost. University of Nebraska Press, 2022.

Frewer, Tim. "What Exactly Do REDD+ Projects Produce? A Materialist Analysis of Carbon Offset Production from a REDD+ Project in Cambodia." *Political Geography* 91 (2021): 1–11. https://doi.org/10.1016/j.polgeo.2021.102480.

Fuchs, Christian. *Social Media: A Critical Introduction*. 2nd ed. Sage, 2017.

Gates, Bill. "Can the Wi-Fi Chip in Your Phone Help Feed the World?" *GatesNotes*, October 9, 2018. https://www.gatesnotes.com/FarmBeats.

Gifford, Lauren. "'You Can't Value What You Can't Measure': A Critical Look at Forest Carbon Accounting." *Climatic Change* 161, no. 2 (2020): 291–306. https://doi.org/10.1007/s10584-020-02653-1.

Gillen, Jamie, Tim Bunnell, and Jonathan Rigg. "Geographies of Ruralization." *Dialogues in Human Geography* 12, no. 2 (2022): 186–203.

Goldstein, Jesse. *Planetary Improvement: Cleantech Entrepreneurship and the Contradictions of Green Capitalism*. MIT Press, 2018.

Greene, Daniel. "Landlords of the Internet: Big Data and Big Real Estate." *Social Studies of Science* 52, no. 6 (2022): 904–27. https://doi.org/10.1177/03063127221124943.

Halpern, Orit, and Gökçe Günel. "Demoing Unto Death: Smart Cities, Environment, and Preemptive Hope." In "Computing the City," edited by Armin Beverungen, Florian Sprenger, and Susan Ballard, *Fibreculture Journal*, no. 29 (July 2017). https://doi.org/10.15307/fcj.29.215.2017.

Halpern, Orit, Robert Mitchell, and Bernard Dionysius Geoghegan. "The Smartness Mandate: Notes Toward a Critique." *Grey Room* 68 (2017): 106–29. https://doi.org/10.1162/GREY_a_00221.

Harvey, David. "Globalization and the 'Spatial Fix.'" *Geographische Revue* 2 (2001): 23–30.

Hawken, Paul, Amory Lovins, and Hunter Lovins. *Natural Capitalism: Creating the Next Industrial Revolution*. Little, Brown, 1999.

Hogan, Mél. "The Data Center Industrial Complex." In *Saturation an Elemental Politics*, edited by Melody Jue and Rafico Ruiz. Duke University Press, 2021.

Huff, Amber. "Frictitious Commodities: Virtuality, Virtue, and Value in the Carbon Economy of Repair." *Environment and Planning E: Nature and Space* 6, no. 4 (2023): 2203–28. https://doi.org/10.1177/25148486211015056.

Jenkins, Jennifer. "A Response to Verra's Decision on Tonne-Year Accounting—NCX." NCX Learning Hub (blog), June 22, 2022. https://

ncx.com/learning-hub/a-response-to-verras-decision-on-tonne-year-accounting/.

Johnson, Alix. "Emplacing Data Within Imperial Histories: Imagining Iceland as Data Centers' 'Natural' Home." *Culture Machine* 18 (2019): 1–12. https://culturemachine.net/vol-18-the-nature-of-data-centers/emplacing-data/.

Klein, Naomi. *This Changes Everything: Capitalism vs. the Climate*. Simon and Schuster, 2014.

LaDuke, Winona, and Deborah Cowen. "Beyond Wiindigo Infrastructure." *South Atlantic Quarterly* 119, no. 2 (2020): 243–68.

Levenda, Anthony, and Dillon Mahmoudi. "Silicon Forest and Server Farms: The (Urban) Nature of Digital Capitalism in the Pacific Northwest." *Culture Machine* 18 (2019): 1–14. https://culturemachine.net/vol-18-the-nature-of-data-centers/silicon-forest-and-server-farms/.

Lin, Cindy. "How to Make a Forest." *e-flux Architecture*, April 20, 2020. https://www.e-flux.com/architecture/at-the-border/325757/how-to-make-a-forest/.

Lovell, Heather, and Donald MacKenzie. "Accounting for Carbon: The Role of Accounting Professional Organisations in Governing Climate Change." *Antipode* 43, no. 3 (2011): 704–30. https://doi.org/10.1111/j.1467-8330.2011.00883.x.

Lukacz, Matt. "The Dashboard Approach to Environmental Governance: Situating Microsoft's 'Planetary Computer' Within the Environmental History of Computing." Paper presented at the virtual annual conference of the Special Interest Group for Computing, Information, and Society, 2021.

Microsoft Corporation. "AI for Earth Grantee Profile: SilviaTerra." 2020. Accessed April 2, 2023. https://ai4edatasetspublicassets.blob.core.windows.net/grantee-profiles/SilviaTerra_US_Ag_AI4E%20Grantee%20Profile.pdf.

Microsoft Corporation. "Microsoft Carbon Removal: An Update with Lessons Learned in Our Second Year." 2022. https://query.prod.cms.rt.microsoft.com/cms/api/am/binary/RE4QOoD.

Microsoft Corporation. "Microsoft Carbon Removal: Lessons from an Early Corporate Purchase." 2021. https://query.prod.cms.rt.microsoft.com/cms/api/am/binary/RE4MDlc.

Mintert, James, and Michael Langemeier. "Ag Economy Barometer—February 2022." Purdue University Center for Commercial Agriculture. https://ag.purdue.edu/commercialag/ageconomybarometer/farmer-sentiment-rises-during-commodity-price-rally-concern-over-production-costs-remains/.

Moore, Jason W. "The Capitalocene, Part I: On the Nature and Origins of Our Ecological Crisis." *Journal of Peasant Studies* 44, no. 3 (2017): 594–630. https://doi.org/10.1080/03066150.2016.1235036.

Morris-Suzuki, Tessa. "Capitalism in the Computer Age." *New Left Review* 160 (1986): 81–91.

Murray, Brian C., Bruce A. McCarl, and Heng-Chi Lee. "Estimating Leakage from Forest Carbon Sequestration Programs." *Land Economics* 80, no. 1 (2004): 109–24. https://doi.org/10.2307/3147147.

Murray, Nick. "Agriculture Wars." *Viewpoint Magazine*, March 12, 2018. https://viewpointmag.com/2018/03/12/agriculture-wars/.

Myhrvold, Conor. "The App Craze Branches into Forestry." MIT Technology Review (blog), June 25, 2013. https://www.technologyreview.com/2013/06/25/177594/the-app-craze-branches-into-forestry/.

NCX (Nature Capital Exchange). "NCX's Natural Capital Exchange." Accessed April 20, 2023. https://story.ncx.com/pennsylvania.

Parisa, Zack, and Max Nova. "This AI Can See the Forest and the Trees." *IEEE Spectrum*, July 30, 2020. https://spectrum.ieee.org/this-ai-can-see-the-forest-and-the-trees.

Reuben, Aaron. "Foresters Now Monitoring Tree Populations from Space." *Scientific American*, August 6, 2014. https://www.scientificamerican.com/article/foresters-now-monitoring-tree-populations-from-space-slide-show/.

Scales, Ivan R. "Green Capitalism." In *International Encyclopedia of Geography: People, the Earth, Environment and Technology*, edited by Douglas Richardson, Noel Castree, Michael F. Goodchild, Audrey Kobayashi, Weidong Liu, and Richard A. Marston. Wiley, 2017.

Schiller, Dan. *Digital Capitalism: Networking the Global. Market System.* MIT Press, 1999.

Schmelzer, Matthias, Andrea Vetter, and Aaron Vansintjan. *The Future Is Degrowth: A Guide to a World Beyond Capitalism*. Verso, 2022.

Schüll, Natasha Dow. "Data for Life: Wearable Technology and the Design of Self-Care." *BioSocieties* 11, no. 3 (2016): 317–33. https://doi.org/10.1057/biosoc.2015.47.

Srnicek, Nick. *Platform Capitalism*. Polity Press, 2017.

Tsing, Anna Lowenhaupt. *The Mushroom at the End of the World: On the Possibility of Life in Capitalist Ruins*. Princeton University Press, 2015.

van Kooten, Gerrit Cornelis, Timothy N. Bogle, and Frans P. de Vries. "Forest Carbon Offsets Revisited: Shedding Light on Darkwoods." *Forest Science* 61, no. 2 (2015): 370–80. https://doi.org/10.5849/forsci.13-183.

West, Sarah Myers. "Data Capitalism: Redefining the Logics of Surveillance and Privacy." *Business and Society* 58, no. 1 (2019): 20–41.

Williams, Raymond. *The Country and the City*. Chatto and Windus & Spokesman Books, 1973.

2

Scenes of Extraction

Mediating Rurality, Wilderness, and Hinterland in Dutch and Chinese Film

EMILY NG AND ESTHER PEEREN

Raymond Williams starts his seminal study of the country and the city in English literature and culture by noting, "On the actual settlements, which in the real history have been astonishingly varied, powerful feelings have gathered and have been generalised. On the country has gathered the idea of a natural way of life: of peace, innocence, and simple virtue. On the city has gathered the idea of an achieved centre: of learning, communication, light. Powerful hostile associations have also developed: on the city as a place of noise, worldliness and ambition; on the country as a place of backwardness, ignorance, limitation."[1] These structures of feeling are not a mere fog clouding rural and urban actualities but rather actively mediate them. What people think *country* and *city* are or should be affects how such places are concretely made and remade. Ruralities and urbanities are, in the words of the editors of this volume, "plural, dynamic, diverse, situated, and emergent." Certain (infra)structures, activities, and medialities are welcomed and attended to because they appear to belong; others are resisted or overlooked because they seem out of place. As Williams puts it, "What is knowable is not only a function of objects, of what is there to be known. It is also a function of subjects, of observers—*of what is desired and what needs to be known*."[2] While this mediation is reciprocal, and radical transformations in actual ruralities or urbanities can also shift the wants involved, Williams emphasizes how "certain images and associations persist" and come to "stand as themselves; not in a living but in an enamelled world."[3] The term *enamelled world* suggests a process of fixing, turning a relation of mutual mediation into one of determination.

The Russian literary theorist Mikhail Bakhtin conceptualizes genres as *chronotopes*, which are particular sociohistorical configurations of time-space assimilated into literature that reflect and refract social reality. According to Bakhtin, literary genres may linger long after they cease to articulate with the actualities from which they sprung, and they do so precisely as affective structures because "in literature and art itself, temporal and spatial determinations are . . . always colored by emotions and values."[4] Both Williams and Bakhtin emphasize the difficulty of apprehending how feelings, emotions, and values mediate social actualities in literature and in life; this mediation has to be *read for*, something that is particularly tricky when one is oneself participating in a chronotope or structure of feeling taken as unmediated reality.

Williams's unfixing of the enameled English country–city dichotomy in *The Country and the City*, which he describes as a slow, laborious process, proceeds primarily through a consideration of the genres of the pastoral and counterpastoral.[5] Such genres persist despite having "waned," in Lauren Berlant's sense, not just because of the widespread affective investment in the "good-life fantasies" they offer, but also because of the strategic use of the pastoral by the land-owning class as "a self-consciously rural mode of display" to consolidate, legitimate, and naturalize capitalist and (neo)colonialist expansion on a national and global scale.[6] Bakhtin, on his part, marks out the chronotope of the idyll as a main driver of Western literary history, emphasizing how its many forms are all grounded in "the special relationship that time has to space in the idyll: an organic fastening-down, a grafting of life and its events to a place, to a familiar territory with all its nooks and crannies, its familiar mountains, valleys, fields, rivers and forests, and one's own home."[7] This "organic fastening-down" in distinctly ruralized environments is associated with an affective atmosphere of safety and comfort. What this renders illegible is what Williams lays bare, namely that rurality—and, may we add, wilderness—is far from offering a place insulated from urban life and with that, from capitalist and (neo)colonialist development. Such formations are in fact deeply implicated in the great world of globalization and its extractive relationship to the earth and large sections of its human population.

In this regard, Gayatri Chakravorty Spivak points to the "spectralization of the rural" that globalization has wrought, most notably in the global south, involving ruralities—and wildernesses—converted "into a database for pharmaceutical dumping, chemical fertilizers, patenting of indigenous knowledge, big dam building and the like."[8] Williams and Bakhtin help explain why these destructive (infra)structures, materialities, and epistemologies remain spectral—invisible—for the many who continue to see rurality and wilderness through the lens of genres like the idyll and pastoral that reify urban–rural–wilderness distinctions. Critical attention to media rurality, then, allows for reconsiderations of spaces often neglected in urban-centered analyses of global capital.

One way to open Spivak's database up to scrutiny is by thinking spectralized rurality or wilderness as also hinterland. This notion has recently been taken up by scholars like Phil A. Neel, Neil Brenner, and Nikos Katsikis to address the economic exploitation of geographies—urban, rural, and wilderness—in service of global systems of extraction, production, storage, distribution, and consumption. Affectively, the hinterland is linked to feelings of loss, abandonment, and sacrifice.[9]

Coined in the nineteenth century, the term *hinterland* is closely connected to colonial expansion.[10] In Williams's rendering, the exploitative metropole–colony relation of the British Empire is an upscaling of the city–country one, while the hinterland tellingly enters this analogy as a mediating figure: "The 'metropolitan' states, through a system of trade, but also through a complex of economic and political controls, draw food and, more critically, raw materials from these areas of supply, *this effective hinterland*, that is also the greater part of the earth's surface and that contains the great majority of its peoples. Thus a model of city and country, in economic and political relationships, has gone beyond the boundaries of the nation-state, and is seen but also challenged as a model of the world."[11] Much like the hinterlands of historical colonial expansion, contemporary hinterlands are designated by global capitalism and neocolonialism for extraction and exploitation of natural, human, and nonhuman resources in service of just-in-time flows of logistics.[12] Brenner and Katsikis aptly note that the hinterland long remained a "black box" or "ghost acreage," not deemed of interest in and of itself, noting, "While studies of urban metabolism have exhaustively quantified the material and energetic flows that mediate city/hinterland relations, they have tended to bypass the question of how noncity spaces are reconfigured through these mediations."[13] In addition to addressing this question, another necessary step involves contesting Brenner and Katsikis's definition of hinterlands as "the diverse non-city landscapes that support urban life," which continues to center the urban as what rural landscapes are geared toward and thus affirms planetary urbanization's "notions of the urbanization of everywhere in a zero-sum relationship with a residual rural."[14] In fact, not all rural or wilderness areas operate as hinterlands, and as Neel makes clear, there are also urban and suburban hinterlands.[15]

In this chapter, we explore how rurality, wilderness, and hinterland, in terms of their actualities and the affectivities that have gathered on them, mediate each other, and what critical potentialities such mediation has for moving beyond stale genres and their associated structures of feeling. We do this through a comparative close reading of two films highlighting this mediation—as well as the specific medialities or media forms that bring it into view—at sites of fossil fuel extraction in, respectively, the Netherlands and China.

Piet Hein van der Hoek's 2017 documentary *The Silent Quake* (*De stille beving*) tracks a middle-class family seeking to register, by using surveillance and handheld cameras, the ruinous impact of earthquakes caused by gas extraction in the northern Dutch province of Groningen on their renovated farmhouse. Here, rurality and hinterland are conceived as incompatible. The gradual destruction of a rurality thought to be idyllic—peaceful, wholesome—by a hinterland erupting violently from underground is presented as a sudden aberration to be remedied by financial compensation.

Wang Bing's 2008 video installation/film *Crude Oil* (*Caiyou riji/Yuanyou*), running fourteen hours in length, follows oil field workers throughout their workday in the northwestern Chinese province of Qinghai.[16] Here, rurality and hinterland are not rendered mutually exclusive, as in *The Silent Quake*. Instead, rurality, hinterland, and wilderness seem to act as mutually constitutive, as oil workers relocate temporarily to the remote oil-drilling desert town of Huatugou in order to sustain their lives in their home villages, where small-scale farming no longer offers a secure livelihood. This makes these ruralities far from idyllic, while the wilderness-hinterland of the oil field, in addition to exhaustion and despair, also fosters fragile modes of survival, care, and hope.

In our comparison, we focus on how narrative and formal aspects of the two films foreground the way rurality/wilderness-hinterland mediation—whether conceived as aberrant or recognized as inherent to the workings of present-day global capitalism—involves oscillations between above- and underground; horizontality, verticality, diagonality, and perpendicularity; the mundane and the spectacular; slowness and speed; and constancy and transformation. These oscillations mark different structures of feeling that variously deny (in what Berlant would call a relation of cruel optimism) or acknowledge global (neoliberal) capitalism's human and environmental destructiveness.

The Becoming-Hinterland of the Rural Idyll in *The Silent Quake*

Night-vision surveillance footage shows, up close and from a low angle, undergrowth with wild flowers as rain pours down and stormy winds blow.[17] After a lightning bolt exposes a farmhouse and some trees in the distance, tense nondiegetic piano-dominated music begins to play. A second bolt lights up the same scene but then follows a shot taken from a different camera position much closer to the farmhouse; the building's outlines and an old-model tractor in front of it can now be seen even when the sky remains dark. The rain is so heavy that it creates diagonal white streaks across the frame. Text, also in white, superimposed on the image reads: "Since 1992, the province of Groningen has been hit by more than 1,000 earthquakes as a result of natural gas extraction. This led to 100,000 damage reports."[18] After another night-vision shot showing the surrounding countryside from up high, a camera positioned closer to the house shows, through a window, a couple having drinks in the farmhouse living room as the following text appears: "Albert and Annemarie live right in the middle of the earthquake zone. In and around their farm there are tens of cameras to capture the quakes." A high-angle interior surveillance camera shot of a room with children's toys strewn across the floor is followed by an exterior shot showing a side wall of the farmhouse reinforced with diagonally placed wooden struts. After the documentary's title appears on screen, the image lightens and morphs into a color image of the farmhouse now bathed in sun, chirping birds audible beneath the continuing music.

This sequence establishes the farmhouse and the family living in it as under threat, assailed by the storm and surveilled in a way reminiscent of the disturbing start of Michael Haneke's 2005 film *Caché*, where a man receives video recordings of the family home made by someone unknown. At multiple points, the documentary translates what Nixon calls the "slow violence" of the intermittent underground rumblings into the visually and auditorily imposing spectacle of a thunderstorm captured in night vision.[19] It is not that the earthquakes cannot be portrayed more directly;

in other scenes, the surveillance cameras and various other contraptions—including an aquarium filled with water and Ping-Pong balls—allow the vibrations to show up in their immediate small-scale effects, like knickknacks stored in the attic shaking and falling over as the balls bob in the water. These images, however, are too mundane to fulfill the documentary's aim of underscoring the severe, lasting material and mental toll the earthquakes took on people in Groningen. The Dutch Oil Company (NAM, a partnership between Shell and ExxonMobil, with profit sharing by the Dutch government) long minimized this toll, first denying any link between gas extraction and the earthquakes, then drawing out the processing of damage claims. The spectacular horror aesthetic of the opening sequence, reprised several times in the documentary, is designed to convey the seriousness of the threat posed by the earthquakes and the family's growing fear, anxiety, and frustration, as well as to make the audience share these affects.

What interests us here is how this aesthetic presents a particular mediation of rurality and hinterland, where the hinterland is seen as terrifyingly erupting into an idyllic rurality to destroy its entrenched affective association with "peace, innocence, and simple virtue."[20] The specific mediality of the surveillance cameras matters because they do not simply register the horror but are themselves part of it. Surveillance evokes dread anywhere (as *Caché*, set in Paris, shows), but in a rural context, its disembodied perspective is particularly unsettling. If rurality is dominantly thought of and felt as idyllic, and as such is expected to appear as completely "familiar territory," then the need to surround and fill one's home with cameras to capture an elusive threat suggests a rurality despoiled.[21]

The Silent Quake evokes the good life of an idyllic rurality that was (before the earthquakes) and ought to be restored (after adequate compensation is received) in scenes filmed with handheld cameras, often by one of the family's two young daughters.[22] This introduces a different mediality, more attuned to the rural idyll than the surveillance images because of these cameras' association with home videos. In scenes showing the family eating breakfast in the garden and putting up Christmas decorations, the damage done by the earthquakes is not visible. In others, damage is deliberately pointed out in a direct address to the audience: "If you look over here, you can also already see a tear, a crack." The juxtaposition of these two types of scenes implies that the tears and cracks are not just weakening the house's structural integrity but also unsettling the rural idyll as a time-space of leisurely (childhood) enjoyment governed by a cyclical temporality (the documentary is structured according to daily and seasonal rhythms) that renders life predictable. The physical shocks of the earthquakes and the mental shock of not receiving compassion and prompt compensation from the NAM mark the transformation of the safe, idyllic rural home into a hinterland space of abandonment and exposure. This transformation exposes the family's belief in the attainability of the good life promised by the rural idyll in the context of "a spreading precarity . . . cutting across class and localities" that offers "no guarantees that the life one intends can or will be built" as a form of cruel optimism.[23]

What the plot of a suddenly encroaching hinterland obscures, however, is that Groningen's rural areas have for centuries been assigned the hinterland function of being "a supplier to the national ambitions managed from the Randstad."[24] In the eighteenth and nineteenth centuries, Groningen had many peat colonies; the so-called peat lords lived mostly in the cities, offering those cutting the peat meager wages and ignoring the way peat removal caused the land to subside and the water to rise.[25] In the twentieth century, shortly after gas was discovered in its depths, the laying of gas pipelines between Groningen and the west of the country caused significant damage to agricultural land, while in the 1970s the province saw several cases of chemical industries polluting groundwater.[26] Thus, far from having been an idyllic countryside until the earthquakes of the 1980s, rural Groningen has long been a hinterland "sacrifice zone," underlining Spivak's point that rural areas are not peripheral but central to absorbing the harms caused by globalized capitalism.[27]

Importantly, the large-scale intensive agriculture that came to dominate the Netherlands after World War II also constitutes an extractive, sacrificial hinterland practice. Farming—in a generalized sense that glosses over distinctions between agricultural modes and their divergent human and nonhuman costs—is often considered an integral part of a countryside, thought of and felt as idyllic. It is opposed to fossil fuel extraction in a valuated schema that, Jennifer Wenzel explains, puts the horizontality of farming, as on the surface (and therefore *on the level*—honest, honorable) over against the invasive verticality of mining.[28] Whereas mining is readily recognized as destroying landscapes, often agricultural ones, "processes of rural exploitation . . . have been, in effect, dissolved into a landscape" perceived as attractive and even natural.[29] Disentangling horizontality and surface from farming—which, in intensive forms, penetrates and exhausts the soil—and verticality and subsurface from resource extraction—which may proceed along horizontal planes underground (for example, through pipelines) and requires aboveground (infra)structures—makes clear that rurality and hinterland, idyll and sacrifice zone, may well converge.

The cameras in *The Silent Quake* do not venture underground; the infrastructures of gas extraction never appear onscreen. Formally, the hinterland's eruption into the rural idyll is expressed not through verticality but diagonality. It is the sharp white diagonals of the lashing rains (figure 2.1) and the wooden struts propping up the outside wall in the opening sequence (figure 2.2) that visually convey the extractive hinterland's unhomely crossing (out) of the homely idyll.

Besides appearing formally in the image as diagonality, the idyll's destruction by the hinterland is also conveyed by the image's form, resulting from the specific media through which rurality is conveyed. As noted, the surveillance cameras, by their very presence, disrupt the way rurality is dominantly felt as a safe, familiar space by invoking a sense of horror, while the handheld camera, against its association with happy home movies, documents the falling apart of the family's everyday life, adding an affective charge of fear and sadness. *The Silent Quake* also features several sequences filmed using drones. Bird's-eye perspectives on rural areas used to be rare in visual media because of the cost involved in taking a person wielding a camera up in the air, but the affordability of drones has made it commonplace to capture these areas as landscapes seen from above, stretched out horizontally on the surface. In the documentary's first drone sequence, rural Groningen is traversed at a leisurely speed from up high, appearing beautiful, knowable, and controllable—in other words, idyllic. However, the sequence ends with the camera zooming in on a collapsed farm, subsequently also filmed from the ground and identified as an earthquake casualty. A later scene begins with the drone-mounted camera tracing a narrow country road running in between rows of green trees to its end, only to find an empty piece of land. The gray patch left by the foundations of a demolished house appears as a stain put on the verdurous idyll by the hinterland rising up from below.

Toward the end of the documentary, the family decides to accept a compensation offer from the NAM that will allow them to replace the farmhouse with a new, earthquake-proof home. In the scene showing the demolition of the farmhouse, the diagonal struts marking the hinterland's crossing (out) of the idyll are the final structures standing as the excavator's arm forms another diagonal on the screen.

With the new home yet to be built, the family's hopes that the deal with the NAM will restore their idyllic good life remain unrealized. The

Sinds 1992 is de provincie Groningen getroffen door
meer dan 1.000 aardbevingen als gevolg van aardgaswinning.
Dat leidde tot 100.000 schademeldingen.

2.1 Still from Piet Hein van der Hoek's documentary *The Silent Quake* (2017).

2.2 Still from Piet Hein van der Hoek's documentary *The Silent Quake* (2017).

cyclical temporality of the idyll, evoked by the way the documentary moves through the four seasons (even though filming went on for longer than a year) and promising regeneration, is undone by the hinterland's temporality of simultaneous speed and stagnation. This temporality, which makes the family feel both like "we have jumped on a train and do not know yet where it will end" and as if "for three years now we have been standing completely still," configures the future, one situated within "the neoliberal feedback loop, with its efficiency at distributing and shaping the experience of insecurity throughout the class structure and across the globe," as incalculable.[30]

In the end, however, no systemic critique of this generalized experience of insecurity, tied to the proliferation of hinterlands across urban, rural, and wilderness spaces, is articulated. While the documentary powerfully conveys how destabilizing it is to face, on an everyday basis, a violence that may be slow but nevertheless imposes itself as a looming threat difficult to make tangible for those not living with it, its insistence on presenting the rural, properly idyllic, and the hinterland, properly sacrificial, as mutually exclusive disregards how this particular rural area was historically a hinterland for most and only ever an idyll for a small elite. Instead, the focus remains firmly on one middle-class family (with less well-off neighbors only making sporadic appearances) being thwarted while in pursuit of a rural idyll. Crucially, where Bakhtin sees the idyll as characterized by "a grafting of life and its events to a place" that includes "one's own home" but is fundamentally a collectivity, a commons, the good life pursued in *The Silent Quake* centers on the *owned* home.[31] It is the damage the earthquakes have caused to private property, the documentary insists, that makes the rural idyll's ruination by the hinterland particularly horrifying, and it is fairly calculated, quick compensation that should restore it. In this way, the documentary reaffirms a capitalist logic in which certain people can legitimately be sacrificed while others—especially those accumulating capital—should and can be protected. A very different picture of the hinterland and its mediating relations with the rural and the wilderness appears in *Crude Oil*, where the protagonists are not middle-class homeowners but migrant workers.

Crude Oil: Hinterland and Rurality as Mutual Elsewheres

The camera sits still, gazing slightly downward at the slumped body of an oil worker in a stained orange jumpsuit, nestled by the heater. Voices murmur from beyond the frame, the chitchat of other workers. The viewer follows the drift of their conversations, from the timing of work shifts to the depth of the drill pipes. The shot lasts nearly two minutes before cutting to a broader view of the room as they enter an extended discussion of sleep and health.

This opening shot of *Crude Oil*—its lengthiness, framing, and attention to a moment of rest while signaling labor through the conversation and uniform—foretells the tempo and sensibility of the film's remaining hours. Wang Bing's cinematic works, known for their extreme length, have been described as a "cinema of labor," often gesturing to the limits of representation both in depictions of historical violence and the sense of temporal excess.[32] Here, we focus on how *Crude Oil* creates a critical aesthetics of the hinterland through its use of the extreme long take in a time of shrinking attention economies and its intimate portrayals of labor through the juxtaposition of vertical, horizontal, perpendicular, and nonangular visual elements. We also consider how rurality/wilderness-hinterland mediations are evoked in the way the film touches on themes and temporalities of risk, protection, and calculability. Unlike *The Silent Quake*, which renders the effects of capitalist extraction and sacrifice external and intrusive to the rural idyll, *Crude Oil* dwells on their emplacement in nonurban spaces by construing its setting as hinterland, without recourse to an idyllic counterpart. And where risk is portrayed in *The Silent Quake* as capable of being warded off by proper financial compensation for middle-class property owners, *Crude Oil* gestures toward a more totalized risk for the rural working classes, whose very bodies and lives are at stake in the hinterland as sacrifice zone.

The long take has been a technique of interest and debate in film scholarship. Some have emphasized its realist potential. André Bazin, for instance, defends the long take on the grounds that "the long take's time is the event's time."[33] But given the structure of temporal expectation habituated by histories of cinema, in which narrative meaning often relies on the tempo of the filmmaker's cuts instead of an uninterrupted flow of time (Eisenstein's montage being a classic example), others have taken up the long shot precisely to unsettle expectations. For those like Warhol and the makers of cinema verité, long continuous shots were aimed at disrupting normative tempos of spectatorship, shifting the onus of signification from filmmaker to viewer.[34] Moreover, the divergence in the experience of cinematic time created by the long take has grown increasingly pronounced as the length of time between cuts in Hollywood film has shrunk across the decades, with optimal video lengths on contemporary social media platforms now measured in seconds.[35] To build a fourteen-hour film out of shots that average thirteen minutes and thirty-three seconds in length, as in the case of *Crude Oil*, then, is to disturb structures of anticipation common in the historical present. Rather than conveying objective time, such works are described by Michael Walsh as "durational cinema," in which cinema formally foregrounds the passage of time.[36]

In *Crude Oil*'s cinematic staging of hinterlands, the use of the long take also evokes the question of labor time. As Mary Ann Doane sketches out, the problem of the representability of time arose at the turn from the nineteenth to the twentieth century and was linked to industrialization.[37] Attempts to visualize time using new technologies of photography and film accompanied efforts to standardize and rationalize time in the service of capitalist production. A parallel is found in Taylorism-inflected Chinese socialist films of the 1950s, where human movement is juxtaposed with the sounds and imagery of ticking clocks and wristwatches, gesturing toward a rationalization of time. The syncing of the laboring body with clock-time exhorts dreams of modern industrialization, founded on temporal notions of speed, efficiency, and progress.[38] If the misrecognition of value in commodity fetishism is indeed enabled, as in Marx's rendering, by the concealment of unequal exchanges of value—a value fundamentally hinged on the labor-time it takes to produce a commodity—then an aesthetic rendering of labor-time in the hinterlands that disturbs common genres might open up to reconsiderations of capitalist value, especially amid the just-in-time temporality of global logistics.[39] While certain genres portraying labor can at times risk a romantic aestheticization of labor—such as what Salomé Aguilera Skvirsky calls the "process genre," in which labor is depicted sequentially, thereby explicitly or implicitly producing the sensation of a how-to—Wang Bing's works at once evoke and exceed them.[40] As with many of his other works, *Crude Oil*—unlike *The Silent Quake*, with its more normative shot length and narrative tempo—makes an impression in feeling nearly unwatchable within the viewing habits of its own historical time. Such a sense of unwatchability (especially strong in the context of an installation) disturbs the pretense to mastery that might be offered by the experience of watching a filmic work to completion.[41]

With the above in mind, the scenes of rest and leisure so prominent in *Crude Oil* become noteworthy, both in contrast to other cinemas of labor and in Wang Bing's own oeuvre. Unlike Wang's *15 Hours* (2017), which leaves hours of leisure and rest offscreen in spite of a comparable total length, in *Crude Oil*, off-duty time is as—if not more—prominent than on-duty time. The majority of the film consists of shots away from the oil rig. Workers eat from temporary canteens, take smoke breaks, watch DVDs sitting on dorm beds, and, most frequently, chat or nap in the converted containers. Each of these scenes is imbued, by the film's overall temporal structure, with a certain endlessness. While this brings with it an air of drudgery and boredom, there is also a sense of warmth, conviviality, and care, of informal modes of solidarity. The supposedly linear, accelerating time of capitalist progress is disturbed and stalled by a stagnant temporality of the hinterland, filled here not only with the slog of undercompensated labor but also with other rhythms of life that, when defamiliarized through the long take, do not merely cave into the fold of capitalist time.

Moreover, in contrast to Wang's *West of the Tracks* (2003), which features state-run housing visually reminiscent of Maoist and post-Mao villages, *Crude Oil* brings an aesthetics of the frontier to the forefront by integrating wilderness with the industrial visuality of hinterland. Through long shots that highlight the bareness of the flat, horizontal landscape of the Tibetan Plateau (on which Qinghai is mostly located), against which the oil rig rises up vertically, wilderness and hinterland infrastructure are paired, heightening a visual sense of hinterlands as sites of extraction and uninhabitability (in spite of, precisely, the lives temporarily present there). In the context of Qinghai, this impression of empty land seemingly lacking human life other than the oil workers may occlude previous state-implemented policies of rural resettlement and ethnicized displacement. Across Republican-era thinking on colonizing the northwest, the Chinese Communist Party's Supporting Frontier Construction, or Zhibian (*zhichi bianjiang jianshe*), policies of the 1950s, and the more recent Develop the West (*xibu da kaifa*) policies of the 2000s, imaginations of Qinghai as an empty frontier zone in need of relocated human labor for resource extraction—including oil, natural gas, and minerals—were central to the region's territorial and political incorporation. The multiplicity of at times ethnically marked political powers contending and shaping the region historically and into the present—Tibetan, Mongolian, Hui, Han, British, Anglo-Indian, among others—are erased in the discourse of emptiness and "need" for (Han) labor and governance from elsewhere.[42]

Meanwhile, in *Crude Oil*, classic visual signals of rural life and community remain minimal. While the workers at times mention their rural hometowns, the seasonal cycles quintessential to depictions of the rural idyll (so vividly invoked in *The Silent Quake*) are absent. Instead, time moves through the tempo and schedule of the workday. Between time spent in windowless containers and shifts that—as the workers note—disregard normative sleep hours, bifurcations of day and night, time spent on and off duty, blur into a generalized flow of work life, disregarding bourgeois distinctions between work and leisure that previously replaced their relative indistinction in rural life.[43] In the workers' conversations, time is marked by ponderings on the one hand about the effects

2.3 Still from Wang Bing's video installation *Crude Oil* (2008).

2.4 Still from Wang Bing's video installation *Crude Oil* (2008).

of disrupted sleep on health, and on the other about the attractiveness of the extra pay offered for night shifts.

When showing workers on duty at the oil rig, there is an aesthetics of bareness and seriousness paired with perpendicular lines. Long shots are more frequently used, displaying the horizontality of the desert landscape while formally aligning the vertical body of the worker with the verticality of the rig (figures 2.3 and 2.4), evoking Wenzel's speculations on what "the

2.5 Still from Wang Bing's video installation *Crude Oil* (2008).

relationship between the horizontal and the vertical means for the material and cultural politics of hinterlands."[44] Here, though, rather than the northern Canadian marshland seen as "overburden" covering the valuable materials beneath, the dry, bare desert landscape displays a seemingly unburdened potentiality for extraction.

In the containers, whether fully off duty or pausing during active shifts, the scenes feel intimate. Instead of echoing the colder visuality of abstract labor expressed by media and documentary images of Chinese labor in an era of global trade, with factory workers lined up in rows and uniforms, signaling sameness, quantifiability, and substitutability, the workers' bodies and material objects on view exude a sense of warmth, mutuality, and care. Bodies lean on one another in moments of rest (a head on a shoulder) and huddle near one another in moments of leisure (sitting in close proximity on a single bed), making shapes that diverge from the straight, perpendicular lines of the oil field (figure 2.5). Objects in recent use like half-filled tea thermoses and gloves drying on the heater signal life as lived, even as the room turns into a filmic still life during work hours (when the camera continues filming after all workers leave for the rig). Although the workers are in uniform at the rig and often (though not always) also in the container, the postures of their bodies—curled up, leaning, slouching—are distinct from the verticality and angular movements (squatting, moving the levers) at the rig. Yet distinctions between work and nonwork remain ambiguous as workers live on site during their temporary employment, and rest might be interrupted by calls to the rig. This sense of unpredictability points to a broader question of calculability in the hinterland, which is answered very differently here than in *The Silent Quake*.

In their conversations, the workers render absurd the logics of calculability supposedly governing their lives as temporary workers. This comes through in an extended discussion of wages in part 2 of the film:

> WORKER 1: I worked overtime on May Day. What about others?
> WORKER 2: The salaries are not fair. Two years ago it was more.
> WORKER 1: That's it. We're all going crazy. We make less. . . . Nobody's stupid. For two months, some make 999, some 800. Some only 700. If you calculate, it's 1,100 to 1,200, but you only get 700 *[group laughter]*. At this rate, by the end of the year, we'll make 100, 200 yuan. There's nothing you can do. . . . These people keep our money. They gamble with our money. These bosses. They don't give a fuck. . . . The more you calculate, the more irritated you get.

In contrast to the middle-class family in *The Silent Quake*, who appear convinced that the damage to their house can be accurately assessed and compensated for, the workers' positionality as rural temporary workers does not afford them such an assumption. The precarity associated with the style of temporary wage labor shown in *Crude Oil* cannot be fully disavowed through cruel optimism, perhaps because of the more drastic gap between one's life and a good life fantasy centered on economic stability and various forms of state-provided care, and perhaps also because of the disruptive visibility of oil drilling compared with gas extraction.[45] Since the market reform era of the late 1970s, the vast rural population of China has become a crucial, if spectralized, labor force driving China's economic expansion and capital accumulation.[46] In the scene above, the purported calculability associated with precarious wage labor is revealed in its lived absurdity as numbers seem to move arbitrarily, though unsurprisingly, to the detriment of the workers. In its contrast to the bafflement and indignation expressed by the middle-class family in *The Silent Quake* about the instability they face, the clear-eyed, tragic-yet-jovial affect surrounding discussions of instability among the oil workers indicates a recognition of both their precarity and its irremediability within global capitalism's logics.

Notably, the workers' acknowledgment of their subjection to an unjust unpredictability comes after a reference to a shared sociality of the rural, which had signaled a sense of safety and mutual protection:

> WORKER 1: Both of them have the same accent. . . .
> Little Old Man, you're so happy *[laughter]*.
> LITTLE OLD MAN: We're from the same village. We look out for each other. We use the same tone of voice, and stay calm. Here's what I'll do: I'll rest one month and work the next two.
> WORKER 2: Don't count your chickens.
> LITTLE OLD MAN: You don't believe me then. You haven't done your figures yet. I've done mine.

Here, rurality and hinterland are not held to be mutually exclusive, as in *The Silent Quake*. Instead, one social and material condition exists as the spatial elsewhere of the other while their inhabitants alternate between them, bringing the markings of the former into the latter. After the pronouncements of mutual protection, calm, planning, and calculability on the part of Little Old Man comes the devolving of numbers into madness described above. The capacity of the rural to offer a sense of security—evoked here in terms of shared linguistic inflections and mutual protection, as a realm of familiarity—comes up against the more exploitative logic present in the hinterland. The sense of sameness and continuity affectively attached to imaginations of the rural hometown confronts that of disposability and replaceability associated with abstract labor and the hinterland. While the former is evoked here as an antidote to the latter, deep skepticism abides. The predictability posited by Little Old Man is soon disrupted by a numbers game in which victory belongs solely to the figure of the boss. With the space of the rural hometown rendered a distant geographic elsewhere, but one to which a return is nonetheless envisioned, what seems to linger is a sense of shared identity and promise of mutual protection that mediates and ameliorates, if only weakly, the rigged game of labor-time in which any sense of rationality and security is ultimately deemed impossible.[47]

The sense of risk evinced by the hinterland goes beyond the question of numbers. As Deborah Cowen describes, the body of the worker is central to the global flow of commodities and the supposed seamlessness of logistics systems. Cowen notes that "disruption in a world built on fast flows takes on epic proportions," both for the systemic flow of goods and in the life of the worker.[48] In *Crude Oil*, the specter of major risk enters a moment of filming when one worker slips and falls, causing the camera person to stumble in tandem, leading to a notable quake in the footage. Unlike the individually minor earthquakes rendered spectacular through the horror aesthetic in *The Silent Quake*, in *Crude Oil*, we are told, in a mundane and matter-of-fact voicing, that when mishaps occur at the rig, they tend to be catastrophic. One worker speaks of his younger brother, who, while working at the rig, was collided into, and woke up in the hospital four days later with half his body paralyzed: "When everything is fine, no problem. No small accidents here. Only big ones (*yi chu shi, quan shi da shi*)." His fellow worker describes another case of severe injury, relaying what the company leadership was rumored to have said: "Why didn't he just die? One payment and it's over. Now it's dragging on." The scene closes with one of the workers pondering aloud the notion of "mutual assistance" (*huzhu*).

Looming behind the workers' talk of mutuality and protection, then, is the specter of major risk and substitutability when it comes to their bodies. Workers rendered unable to work vanish from the hinterland, winding up at the hospital or returning to the rural hometown. While the rural hometown is portrayed as a site that may offer shelter from the precarity of the hinterland, it is far from an idyllic site of normative family life, as in *The Silent Quake*. Aside from overlapping with hinterland spaces, rural areas at times absorb the catastrophes and excesses of contemporary labor in the hinterlands. Rather than positing a single sacrifice zone, in Wang's film the theme of sacrifice runs across the two conceptual-material landscapes as counterparts. Those who leave the home village may experience their departure and labor as sacrificing for the sake of the family, in and through entering a capitalist economy of sacrifice.

Conclusion

The Silent Quake, by suggesting that rural Groningen ought to offer the propertied middle-class family at its center an idyllic good life and that the hinterland is no more than a recent intruder that can be expelled through fairly calculated compensation, elides the complex rural–hinterland mediations that have long shaped this area along classed (and gendered and racialized) lines. Rendering rurality and hinterland as mutually exclusive leaves unilluminated how extractive industries and private land and home ownership are pillars of a globalized capitalist system whose relentless expansion is facilitated by a perverse calculative logic that counts certain people, animals, and other life-forms as expendable. In *Crude Oil*, rather than evoking a hinterland alien and distinct from the rural, rurality and hinterland are thought together. The hinterland is infused by rural sociality, and the rural hometown is imagined more as the other side of the rural hinterland sacrifice zone (often entailing short- or long-term outmigration and undercompensated labor to eke out survival) than as rural idyll (offering a secure good life of rootedness in place). Nonetheless, potential rural aspects of the Qinghai plateau—other communities that reside in the region aside from the temporary oil workers, or histories of rural resettlement—may be occluded by the film's invocation of the relationship between coloniality and rurality as reliant "on the projection and production of emptiness" (see this volume's introduction) in its portrayal of the plateau as an empty wilderness frontier. This suggests that just as there is a danger in insisting on the separation of rurality and hinterland, there is also a danger in conceiving certain wildernesses as destined to become hinterlands. Wilderness, rurality, and hinterland should be seen as distinct but capable of mediating each other in complex, changing ways specific to particular contexts.

The Silent Quake's mobilization of surveillance, handheld, and drone-mounted cameras produces an eminently watchable and empathy-arousing spectacle of a hinterland interrupting, via a disorientating concurrent speeding up/slowing down, a rurality thought and felt as properly idyllic, as structured by a predictable cyclically regenerative temporality. It takes aim, sharply but narrowly, at the NAM and national politicians for allowing *this* hinterland to wreak havoc on *this* middle-class family; there is no indictment of hinterlandization as such. *Crude Oil* mobilizes a broader, more pensive gesture of critique toward contemporary conditions of extraction—of human labor and material resources—through a staging of unwatchability and stagnation paired with an intimate aesthetics and focus on downtime uncommon in cinemas of labor. Moreover, questions of risk, disruption, and calculability enter the two films differently. Whereas *The Silent Quake* charts the efforts of a propertied family to pursue what they expect to be rightful compensation for the fissures brought on the idyllic infrastructure of their farmhouse by the hinterland, *Crude Oil* registers the specter of calamitous accidents that take life and limb, paired with a deep distrust in claims to calculability in economies of precarious labor that straddle and entangle rural and hinterland spaces as sacrifice zones.

Notes

1 Williams, *Country*, 1.
2 Williams, *Country*, 165; emphasis added.
3 Williams, *Country*, 1–2, 18.
4 Bakhtin, "Forms of Time," 243.
5 Williams, *Country*, 306.
6 Berlant, *Cruel Optimism*, 6, 2; Williams, *Country*, 6, 2, 282.
7 Bakhtin, "Forms of Time," 225.
8 Spivak, *Death of a Discipline*, 92–93.
9 Peeren et al., "Introduction."
10 Unangst, "Hinterland"; Uzoigwe, "Spheres of Influence."
11 Williams, *Country*, 279, emphasis added.
12 Cowen, *Deadly Life*.
13 Brenner and Katsikis, "Operational Landscapes," 26.
14 Brenner and Katsikis, "Operational Landscapes," 24; Gillen et al., "Geographies of Ruralization," 186.
15 Neel, *Hinterland*.
16 Within his oeuvre, Wang distinguishes between filmic works created for theaters and installations to be exhibited in galleries and other contexts. *Crude Oil* falls in the latter category and premiered as an installation at the International Film Festival Rotterdam in 2008.
17 The documentary can be found online in several versions. We refer to the seventy-four-minute-long version available on Vimeo (https://vimeo.com/206711457), which was broadcast in 2017 by NTR, a national television channel, and RTV–Noord, a local one, drawing half a million viewers in total. The film also won the audience award at that year's Netherlands Film Festival.
18 All text and dialogue from *The Silent Quake* are translated from the Dutch by Esther Peeren.
19 Nixon, *Slow Violence*.
20 Williams, *Country*, 1.
21 Bakhtin, "Forms of Time," 225.
22 Berlant, *Cruel Optimism*.
23 Berlant, *Cruel Optimism*, 192.
24 Mayer, "From Peat to Google Power," 904. The Randstad is the economically, politically, and culturally dominant region in the east of the Netherlands in which the cities of Amsterdam, The Hague, and Rotterdam are situated. Although it also comprises rural parts, it is strongly associated with urbanity (and, increasingly, with urban elites seen as blind to the plight of ruralites).
25 Mayer, "From Peat to Google Power"; Noordhoff, *Ontaard land*.
26 Boersema, *Gronings goud*.

27 Brenner and Katsikis, "Operational Landscapes," 28; Lerner, *Sacrifice Zones.*

28 Wenzel, "Hinterland, Underground."

29 Williams, *Country*, 46.

30 Berlant, *Cruel Optimism*, 192–93.

31 Bakhtin, "Forms of Time," 225.

32 Balsom, "Chimera of Endlessness"; Pollacchi, "Wang Bing's Cinema"; Veg, "Limits of Representation"; Walsh, "Wang Bing."

33 Henderson, "Long Take."

34 McDougall, "When Less Is Less."

35 Cutting et al., "Attention"; Bhatti, "Ask Buffer."

36 Walsh, "Wang Bing," 241.

37 Doane, *Emergence.*

38 Qian, "When Taylorism Met Revolutionary Romanticism."

39 Cowen, *Deadly Life.*

40 Skvirsky, *Process Genre.*

41 Balsom, "Chimera of Endlessness."

42 Rohlf, *Building New China.*

43 Lefebvre, *Critique.*

44 Wenzel, "Hinterland, Underground," 301.

45 Berlant, *Cruel Optimism.*

46 Yan, "Spectralization."

47 Zhang, *Strangers.*

48 Cowen, *Deadly Life*, 96.

Bibliography

Bakhtin, Mikhail. "Forms of Time and of the Chronotope in the Novel." In *The Dialogic Imagination: Four Essays by M. M. Bakhtin*, edited by Michael Holquist. University of Texas Press, 1996.

Balsom, Erika. "Wang Bing's *15 Hours* and the Chimera of Endlessness." In *Taking Measures: Usages of Format in Film and Video Art*, edited by Fabienne Liptay, Carla Gabrí, and Laura Walde. Scheidegger and Spiess, 2023.

Berlant, Lauren. *Cruel Optimism*. Duke University Press, 2011.

Bhatti, Umber. "Ask Buffer: What Is the Ideal TikTok Length?" Buffer.com, accessed July 26, 2023. https://buffer.com/resources/best-tiktok-video-length/.

Boersema, Wendelmoet. *Gronings goud* [Groningen gold]. Ambo/Anthos, 2021.

Brenner, Neil, and Nikos Katsikis. "Operational Landscapes: Hinterlands of the Capitalocene." *Architectural Design* 90, no. 1 (2020): 22–31. https://doi.org/10.1002/ad.2521.

Cowen, Deborah. *The Deadly Life of Logistics: Mapping Violence in Global Trade*. University of Minnesota Press, 2014.

Cutting, James E., Jordan E. DeLong, and Christine E. Nothelfer. "Attention and the Evolution of Hollywood Film." *Psychological Science* 21, no. 3 (2011): 432–39. https://doi.org/10.1177/0956797610361679.

Doane, Mary Ann. *The Emergence of Cinematic Time: Modernity, Contingency, the Archive*. Harvard University Press, 2002.

Gillen, Jamie, Tim Bunnell, and Jonathan Rigg. "Geographies of Ruralization." *Dialogues in Human Geography* 12, no. 2 (2022): 186–203. https://doi.org/10.1177/20438206221075818.

Henderson, Brian. "The Long Take." *Film Comment* 7, no. 2 (1971): 6–11.

Lefebvre, Henri. *The Critique of Everyday Life*. Verso, 2014.

Lerner, Steve. *Sacrifice Zones: The Front Lines of Toxic Chemical Exposure in the United States*. University of Minnesota Press, 2012.

Mayer, Vicki. "From Peat to Google Power: Communications Infrastructures and Structures of Feeling in Groningen." *European Journal of Cultural Studies* 24, no. 4 (2021): 901–15. https://doi.org/10.1177/1367549420935898.

McDougall, David. "When Less Is Less: The Long Take in Documentary." *Film Quarterly* 46, no. 2 (1992): 36–46. https://doi.org/10.2307/1213006.

Neel, Phil A. *Hinterland: America's New Landscape of Class and Conflict*. Reaktion, 2018.

Nixon, Rob. *Slow Violence and the Environmentalism of the Poor*. Harvard University Press, 2011.

Noordhoff, Ineke. *Ontaard land: De strijd van een Groninger tegen de gasregenten* [Degenerate land: The struggle of a Groninger against the gas regents]. Atlas Contact, 2022.

Peeren, Esther, Hanneke Stuit, Sarah Nuttall, and Pamila Gupta. "Introduction: Conceptualizing Hinterlands." In *Planetary Hinterlands: Extraction, Abandonment, and Care*, edited by Pamila Gupta, Sarah Nuttall, Esther Peeren, and Hanneke Stuit. Palgrave Macmillan, 2024. https://doi.org/10.1007/978-3-031-24243-4.

Pollacchi, Elena. "Wang Bing's Cinema: Shared Spaces of Labor." *Working USA* 17 (2014): 31–43. https://doi.org/10.1163/17434580-01701004.

Qian, Ying. "When Taylorism Met Revolutionary Romanticism: Documentary Cinema in China's Great Leap Forward." *Critical Inquiry* 46, no. 3 (2020): 578–604. https://doi.org/10.1086/708075.

Rohlf, Gregory. *Building New China, Colonizing Kokonor: Resettlement to Qinghai in the 1950s*. Lexington Books, 2016.

Skvirsky, Salomé Aguilera. *The Process Genre: Cinema and the Aesthetic of Labor*. Duke University Press, 2020.

Spivak, Gayatri Chakravorty. *Death of a Discipline*. Columbia University Press, 2003.

Unangst, Matthew. "Hinterland: The Political History of a Geographic Category from the Scramble for Africa to Afro-Asian Solidarity." *Journal of Global History* 17, no. 3 (2022): 496–514. https://doi.org/10.1017/S1740022821000401.

Uzoigwe, Godfrey N. "Spheres of Influence and the Doctrine of the Hinterland in the Partition of Africa." *Journal of African Studies* 3, no. 2 (1976): 183–203.

Van der Hoek, Piet Hein, director. *De stille beving* [*The Silent Quake*]. Documentary film, 2017.

Veg, Sebastian. "The Limits of Representation: Wang Bing's Labour Camp Films." *Journal of Chinese Cinemas* 6, no. 2 (2012): 173–88. https://doi.org/10.1386/jcc.6.2.173_1.

Walsh, Michael. "Wang Bing: Duration, Deindustrialization, and Industrial Work-Discipline." In *Durational Cinema: A Short History of Long Films*. Palgrave Macmillan, 2022.

Wang Bing, director. *Caiyou riji/Yuanyou* [*Crude Oil*]. Documentary film, 2008.

Wenzel, Jennifer. "Hinterland, Underground." In *Planetary Hinterlands: Extraction, Abandonment, and Care*, edited by Pamila Gupta, Sarah Nuttall, Esther Peeren, and Hanneke Stuit. Palgrave Macmillan, 2024. https://doi.org/10.1007/978-3-031-24243-4.

Williams, Raymond. *The Country and the City*. 1973; reprint, Spokesman, 2011.

Yan, Hairong. "Spectralization of the Rural: Reinterpreting the Labor Mobility of Rural Young Women in Post-Mao China." *American Ethnologist* 30, no. 4 (2003): 578–96. https://doi.org/10.1525/ae.2003.30.4.578.

Zhang, Li. *Strangers in the City: Reconfigurations of Space, Power, and Social Networks Within China's Floating Population*. Stanford University Press, 2002.

3

Imperial Wireless

Energetic Mediation at Marconi's Connemara Station

PATRICK BRESNIHAN AND PATRICK BRODIE

From Colonial Gothic to (Techno)Science Fiction

> In a remote bog in Connemara, Marconi
> built a radio transmission station.

Indeed, radio pioneer Gugliemo Marconi's wireless stations scattered around the world have long shared a historical imagination of remoteness, tucked away as they were in the farthest corners of imperial territories and relaying messages across vast distances. Fulfilling a communicative and migratory longing from empire to colony, his stations on the west coast of Ireland were built on patches said to be as close to North America as you could get in Europe. Similar to his other installations in places like Hawai'i, the strategic location of the west of Ireland mapped its place within a geography of empire that produced remote imaginations of what would later be classified as rural areas, far from the technological and industrial concentrations of urban metropoles.[1] In Ireland, these places, especially the peat boglands, were characterized by the attraction of what literary scholar Derek Gladwin refers to as "postcolonial gothic," where the wild, sparsely populated landscapes of Ireland's western regions awaited discovery and consumption by adventurous industrialists, constructed as much by the depopulation brought on by the holocaust of the Great Famine (1845–52) as the natural features of the landscape itself.[2] His station built in Derrigimlagh Bog in the early 1900s, near Clifden, County Galway, in the heart of Connemara, became operational little more than half a century removed from this episode of mass death and migration. The station promised to connect this impoverished, remote location to the world via the technological marvel of radio, delivered by the free market pioneer Marconi. As Mike Davis describes the colonial famines of the 1800s in *Late Victorian Holocausts*, colonies like India and Ireland became, in the free market policies that amplified the impacts of the Famine, "utilitarian laborator[ies] where millions of lives were wagered against dogmatic faith in omnipotent markets overcoming the 'inconvenience of dearth.'"[3] In the intervening years, these policies intensified through the technological forges of industrial modernity.

As the sharp contrast between the abjection of the Famine and the technological marvel of radio demonstrates, the colonial ideology of remoteness masks the violent materiality of imperial presence. Reports and descriptions of the Derrigimlagh site, and any existing photographs, conjure a science-fictional scenario of stark, confounding contrasts. Officially opened in 1907, the facility entangled the traditional landscapes and cultures of the region, including the subsistence practice of cutting turf for fuel, with the scale of an anachronistic technological modernity that would otherwise evade Ireland's rural reaches for decades. In the blustery, rainy west, shrouded by fog, landscape faintly swaying in the wind, Clifden residents could see and hear sparks flying from 210-foot-high masts in the distance.[4] The small train that brought peat briquettes cut from the surrounding bogland chugged its way across tracks built into the wetlands to deposit its haul at the generating station. Steam from the engines and the turf-fired turbines puffed out and dissipated in the sky. The noise of the condenser station sent signal operators across the bog to a quieter place to work. The surrounding area, much of which would not be serviced with electricity until the mid-1900s (and the town of Clifden not even until 1927, some years after the arrival and departure of the Marconi Company, with a local hydroelectric dam), remained shrouded in darkness at night, the light of the station buildings, we can imagine, starkly illuminated against the landscape. Imperial location was thus experienced as a spectacular materiality of emplaced cultures and ecologies. The paired imposition of technoscience and free market capitalism has always navigated these messy specificities, infrastructuralizing the extractive imbalances of rural remoteness and metropolitan connectivity. This "colonial sublime," as Brian Larkin describes the sensory experience of energy modernity in colonial Nigeria,[5] explodes with the contradictions of distance and proximity embedded within the spatial and technical politics of imperial capital.

3.1 The Marconi Company's Derrigimlagh facility in Connemara, Ireland, ca. 1907. Colorized. Source: https://connemara.net/the-marconi-station.

3.2 Clifden Marconi telegraph station at Derrigimlagh, Ireland, with its fuel, turf briquettes. Source: http://viemagazine.com/article/guglielmo-marconi/.

3.3 Oweninny Wind Farm in Mayo, Ireland, built on hectares of cutaway bogland. Source: Bord na Móna.

Today, Ireland's rural boglands host different science-fictional landscapes, but they are similarly bewildering and anachronistic. The lunar landscapes of cutaway bogs stretch for miles across the Ireland's Midlands, where half a century of industrial turf extraction has stripped peat down to the bedrock. Boneyards of industrial equipment from the former state-run peat industry sit beside solar arrays and meteorological monitoring towers beaming turbulent weather data back to energy companies and grid managers. Further west, unstaffed wind farms tower over mountain blanket bog transected by twirling lattices of access roads, overhead wires, and substations. In the suburban and semirural hinterlands of Dublin, Ireland's metropolis, data sheds and logistics warehouses hum with the pulse of consumer activity and supply chains infrastructurally entangled with these landscapes, their subterranean and atmospheric resources and connections circulating materials and beaming information worldwide. Constructed in landscapes struggling with different, albeit remarkably analogic, problems of viability and sustainability to the Connemara encountered by Marconi, you may think that a smart, ecomodern future has already arrived, postindustrial and carbon neutral, with new technological forms harnessing air and ether into harmless circulations of electricity rather than spewing black clouds of carbon into the atmosphere.

Responding to these anachronisms, author Mike McCormack, in his futuristic scenarios set in imagined, overdeveloped Irelands since the 1990s, claims to pursue a vision of the rural west of Ireland as a science-fictional landscape.[6] McCormack emphasizes the contradictions and juxtapositions between the real and the imagined—for example, the long-standing historical assumption that rural Ireland is not industrialized or even fully modernized—when in fact these ideologies are formed across contested geographies of uneven development, and most of the country's most vital infrastructures require rural land and resources as a basis. In doing so, he conjures a rural Ireland of simultaneity and coexistence, of over- and underdevelopment, of rupture and surprising entanglements. As much contemporary research has meticulously demonstrated, the cloud is nothing if not material, and rural places remain underserved but overburdened with its material impacts, mapping onto the much longer imperial histories already gestured toward here.[7] To keep sight of the contradictions and ambivalences embedded in such industrial-scale activities and infrastructures, we must pay close attention to the places their supply chains cross and operate through, the resources they extract and require, and the natural and global histories they entail. Apart from their construction across preexisting colonial and postcolonial routes and geographies, digital infrastructures have inherited a logic of global connectivity built along the supply lines and ecological dynamics of imperial capitalism. The material prehistories of these activities continue to shape the directionality of their flow and the geographies of their construction.[8]

Contemporary rural Ireland is a place where periodizations and characterizations of modernity and development short-circuit in the muddy waters of emplaced histories, especially in relation to Ireland's economic development policies. Since the 1960s, state economic policy has primarily served to attract multinational investment at the expense of more carefully and equitably managed industrial and environmental development policies.[9] Marconi's facility in Derrigimlagh represents a site through which to understand how rural land and territory in Ireland have been contested and navigated in relation to this global connectivity and Ireland's role within empire. Through this rural, colonial-era industrial installation, we can speculate how and why these politics materially persist in spite of the historical transformations of the state and its economies in the intervening century.

What we find especially crucial is the role of private capital. A corporate technological installation infrastructuralized the harnessing of heat from turf and the creation of electricity for this distinct form of "development." It utilized a rural location and local materials to enable ethereal transoceanic communication. We propose the concept of energetic mediation as a relation that made possible the forms of circulatory, communicative capitalism embedded within coming financial and logistical systems at a global level. As media scholars Anne Pasek, Cindy Kaiying Lin, Zane Griffin, Talley Cooper, and Jordan Kinder argue, by focusing on the "energetics" of materials and methods of technical processes, "energy and media . . . become ways of organizing and qualifying relations between entangled objects, processes, and systems."[10] What can we learn about how energy and mediation were developed in Ireland through the entangled relations at this rural site, one historicized as pioneering but also materially disconnected from the infrastructural developments of modern Irish history? What did the intersections between harnessed energy and the technoscientific marvel of ethereal communication require and make possible, in both its forms of energy extraction and generation as well as the communicative apparatus itself? What can the ways in which this site is remembered tell us about conflicting imaginaries of Irish ecomodernity today? In short, what can the logics, histories, and infrastructures that make up this site *do* for a wider analysis of the rural construction of industrial and telecommunications development on the island of Ireland? As we will suggest, energetic mediation does not need to be privately owned or extractive. Rather, in studying the relations of imperial location, resource materialities, place-specific cultures and ecologies, and the struggles and frictions that arise then and now, we can start to see and recover paths not taken toward more just and even radical ruralities.

Location, Materiality, and Place: Peat Through the Ether

> Through that remote bog in Connemara,
> Marconi enabled modern connectivity.

He was not just an inventor, as popular accounts would lead you to think, but a businessman; the Marconi Company was one of the first communications companies to incorporate, formally seek public investment, and sell shares on the London stock exchange.[11] Media scholar Greg Elmer argues that we need to understand Marconi not as an individual but "as a discrete limited and public company—a disruptive 'new media' business—indeed one that foreshadows the initial public offerings . . . of new or digital media corporations like Apple, Google, Amazon, and Facebook, among others."[12]

Across the historical literature, we see an emphasis on the communicative potential and benefits of Marconi's wireless technology. In these narratives, Marconi's technologies perform a universal service to mankind, enabling progress via technological feats of achievement.[13] But the Marconi Company was set up in a moment of aggressive imperial expansion, when communications technologies had significant strategic and commercial applications—and Marconi knew it. His own publicity efforts focused on the potential of his technology for enabling the reach of imperial control and supply chains into new, difficult terrains, or what media scholar Tyler Morgenstern refers to as Marconi's role in energizing "potent fantasies of scalar extensibility and global connectivity."[14]

These points are made in similar terms in other remote outposts of imperial territories. Media historian Marc Raboy notes that the establishment of the largest Marconi station in the world in Hawai'i in 1914 points to the strategic role such developments played for Marconi's company at this time of public criticism and financial challenges.[15] The value of such technology in drawing the most remote islands into a global community, defined by integration into networks of imperial capitalist trade and communication, was promoted by a boosterism similar to what we see circulating around Clifden and other Marconi stations in Ireland (particularly in the national press). What these projects manage to do is simultaneously enroll rural geographies within imperial state governance and produce them as distinct and remote—or rather, they do the former through the latter, a recurring trope within (especially British) colonial practice. The sites of Marconi stations were not so removed at this point in imperial modernity that they could not enable the activity of empire directly—these places were often able to visited by wealthy tourists, after all—but their materiality was distant (and mediated) enough to make the place of their operation a marvel or an infrastructural spectacle rather than a place with people and relations worth considering in the metropole.

When Marconi's station arrived in Connemara in 1907, it was in the aftermath of Ireland's turbulent land politics of the late 1800s. These politics had centered around colonial land use and valuation, to deploy toward imperial strategies meant to enroll Irish territory into the previously mentioned networks of trade and production. The most prominent event in this history remains the Great Famine—less than half a century before the station's construction, Connemara had been decimated by this engineered crisis, by which the British state created and amplified starvation conditions via colonial trade policies.[16] With the majority of paying tenants and cottier farmers either starving or fleeing on ships, the region of Connemara was sold as a single estate of 200,000 acres of bog and mountain. When put up for sale by the London Law Life Insurance Society in 1849, the vendors described the estate as having vast potential, just lacking in capital investment.[17] As scholars of improvement within capitalist accumulation strategies have made clear, there is a long history of material and ideological processes that produce certain places and people as waste(d) and in need of development via capital investment and new rounds of private accumulation, traversing colonial and postcolonial states of abandonment.[18] In the case of Ireland's peat bogs, British colonial surveys which mapped the country's land and resources in the eighteenth and nineteenth centuries treated the bogs as wastelands, which must be civilized and made productive for agricultural markets in the metropole.[19] The strategies of colonial administration and governance that followed the Famine on similar guidelines did little to improve the situation for the majority of peasant farmers and laboring rural poor. The large estates, including the Martin estate, were sold to wealthy industrialists as gothic holiday homes.[20]

The history of under- or dedevelopment in Connemara thus also created an accidental point of synergy for Marconi's operations, especially in the eyes of colonial governance. The intertwining of a strengthening imperial free market that required the subordination of colonial territories to capital was thus the primary policy for the management of land and population, a market that would come to be increasingly organized by emerging networks of global communication. Thus, by the time Marconi arrived, the ground was set for a free market savior of this apparently abandoned region.[21] A total of 140 local seasonal turf cutters came to work under Marconi's employ, thereby drawing significant employment into the area, which, once operational, separated it from the otherwise extremely poor Connemara region surrounding it. By historical reports, Clifden became a jewel of Connemara, and people were able to stay and work, without having to leave to seek opportunities elsewhere. This is hugely significant in the historical perception and development of the area.[22]

The purpose of these stations, though, was Marconi's commercial and territorial strategies as embedded and supported by imperial powers, whatever the secondary effects of their operation, including improvement for the surrounding areas. If local development was embedded within this energetic process and locational dynamic, then the benefits become clear, "overcoming the 'inconvenience of dearth.'" If not, there was still the technological marvel and its benefits to imperialist development. Like the telegraph before it, Marconi's technological invention, at its aspiring scale of use, was about annihilating space via managing supply chains, permitting immediate communication, expanding frontiers, and standardizing trade. Indeed, his other stations were even more explicitly about trade and movement—for example, his stations at the sites of lighthouses and the transition from visual to auditory communication for shipping.[23] To do so, radio needed infrastructure and management across locations. The purpose of the technology and its application was not to develop a region, a potential promoted far more by colonial state authorities (and later Irish state revisionism for touristic narratives, as the final section will describe) than by Marconi himself. Location in this sense, as a discursive and geographical mechanism for imperial activity, envisions a dot on the map as it charts an effective pathway across points (whether through cables or ether). Other elements of local economies and ecologies must be considered before installation, of course, but only insofar as they can either be exploited or need to be smoothed over for efficiencies. In this context, benefits for local workers and economies were secondary to the scalar ambition of wireless communications.

Marconi's wireless technology, by the ideological construction promoted by the company, promised also to elevate global communication above the messy particularities of place and ecology—resonant with the ideologies and metaphors of the internet's cloud today.[24] To harness the power of the ether, radio had to harness heat to produce electrical power and signals. Similar to earlier heralding of the conquest of space by wired telegraphy,[25] Marconi's stations promised to overcome the tyranny of wires with ethereal circulations of communicative matter, thus accomplishing for water what the telegraph had accomplished for land. These accounts tend to neglect the significance that was attached to the ethereal qualities of radio transmission in the context of the increasingly carbon-polluted atmospherics of industrial cities. Ether was not only not-wires but also not-smoke. Wireless telegraphy was emerging as a response to the material intransigence and costs of both wired communications systems and industrial modernity more generally. This perception served to occlude the energetic materiality of radio transmission itself, which required significant quantities of carbon fuel extraction and burning, generating exhausts associated with combusting it.[26] Radio was thus both smoke and wires, but only at its individual sites of receiving and transmission. The material abstraction across points was achieved by the transport of electrically produced signals through the air.

Thus even as wireless did away with some kinds of material resources and infrastructures for transport and communication, *other* resources and infrastructures became necessary, in particular the quantities of energy required to shoot a message over the Atlantic, harnessed and electrified on the Irish end of the signal via turf cut from a bog. Take this vivid description in the *Daily Mail* from 1907, subtitled "A Wireless Thunderstorm," which resolutely positions the "technological marvel" of radiotelegraphy within its material operation while promoting its affective and circulatory atmospherics:

> An entire room is given up to strange sheets of steel, which are hung from roof to floor, like washing on a line, until only narrow alleys are left. Queer brown earthenware jars, like old-fashioned receptacles, and all manner of outlandish electrical apparatus now confront the visitor. The plates are for acting as a reservoir to store electrical energy. The jars are transformers. The engineer gave a few directions to his assistant, who, seated before an ordinary Morse telegraph instrument in the operating room, placed a telephone headpiece to his ears, and began to fumble with the key, hastily bidding me to stuff cotton wool in my ears and don a pair of blue-glass spectacles. The engineer beckoned me to the connection room on the floor above, which is equipped with a medley of strange electrical contrivances. The use of the cotton-wool and smoked glasses became at once startlingly apparent. From the "interrupter" instrument corresponding exactly in duration to the assistant's touch of the key below, came three blinding flashes of blue-white flame, followed by a short flash, and then three more short flashes. The two side-mouths of the instrument likewise spout eye-blinding flame of the same colour and intensity. Simultaneously, the discharger, a few feet across the room, emitted similar blinding flames, and there came a wearing, tearing boom like the deep bass of some gigantic organ, but immeasurably cruder and louder. The duration of each note again corresponded exactly with the assistant's dot or dash on the instrument below. This was the electrical discharger, which sends oscillating electrical currents from the building into the aerial wires outside. These at once begin to set up vibrations of the ether, which in loops and waves travel with inconceivable rapidity across the sea.[27]

In this description, radio is a "crude" apparatus of harnessed fire and electricity, with communication reduced to patterned crashes and thunderous booms sent and received via a worker below. This mechanism was powered by the 300-kilowatt turf-fired generator and 15,000-volt battery, powered by up to six thousand tons of turf extracted on site per year as well as supplementary coal shipped into the small port in Clifden town (to be transported via Marconi's proprietary railway).[28] This turf was seasonally cut and transported by hundreds of local turf cutters, whose artisanal knowledge of this practice was key to the viability of the station.

The "crude" materiality of radio, however, also describes its core productive relations, coming at a point in the development of energetic modernity where the infrastructuring of electricity and utility delivery was not only the imaginative horizon but was also being planned and negotiated by states and scientists[29] in arenas beyond transport and mobility. Undoubtedly, modes of transport before Marconi's signals (and the telegraph cable before it) also required harnessing and using energy, whether from the flow of wind captured by sails that moved ships or from the legacies of colonial plunder, migration, and enslavement that carries across the Atlantic.[30] Steamships brought coal to port to supplement the peat, and trains were presumably powered by coal or by forms of human power. But what the Marconi signal represents is the imaginative obliteration of space by electricity, six thousand tons of turf per year. The near-instantaneity of the communication, which occurs via the powering and then harnessing of electrical signals across the airwaves, is what introduces the landed implications of energetic mediation and its entanglement with local processes, which then get upscaled through the spatial extensivity and intensification of supply-chain capitalism. Harnessed energy obliterated time and space via electrical signals. Energetic mediation thus provides a framework through which to understand the productive relations at the infrastructural sites of advanced technological development, in this case the landing point of distanced communication. In centering energetic mediation as a core relation of modernity, land (and rurality) can be recentralized in the understanding of the energetic bases of technological communication.

With the first radio transmission from Derrigimlagh to Nova Scotia in 1907, the intersecting projects of colonial modernity and imperial capitalism manifested in the west of Ireland. The harnessed local energy source (peat) and the communication it enabled—via what was ultimately a hugely advanced infrastructural apparatus, built in vital relation to ongoing environments and communities *through* this unique intersection of private pioneering entrepreneurialism and global imperialism—formed an experimental template for the sorts of rural relations that would sustain Irish modernity both before and after formal independence.

The specificity of Ireland's carbon culture—and its relationship to the modernization and industrialization of Ireland's infrastructural systems—is worth lingering over, especially as a transition to the final section of this chapter. Turf cutting in Ireland is demonstrative of the ruptures and continuities of energetic development in the country across the colonial, postcolonial, and contemporary neoliberal period. The practice of turf cutting specifically and its significations around Irish modernity are complex and multiscalar. Turf has long been a fuel source for domestic and subsistence use. In postcolonial Ireland, the state company Bord na Móna industrialized the process of turf cutting to power electrification in a process of limited national infrastructural modernization intended to reduce dependence on coal supplies from Britain, develop domestic industries, and provide social benefits.[31] Thus, turf cutting to power Marconi's private, imperial venture is not the same as turf cutting to power a state-owned power station electrifying rural homes. While scholars, including ourselves, have frequently commented on the inheritance of improvement planning in the industrialization of turf extraction and burning in the Irish Midlands and the west of Ireland in the 1940s and 1950s under the newly independent state, the logics were crucially shifted under postcolonial strategies of national development.[32] Under an administration that was committed, at least partially, to ideas of economic independence and social development, the boglands were viewed as untapped national resources for the electrification and industrialization of the country.[33] They became central to the modern imaginary of Ireland. How could the country become self-sufficient via its own land? How could its resource base sustain its population in ways that broke from its dependency on Britain and the relations of empire?

In the Midlands, this involved the construction of an entire industrial infrastructure, one that was once the largest industrial railway in Europe, spanning turf extraction sites to transport the resource to scattered power stations. Centrally built towns were constructed to draw in and house workers. Modern life came to the Midlands, and parts of the west like Donegal and Galway, which had their own power stations and extraction sites, in ways that appeared relatively anachronistic to otherwise underdeveloped Ireland.[34] This partial decolonization appeared to make good on the unfulfilled plans of two hundred years of colonial improvement, which in the end imagined the country's bogs as undeveloped wastelands

3.4 Screenshot of Screebe Power Station, from RTÉ, "Save Screebe Power Station" (1985).

unfit for purpose. Of course, these policies adopted similar relations to (and technologies for surveying) the bogs—as unproductive landscapes in need of large-scale engineering and control—and to rural populations—as backward and in need of external development. However, these imperfect and incomplete mid-twentieth-century state-led industrial activities were made possible by a political culture motivated by national self-determination and social development.[35]

It is instructive, then, to trace the decline of this political culture to the 1970s and the rise of a new Irish state caught in the embrace of the European Economic Community and US multinationals, one open to investment and seeking financial returns from state companies in a new, market-friendly context. A turf-fired power station, for example, was built and brought into operation in Screebe, on the shores of Galway Bay, in 1951. It was never profitable, but it provided jobs and energy for a part of Connemara that had little other industry. It closed in 1989. In a remarkable RTÉ documentary from 1985, the local parish priest talks about the struggles of the locals to secure commitments from the Irish state. "If you haven't got people in this area you've got a museum, nothing else," he says.[36] Similar arguments arise in the context of rural depopulation and development logics in Ireland today. A lack of development promises ruin, while neoliberal development centered on investment capital suggests unpeopled landscapes, be it attempts to create carbon sinks for offsets that require local activity (including turf cutting) to cease or the largely automated facilities of wind farms and data centers. In an emerging postindustrial and postcarbon culture, troubling memories reemerge as the global market grafts onto the enduring geographies of rural abandonment.

The entanglements of peat's carbon culture and Irish colonial and postcolonial modernity can be found throughout its infrastructural histories, whether in Marconi's station in Connemara or whatever was being powered by the turf-fired power plants scattered throughout the Midlands and the west of Ireland. These are simultaneously sites of foreclosure for many rural populations who are still waiting on the "just" part of the transition away from fossil fuels, as well as potent reminders of the promise and scale of postcolonial development, its landed specificities in Ireland, and its bargains (or buy-ins) with the reformed empire of neoliberal capital. As Bord na Móna plans to develop paired energy/data facilities in the midlands on cutaway peat bogs with multinational tech companies as tenants, the networking of cloud communication and renewable energy responds, in a statist entrepreneurial way, to the vagaries of an emerging low-carbon market.[37]

Paths Not Taken

> Through a remote bog in Connemara,
> Marconi enabled imperial modernity.

Today, at the Marconi site in Derrigimlagh, near the town of Clifden in the region of Connemara, there is ruin but also activity, including that of the sort pursued by British travelers to the west of Ireland in the 1800s: leisure and landscape tourism. Sheep and lambs roam among and rest on the old foundations, which fell into ruin during the twentieth century after being stripped for parts on Marconi's abrupt departure in 1925. These ruins are today augmented by recently installed placards and artworks celebrating Marconi's achievement, put in place with the support of Bord Fáilte, the Irish tourism body, and the Wild Atlantic Way tourist development campaign, rehistoricizing this site of imperial experiment as a global technological triumph, heroically aided by Connemara's unique location and ecologies. Around the exhibition and the rocky foundations, bog life is vibrant, as the area is part of one of the most intact blanket bogs in all of Europe, which has seen no heavy industrial activity since Marconi abandoned the facility in 1925. Tourists can wander the marked Marconi Trail, complete with weather refuges to seek cover from the unpredictable squalls that can arise on the otherwise unsheltered blanket bogs. Depending on the season, stacks of sod dry in the wind, the rows of hand-cut peat demonstrating the lively culture of small-scale turf cutting that persists in largely fuel- and energy-poor rural communities.

There is less detail at the Derrigimlagh site about the events that led to its closure and partial destruction in 1922. Six years after the 1916 Easter Rising, which arguably set off the revolutionary period, Republican forces occupied the station as one of its final strongholds. They were not fighting the British by this stage but rather Irish Free State forces during the Civil War (1921–23), fought between the anti-treaty IRA, which opposed the partition agreement with Britain, and the pro-treaty (Free State) army. The anti-treaty forces, labeled the "irregulars," disabled Marconi's Connemara station under the impression that it would be used to communicate with the British Royal Navy; they also destroyed his station in Crookhaven, Cork, that same year, for the same reasons.[38] Reported in the national press as acts of vandalism by a desperate minority of dissidents, a language of base criminalization that predated the characterizations of the IRA during the later Troubles in the north, the real motivation for the sabotage seems more obvious. While the Marconi Company was privately incorporated, this site was recognized by Republicans as imperial infrastructure. This point was emphasized by the fact that the station's final transmission in 1922 announced the recapture of the station by the Free State and the defeat of the irregulars, becoming a garrison for Free State soldiers.[39]

What does the memorialization of the Derrigimlagh site as a shining testament to Marconi's technological marvel, rather than a site of anti-imperial struggle, say about the relationship between media and rurality in contemporary Ireland, and more broadly in postcolonial contexts? In responding to this question, it is important to emphasize the material and ideological imprints of colonialism on contemporary life in Ireland as well as Ireland's position in ongoing networks of imperialism more broadly. The context for this historical account is, after all, our scholarship's situation within contemporary and ongoing debates about the role of energy, global tech corporations, and cloud infrastructures within Ireland's economy. Ireland's national facilitation of the global operations of tech capital continues to underwrite the imperial networks and technologies that structure the world economy and dictate imaginations of what technological and infrastructural futures look like.[40]

Morgenstern writes of the similarly short-lived Marconi station in Hawaiʻi in the same period: "While the facilities themselves were ultimately fleeting, the models of scale and connectivity they had helped to cultivate would prove far more durable."[41] We share this argument, but we extend it to other enduring legacies mediated by the Marconi station and what it continues to represent about energy systems and communicative capitalism—via the energetic resources and ecologies of immaterial media and colonial relations to rural hinterlands, as well as what it offers as a viewpoint of ongoing struggles surrounding Ireland's present and historical role in imperial capitalist networks. Ireland has once again become a test case for particular kinds of energy infrastructure as data centers and other smart infrastructures increasingly shape energy futures on the rural landscapes encountered and operationalized by Marconi.[42] Ireland's weather, location, existing infrastructure, land resources, and labor force are still offered as reasons why these energy-intensive infrastructures should be built there.[43] This facilitation of extractive multinational capital illustrates perhaps more than anything Ireland's continuously ambiguous role in imperial networks and infrastructural footprints of capitalist modernity.

Another way of responding to these endurances is to consider paths not taken—or, rather, taken and then foreclosed by the reorganization of global capital in the late twentieth century. It was the anti-treaty Republican party of Fianna Fáil that took power in 1932 and sought to formalize an Irish republic, delinking from British domination via the existing Irish Free State dominion and developing indigenous resources for the social and economic development of the Irish people.[44] As in all postcolonial contexts, these efforts were fettered by internal class interests and external imperial interests, but it is still significant that a different relationship to rural geographies, modernity, and energy was experimented with beyond the imperial/private development model wherein huge infrastructures are built that visibly and materially do not directly benefit the people who live with them.

The difference between the Screebe power station and Derrigimlagh radio installation was intent and purpose. Marconi's peat-powered radio station embodied the energetic form of imperial ambitions and its model of infrastructural and ecological mediation. Derrigimlagh was a useful location for a potentially profitable, experimental extractive project that mobilized electrified airwaves to shore up imperial networks of trade and communication. Jobs and development were secondary to this aim and thus precarious and dependent on ongoing viability, with no necessary remit to provide for the place or people living there. In contrast, Screebe, however imperfect and embedded within a model of carbon-fueled national modernity, as a technological infrastructure did bring jobs, development, electricity, and economic activity to the area. However, it did so while relying on the destructive activities of turf cutting and burning. The impacts of this were not necessarily offset by these benefits, but people in this area would at least experience how this version of modernity benefited them in their everyday lives (i.e., rural electrification), not simply the number of transient, poorly paid jobs created in the process of capital accumulation. The models of public activity that this enabled were related to the modern economic and cultural project of nation-building. One example, in the area of media, was the development of a national public service broadcasting service available in the area, which sparked a lively and largely Irish-language media industry from the 1970s onward.[45] Early Electricity Supply Board documentaries from the 1950s and 1960s on the rural electrification scheme make clear that this form of connectivity—broadcast radio and television entering rural households—was central to the national modernization. Thus, however indirectly or rippling these effects seem to be related to electrification, and however much they were also the products of contestation and democratic strife and debate, the relative recency of electrification meant that Screebe remained a crucial infrastructure of enablement. This infrastructure of public life did not require the invention or innovation of a multinational investor but rather the political will to invest directly in the social needs of rural communities.

In this chapter, we have illustrated that the logics surrounding the development and operation of the Marconi station demonstrate not a historical continuity but rather a cycle: imperial capitalism to postcolonial resource independence to multinational-led development. With the advent of Ireland's foreign direct investment–led model of development since the 1970s, along with its more recent convergence with multinational tech companies seeking access to cheap land, energy, and water to host data infrastructures, we have seen a return to colonial-era entrepreneurial rurality. While a data center is not the same as Marconi's radio facility, the ways in which what science and technology scholar Patrick Carroll identifies as a triad between "land, people, and the built environment" are mobilized within technoscientific approaches to economic development shares much more between these eras than it may first appear.[46] Mass, distanced communication, in this lineage, has always been the result of private technological and industrial innovation, and required enormous amounts of resources, particularly energy. The distillation of consumer communication into smaller and smaller technologies continues to hide the fact that simultaneously distributed and larger-scale private infrastructures are necessary to transmit and store information.[47] Whether a cell phone or radio mast on top of a hill, a 5G transmitter on a telephone pole, or a data center in a city's suburbs sustaining the cloud, the activation of technology across distance through material substrates remains central to communicative capitalism, whether wired or wireless, and that includes the energetic processes and resources required to power them.[48] The mutual operation of capitalist extraction and production is always activated across different types of activity, materially stitching together resources and infrastructures spread across distances.

But there have always been different models of energetic mediation. In media histories, there is a justified historiographic resistance to simplistically applying lessons from the past to a modified understanding of the present. On the opposite end of the spectrum, contemporary media research, urban focused and on the cutting edge, is frequently allergic to textured historicizing of the contemporary transformations of media systems and economies. Somewhere in between, and with an emphasis on infrastructure, there is an opportunity to confront the contradictions of the present through a more complex engagement with the historical geographies and built environments that still shape the systemic functions of digital media, especially as constructed historically across urban and rural circulations and tensions. A result of this, when looking at Marconi's stations at the edges of empire, is that the rural geographies of media systems become materially central; they present essential understandings in how they map the enduring landscapes of extraction and imperial activity. Marconi's station in Connemara has allowed us to unravel the persistence of these mechanisms of modernization and improvement as they shape environmental and technological relationships. We have also examined and traced our way back to these paths not taken—those blocked pathways, muddy and treacherous, toward decolonization and the formation of a republic orientated against empire.

In turning to Marconi's activities in Ireland and putting them in conversation with the different yet familiar context of contemporary digital media infrastructures, we can come to understand both how these relations are so stubborn and why it is important to build not just a history but to place it within a political debate and reckoning about media, energy, and imperialism. Energetic mediation describes in this context how imperial valuations of land, harnessing of energy, and the enclosure and making of resources become infrastructuralized into the form of distanced communication. These form not only the prehistory of but also the pathways by which capitalist networks function today. This particular formation of energetic mediation came at a formative time for Ireland's colonial and postcolonial politics, representing a point at which to understand how and why these uneven geographies continue to shape communicative infrastructures here and represent a persistent imagination of rural places as remote outposts of modern life. But other avenues were possible. Even in Ireland's imperfectly decolonized republic, alternative energy possibilities have arisen. It remains to be explored whether these offer a pathway toward forms of energy *and* media justice in an era of digital capitalism's amplified global imbalances.

In this chapter, what we offer toward contemporary debates surrounding rurality and the development of energetic alternatives is, in Ireland especially, their rootedness in stubborn spatial practices and ideologies. Where possible, it is essential to connect these issues to broader political questions of unequal global economic development and projects for radical decolonization. More than other technologies, communicative technologies like wireless radio or digital social media platforms revolve around the charismatic figure of the entrepreneur and the perceived universal value of connectivity, obscuring the willful, insidious, and damaging directions that imperial, communicative capitalism has on a world scale. Yet the development of Screebe power station, among other aborted examples of truly public energetic alternatives, with emplaced specific social intent and effects, tells us rural modernity can look otherwise, directed to different ends and activated via resources and landed imaginations that matter to local communities. What would an energetic mediation toward a republic look like, then, in the contemporary context of digital media? Whatever the answer, it is not administered by Marconi's multinational successors but rather, we can hope, by the infrastructures of a radically restructured and redirected democracy that starts and ends with place and more just materialities.

Notes

Acknowledgment: This chapter shares inspiration with the first chapter in our 2025 book, *From the Bog to the Cloud: Dependency and Eco-Modernity in Ireland.*

1 Morgenstern, "Etherealization."

2 Gladwin, *Contentious Terrains.*

3 Davis, *Late Victorian Holocausts*, 31.

4 Sexton, *Marconi.*

5 Larkin, quoted in Pasek et al., "Introduction," 4.

6 McCormack in de Loughry and McCormack, "Tiny Part."

7 For the cloud as material, see Hu, *Prehistory*; Parks and Starosielski, *Signal Traffic*; Starosielski, *Undersea Network.* For material impact, see Bresnihan and Brodie, "Contested States"; Bresnihan and Brodie, "Waste, Improvement, and Repair"; Brodie, "Data Infrastructure Studies." For imperial histories, see Bresnihan and Brodie, "From Toxic Industries"; Morgenstern, "Etherealization."

8 See, for example, Cowen, "Following the Infrastructures"; LaDuke and Cowen, "Beyond Wiindigo Infrastructure."

9 See O'Hearn, *Atlantic Economy*; Tovey, "Environmentalism in Ireland"; Bresnihan and Brodie, "From Toxic Industries."

10 Pasek et al., "Introduction," 9.

11 Elmer, "New Medium Goes Public," 1830.

12 Elmer, "New Medium Goes Public," 1832.

13 As media scholar Rachel O'Dwyer shows, the vision of "imperial wireless" was of a technology that could be used effectively to enhance British expeditions to the North Pole just as much as it could aid military operations in the Colombian Amazon. O'Dwyer, *Wavelengths.*

14 Morgenstern, "Etherealization."

15 Raboy, *Marconi.*

16 Kearns, "Historical Geographies of Ireland."

17 Robinson, *Connemara.*

18 See Bhandar, "Lost Property."

19 In spite of these surveys and plans, Marconi's station in Clifden was the only large-scale infrastructure that would use turf as an energy resource until the mid-twentieth century, with the exception of a few projects in the Midlands to use the bogs commercially for peat moss to supply for livestock bedding, especially for cavalry horses. See Clarke, *Brown Gold.*

20 Under pressure from liberal reformers, the largely ineffective Congested Districts Board was established in 1891 to "improve" the region and reduce rural disturbance via agricultural programs. However, little of this was delivered, and then in largely piecemeal fashion. Ultimately, as the growing anticolonial, nationalist movement argued, little indigenous

development could be expected as long as Ireland remained subordinate to England and an agrarian economy based on landlordism.

21 In his essay about the Valentia Island telegraph station, the site of the first transatlantic telegraph communication in 1853, literary scholar Chris Morash identifies a similar clash of technological modernity and primitive dearth. Juxtaposing this evental moment of global modern communication with the recent events of the Famine experienced in Kerry, reporters and visitors drawn to the area to celebrate this marvel of technology recount being horrified by the destitution and suffering they also witnessed. This produces a jarring effect, as the supposed advance of civilization is confronted with extreme suffering and barbarism. Morash, "Re-Placing."

22 Sexton, *Marconi.*

23 See Morash, "Re-Placing"; Sexton, *Marconi.*

24 Hu, *Prehistory of the Cloud.*

25 Morash, "Re-Placing."

26 Critical scholars of media and communication have frequently reminded us of the layered physical, elemental, and institutional forms that make up infrastructures of communication. See Mattern, *Code and Clay*; Peters, *Marvelous Clouds.*

27 "Marconi Signal Station."

28 Walsh, "From Newfoundland to Ireland."

29 And all the imperial imbalances this crystallized. See Daggett, *Birth of Energy*; Rubinstein, *Public Works.*

30 See, e.g., Sharpe, *In the Wake.*

31 Bresnihan and Brodie, "Waste."

32 See Bresnihan and Brodie, "Waste."

33 Shiel, *Quiet Revolution.*

34 Bresnihan, *Bog Irish.*

35 Clarke, *Brown Gold.*

36 RTÉ, "Save Screebe Power Station."

37 Bresnihan and Brodie, "Data Sinks."

38 Sexton, *Marconi.*

39 Fritz, "Marconi Station." It is interesting, although outside of the purview of this short chapter, to consider the role of radio in anticolonial politics and its criminalization in Ireland. On April 24, 1916, Irish rebels in the General Post Office (GPO), co-led by socialist and international union leader James Connolly, broadcast what is purportedly the first radio communication in Europe. Their hope was that their message would reach the United States and be reported in the American press, thereby mobilizing potent Irish American sympathies and support in the diaspora in the face of British military repression. While this historiography may be disputable, the broadcast was nonetheless

significant and is a key moment for media theorist Marshall McLuhan in his theories of media politics and power. Marconi's radio station in Clifden was taken off the air by the authorities, but to circumvent this, a coded message was sent to the United States from the Valentia telegraph station by a sympathetic operator. As Morash outlines, the technical know-how and infrastructures that allowed these subversive communications to happen were part of the same transatlantic networks established by Marconi as colonial Ireland emerged as an experimental site for imperial wireless. In this case, media technologies designed and applied for imperial ambitions were appropriated by anti-imperialist forces for decolonial ambitions. Subsequent commentaries on the GPO broadcast have emphasized its illegality as well as the perils of mass media technologies in the hands of popular movements for propaganda and misinformation: "Broadcasting was conceived in sin. It is a child of wrath. There is no knowing what it may not get up to." This criminalization of insurgency would inform the mediated politics of the Irish conflict for decades. See McLuhan, *Understanding Media*, 304; Morash, *History*, 125–30; quote from Irish writer and politician Conor Cruise O'Brien cited in Sexton, *Marconi*, 101.

40 Bresnihan and Brodie, "New Extractive Frontiers."

41 Morgenstern, "Etherealization."

42 Bresnihan and Brodie, "Data Sinks"; Bresnihan and Brodie, "New Extractive Frontiers."

43 Brodie, "Climate Extraction."

44 See McVeigh and Rolston, *Ireland, Colonialism*.

45 Power and Collins, "Peripheral Visions."

46 Carroll, *Science, Culture*.

47 Hogan, "Data Center Industrial Complex."

48 See Pasek et al., "Introduction."

Bibliography

Bhandar, Brenna. "Lost Property: The Continuing Violence of Improvement." *Architectural Review*, October 8, 2020. https://www.architectural-review.com/essays/lost-property-the-continuing-violence-of-improvement.

Bresnihan, Patrick. *Bog Irish*. Distinctive Repetition/Irish Research Council, forthcoming.

Bresnihan, Patrick, and Patrick Brodie. "Contested States: Emerging Rural Geographies of Data and Energy in Ireland." In *States of Entanglement: Data in the Irish Landscape*, edited by Annex. Actar, 2021.

Bresnihan, Patrick, and Patrick Brodie. "Data Sinks, Carbon Services: Waste, Storage, and Energy Cultures on Ireland's Peat Bogs." *New Media and Society* 25, no. 2 (2023): 361–83.

Bresnihan, Patrick, and Patrick Brodie. "From Toxic Industries to Green Extractivism: Rural Environmental Struggles, Multinational Corporations and Ireland's Postcolonial Ecological Regime." *Irish Studies Review* 32 (2024): 93–122. https://doi.org/10.1080/09670882.2024.2304946.

Bresnihan, Patrick, and Patrick Brodie. "New Extractive Frontiers in Ireland and the Moebius Strip of Wind/Data." *Environment and Planning E: Nature and Space* 4, no. 4 (2021): 1645–64. https://doi.org/10.1177/2514848620970121.

Bresnihan, Patrick, and Patrick Brodie. "Waste, Improvement, and Repair on Ireland's Peat Bogs." In *Ecological Reparation: Repair, Remediation and Resurgence in Social and Environmental Conflict*, edited by Dimitris Papadopoulos, Maria Puig de la Bellacasa, and Maddalena Tacchetti. Bristol University Press, 2023.

Brodie, Patrick. "Climate Extraction and Supply Chains of Data." *Media, Culture, and Society* 42, no. 7–8 (2020): 1095–114.

Brodie, Patrick. "Data Infrastructure Studies on an Unequal Planet." *Big Data and Society* 10, no. 1 (2023): 1–14.

Carroll, Patrick. *Science, Culture, and Modern State Formation*. University of California Press, 2006.

Clarke, Donal. *Brown Gold: A History of Bord na Mona and the Irish Peat Industry*. Gill and Macmillan, 2010.

Cowen, Deborah. "Following the Infrastructures of Empire: Notes on Cities, Settler Colonialism, and Method." *Urban Geography* 41, no. 4 (2020): 469–86.

Daggett, Cara New. *The Birth of Energy: Fossil Fuels, Thermodynamics, and the Politics of Work*. Duke University Press, 2019.

Davis, Mike. *Late Victorian Holocausts: El Niño Famines and the Making of the Third World*. Verso, 2000.

Elmer, Greg. "A New Medium Goes Public: The Financialization of Marconi's Wireless Telegraph and Signal Company." *New Media and Society* 19, no. 11 (2017): 1829–47.

Fritz, Jose. "Marconi Station at Clifden Connemara and the IRA." *Arcane Radio Trivia*, December 24, 2017. https://tenwatts.blogspot.com/2017/12/marconi-station-at-clifden-connemara.html.

Gladwin, Derek. *Contentious Terrains: Boglands, Ireland, Postcolonial Gothic*. Cork University Press, 2016.

Hogan, Mél. "The Data Center Industrial Complex." In *Saturation: An Elemental Politics*, edited by Melody Jue and Rafico Ruiz. Duke University Press, 2021.

Hu, Tung-Hui. *A Prehistory of the Cloud*. MIT Press, 2016.

Kearns, Gerry. "Historical Geographies of Ireland: Colonial Contexts and Postcolonial Legacies, Part 2." *Historical Geography* 42 (2014): 27–29.

LaDuke, Winona, and Deborah Cowen. "Beyond Wiindigo Infrastructure." *South Atlantic Quarterly* 119, no. 2 (2020): 243–68.

Marconigraph 2, no. 16 (July 1912).

Marconigraph 2, no. 19 (October 1912).

"Marconi Signal Station." *Wagga Wagga (Australia) Advertiser*, November 23, 1907. https://trove.nla.gov.au/newspaper/article/145073841.

Mattern, Shannon. *Code and Clay, Data and Dirt: Five Thousand Years of Urban Media*. University of Minnesota Press, 2017.

McCormack, Mike. "'. . . A Tiny Part of that Greater Circum-Terrestrial Grid': A Conversation with Mike McCormack." Interview with Treasa De Loughry. *Irish University Review* 49, no. 1 (2019): 105–16.

McLuhan, Marshall. *Understanding Media: The Extensions of Man*. 1964; reprint, MIT Press, 1994.

McVeigh, Robbie, and Bill Rolston. *Ireland, Colonialism, and the Unfinished Revolution*. Beyond the Pale, 2021.

Morash, Chris. *A History of the Media in Ireland*. Cambridge University Press, 2010.

Morash, Chris. "Re-Placing 'the Triumph over Time and Space': Ireland, Newfoundland, and the Transatlantic Telegraph." *Canadian Journal of Irish Studies* 43 (2020): 58–79.

Morgenstern, Tyler. "Etherealization in a Racial Regime of Ownership: Marconi in O'ahu, circa 1900." *Media + Environment* 3, no. 2 (2021). https://doi.org/10.1525/001c.23515.

O'Dwyer, Rachel. *Wavelengths*. Verso, forthcoming.

O'Hearn, Denis. *The Atlantic Economy: Britain, the US, and Ireland*. Manchester University Press, 2001.

Parks, Lisa, and Nicole Starosielski, editors. *Signal Traffic: Critical Studies of Media Infrastructures*. University of Illinois Press, 2015.

Pasek, Anne, Cindy Kaiying Lin, Zane Griffin Talley Cooper, and Jordan Kinder. "Introduction: Locating Digital Energetics." In *Digital Energetics*, edited by Anne Pasek, Cindy Kaiying Lin, Zane Griffin Talley Cooper, and Jordan Kinder. University of Minnesota Press, 2023.

Peters, John Durham. *The Marvelous Clouds: Toward a Philosophy of Elemental Media*. University of Chicago Press, 2015.

Power, Dominic, and Patrick Collins. "Peripheral Visions: The Film and Television Industry in Galway, Ireland." *Industry and Innovation* 28, no. 9 (2021): 1150–74.

Raboy, Marc. *Marconi: The Man Who Networked the World*. Oxford University Press, 2016.

Robinson, Tim. *Connemara After the Famine: Journal of a Survey of the Martin Estate by Thomas Colville Scott, 1853*. Lilliput Press, 2012.

RTÉ. "Save Screebe Power Station." 1985. RTÉ Archives. https://www.rte.ie/archives/category/environment/2020/0425/1124827-esb-screebe-power-station/.

Rubinstein, Michael. *Public Works: Infrastructure, Irish Modernism, and the Postcolonial.* University of Notre Dame Press, 2010.

Sexton, Michael. *Marconi: The Irish Connection.* Four Courts Press, 2005.

Sharpe, Christina. *In the Wake: On Blackness and Being.* Duke University Press, 2016.

Shiel, Mark. *The Quiet Revolution: The Electrification of Rural Ireland.* O'Brien Press, 2003.

Starosielski, Nicole. *The Undersea Network.* Duke University Press, 2015.

Tovey, Hilary. "Environmentalism in Ireland: Two Versions of Development and Modernity." *International Sociological Association* 8, no. 4 (1993): 413–30.

Walsh, Shasti. "From Newfoundland to Ireland with Marconi." *Shastified, Continued,* April 7, 2022. https://shastified.com/2022/04/07/from-newfoundland-to-ireland-with-marconi/.

Mediated Extraction

The Production of Dark Ruralities in the Atlantic World

ASSATU WISSEH

The Infrastructure of Black Extraction

In 1818, the US capital was in a state of in between. British troops set ablaze several US federal structures, including the Capitol, the White House, the Treasury, and the Library of Congress in August 1814 as the War of 1812 ensued. The architect B. Henry Latrobe set about the plan of rebuilding the Capitol and, in the interim, US officials assembled at the Old Brick Capitol, the site of the present-day Supreme Court. In this milieu of restoring the imperiled urban infrastructure of American democracy, several statesmen, including Supreme Court associate justice Bushrod Washington, senator and House speaker Henry Clay, president General Andrew Jackson, Colonel Henry Rutgers, and lawyer and poet Francis Scott Key, author of "The Star-Spangled Banner," penned during the War of 1812, gathered in the US House of Representative chambers. These men endeavored to commemorate with an annual report the one-year anniversary of their association, the American Society for Colonizing the Free People of Colour, also known as the American Colonization Society (ACS). The ACS took on the task of restoring the nation by ridding it of a long-standing vice: free Blacks.

The ACS used their organizational media to produce and circulate the idea that free people of color were wretched beings whose extraction from the United States was necessary. In the *First Annual Report*, free Blacks were constructed as menaces that imperiled the nation: "In reflecting on

4.1 George Munter, *U.S. Capitol After Burning by the British*, 1814. Library of Congress, Prints and Photographs Division.

the utility of a plan for colonizing the free people of color, with whom our country abounds, it is natural that we should be first struck by its tendency to confer a benefit on ourselves, by ridding us of a population for most part idle and useless, and too often vicious and mischievous."[1] This passage comes from an extract of a letter written on August 20, 1817, in Baltimore, Maryland, by former US senator General Robert Goodloe Harper, after whom Harper, Liberia, was named. The plan of the ACS, first organized in Washington, DC, in 1817, was to protect the racial hierarchy of the United States by colonizing free Black Americans in West Africa. From the early 1800s to the early 1900s, the ACS sponsored the voluntary emigration of free people of color and freed slaves from the United States across the Atlantic Ocean to a colony named Liberia. Volume 28 of the *African Repository and Colonial Journal*, the official print medium of the ACS, stated that eighty-four ships sailed from various ports along the East Coast of the United States to Liberia from February 1820 to December 1850. These vessels transported 7,160 Black emigrants sponsored by the ACS and the United States government during this period.[2]

This extractive process, whereby physical bodies were moved from the United States to West Africa, was undergirded by a material infrastructure of transatlantic ships. During this time, news, information, arts, and entertainment were transmitted across the ocean by ships. Further, the first permanent telegraph cable wires were laid with the support of naval ships in 1866. Therefore, early nineteenth-century ships comprised both media and transportation infrastructures of nonelectronic and electronic signal traffic long before data centers and cable wires. This chapter explores the discursive level of infrastructure suggested by anthropologist Brian Larkin, who notes, "Infrastructures operate on multiple levels concurrently. They execute technical functions (they move traffic, water, or electricity) by mediating exchange over distance and binding people and things into complex heterogeneous systems and by operating as entextualized forms that have relative autonomy from their technical function."[3] From the oral and written plans of the ACS, its members built first the colony and later the nation of Liberia in West Africa by transforming Black bodies and land into extractable matter. The power to shape Black bodies and land in this way is the prerogative of White men, a term that connotes an ontological category, not a simple designation of race or gender. This ability to turn ideas into form was enabled and supported by the discursive power of the ACS infrastructure to produce what I call *dark ruralities*.

The conceptualization of *rural* is often associated with an image of a place that is unchanging and natural. Geographers Jamie Gillen, Tim Bunnell, and Jonathan Rigg state that such received and entrenched conceptualizations "of a pre-existing rural space as that which is acted upon makes it much more difficult to imagine the rural in terms of transformative geographies."[4] The term *dark ruralities* foregrounds the ways ideas of and experiences with rural West Africa were transformed and brought forth by the ACS's use of various forms of media, from print to an infrastructure of ships to oral speeches. In this way, dark ruralities describe a form of discursive positioning that enabled the extraction of land and bodies. Tarleton Gillespie, in his study of digital media platforms, writes that "discursive positioning depends on terms and ideas that are specific enough to mean something, and vague enough to work across multiple venues for multiple audiences."[5] The ACS made frequent use of terms such as *darkness* and *gloom*, which are abstractions that Gillespie suggests carry semantic weight, enabling the image of West African dark ruralities to reverberate with audiences of their materials. In the 1800s, with a mediascape largely consisting of pages and podiums, discursive staging was an effective device. The term *rurality* foregrounds how these geographic spaces were not always already there but were largely created by the ACS media infrastructure. The production of dark ruralities was a discursive foundation of the enterprise, making it a critical part of the infrastructure.

Infrastructures are often associated with ideas of supporting connection and communication. Mediated extraction centers on the logical processes of critical infrastructures, specifically media and transportation infrastructures, that have historically facilitated transforming Black and/or Indigenous peoples' bodies and lands into extractable matter. In this sense, mediated extraction is about excommunication, the planned exclusion of bodies from the category of the human with protected rights of life, liberty, and land, an ontological position that the cultural theorist Sylvia Wynter calls Man.[6] This integrated exclusion is not a simple inconvenience caused by an accidental gap, break, or malfunction in infrastructural operation that can be remedied by access or the resourcefulness of marginalized persons. Mediated extraction is a matter of technological layered logic, protocol, practices, and material structures working to produce distinct human (White) and subhuman (nonWhite) ecologies. These processes of extraction, including enslavement, mass incarceration, high medical mortality rates, and the school-to-prison pipeline, are ongoing. They do not become obsolete with so-called new media and technologies. I examine the annual reports of the ACS and the *African Repository and Colonial Journal*, official public-facing documents of the ACS, to explore how media and technology mediate the rural and to examine in what ways the ACS's production of dark ruralities was entangled with material and discursive productions of imagined, White urban and rural geographies in the ACS archive. I provide excerpts from these historical documents to illustrate the centrality of White supremacy to the discursive and material production of a geographic region called Liberia. In what follows, I discuss the production of dark and (un)godly ruralities, the entanglement of dark ruralities and what I call the *geographic lite*, and contraction and dark ruralities to suggest that race, media, and infrastructure mediate the human's symbolic and material relation to land.

Dark and (Un)godly Ruralities

Throughout the ACS archive, records of print journal passages and oral speeches by ACS agents constructed Indigenous West African bodies and geographies as dark, representing them as demonic and devoid of Christianity. Conversely, this archive imagined ACS ships bringing multiple metaphoric rays of light, from education to religion to agriculture, to enlighten a purportedly dark Africa. Notable ACS members, including Francis Scott Key, composed a memorial to Congress that used contrasting imagery of light and darkness to support their request that the US House and Senate federally fund the ACS's African colonization efforts: "That such points of settlement would diffuse their light around the coast, and gradually dispel the darkness which has so long enshrouded that continent, would be a reasonable hope, and would justify the attempt, even if experience had not ascertained its success."[7] Here, Africa and Africans were imagined as beset by darkness that enveloped the entire region, while the ACS colony was pictured as light that could illuminate that dark landscape. With this composition, a form of discursive positioning, the ACS makes use of chiaroscuro. Film scholar Hugh S. Manon describes *chiaroscuro* as "the angular alternation of dark shadows and stark fields of light across various on-screen surfaces in film."[8] The term means "light–dark" in Italian, and as a technique, chiaroscuro has been used for centuries in various media, from paintings to film. In the ACS archive, chiaroscuro is the juxtaposition of striking contrast between the imagined light of ACS ships against the darkness of West African ruralities. This compositional technique in print and oral media helped produce a picture of African colonization that framed it as just and necessary. The discursive positioning, a kind of infrastructural process, also worked to conceal the violence of extracting free Blacks from the United States and land from Indigenous Africans.

With chiaroscuro, the ACS actively built an environment that was material and real across the Atlantic world. While very real, the dark ruralities of Liberia were also imaginative. The entanglement of the material and imaginative enabled the violence of extraction to go on in an organized and unremarkable way. This discursive positioning created an aesthetic hierarchy of light over dark in ACS media that coproduced an existential hierarchy: the casting off of free Blacks and the taking of Indigenous African lands to produce the United States as a White cultural landscape. This hierarchy was obscured with pious framing at the ACS tenth annual meeting by Isaac Knapp: "The time was, when the torch of religion, and the lamp of science, shed their mingled rays over the People of Africa. The torch and the lamp have gone out, and darkness has usurped the place of light. But we shall relume them again, and shed on the darkened minds of the People, the renovated lustre of Christianity and Civilization."[9] The ACS extractive project is here imagined as media, a torch and lamp, of symbolic light, religion, and science. The shining away of Indigenous ways of being, constructed as darkness, was positioned as a religious duty that would bring about Christian enlightenment. Knapp, listed as a manager of the ACS Auxiliary Society of Hampden County, Massachusetts, in the *Tenth Annual Report*, was aware of the mortality rate associated with the ACS African colonization mission, for he later stated that "a thousand objections are found which seem important, and which receive far more attention than they merit. It has been said that the Colony have shed blood. It is not denied; and I believe it was justified."[10] Death for the establishment of a colony was deemed a necessary loss. In these passages, there is again a media practice of composing dark against light. Dark here is the death of Indigenous Africans. The juxtaposition of dark with the light of colonial ships helped to veil and downplay the risks that colonial infrastructures bring on nonWhite bodies.

The characterization of African ruralities as dark and demonic implied the existence of disorder in terms of structure and constitution. Against the production of ACS ships as light, the construction of dark ruralities openly marked the West African coast and inhabitants as a "disreputable place inhabited by disreputable people."[11] By *demonic*, the cultural landscapes were constructed as "ungeographic," outside the mappable and knowable bounds of Christianity and reason.[12] Dark ruralities comprised zones of chaotic spaces occupied but left uncultured by supposed illogical and evil natives. The discursive staging of Black bodies, culture, and place as disordered and disreputable aligned with the ACS's characterization of the West as a space of Christianity and reason. The polarities set up a need for an intermediary such as ACS ships to bridge the gap between the divergent sides that were imagined as real and preexisting, although they were created by ACS media. The ACS believed free Blacks and emancipated slaves acclimated to Western ways would bring the lights of culture and civilization back to their dark native homelands.

In reprinted journal passages in the ACS *Second Annual Report*, Reverend Samuel John Mills, an agent of the ACS appointed with coagent Ebenezer Burgess to explore the coast of West Africa, constructed the environments of Sierra Leone and what would become Liberia as shadowy and satanic cultural landscapes in need of salvation by Providence. Mills wrote on March 22, 1818, that "altars on these mountains, which the natives had dedicated to devils, are falling before the temples of the living God, like the image of Dagon before the Ark."[13] Mills's journal excerpts comprise nearly forty-eight pages of the eighty-page annual report. His production of West African ruralities as demonic is a recurring thematic component of the journal excerpts. On April 14, 1818, he composed a picture of rural Liberia as damned by God: "This territory is generally dry, level, fertile, and covered with forests of ancient growth. Soyarrah has only a handful of people. War, slave-trade, red water, and (as Mr. K. says respecting western Africa), 'the curse of God, for their sins and devil worship' have reduced a considerable population to a few scattered relics."[14] This geographic production shaped early images of Liberia and Indigenous Liberians, as Mills's visuals circulated among prominent readers who comprised the audiences of the ACS annual reports. The tethering of Indigenous Blackness to spiritual and environmental deprivation is an extractive logic that makes the displacement of the so-called natives seem divinely ordained. Extraction is associated with nonhuman matter such as ore, gold, and oil and to technologies such as data centers, rigs, and automated drilling. Here, extraction is connected to technologies like printing and ships and, as geographer Kathryn Yusoff suggests, is conceptualized as a more complicated racialized project that delineates the human from the nonhuman while concealing this demarcation.[15] This specific use of media helped transform and position Indigenous African people and lands into entities that not only lack rights but deserve holy retribution.

The removal of Black bodies and land that undergirded African colonization in Liberia can be characterized as an ecocatastrophe. Critical theorist Katherine McKittrick points out that Black studies and human geography are often at odds, as "the black subject and black communities are rarely given any formal academic geographic relevancy."[16] Dual displacement of Black bodies and lands occurred across two continents. However, the ACS's chiaroscuro method seamlessly constructs African bodies and geographies as dark, un(godly), and subhuman, rendering them suitable for a seemingly justified extraction. At the funeral of Colonial Agent J. Ashmun, ACS member Reverend Bacon stated: "At that time, this cape was literally consecrated to the devil; and here the miserable natives, in the gloom of the dark forest, offered worship to the evil Spirit. All this was only a few years ago. And what see you now? The forest that has crowned the lofty cape for centuries, has been cleared away; and here are the dwellings of a civilized and intelligent people."[17] The produced spiritual and spatial darkness justified the geographic "clearing away" of ecologies. In a different media era, infrastructural connection narratives of twenty-first century fiber-optic cables have been marked by what media scholar Nicole Starosielski calls "developmentalist ideologies" that construct nonWhite countries as ripe for capitalist investment by more developed nations.[18] In this discourse, cables work to connect disconnected users in third world countries to the advanced West. The idea is that infrastructures open up these geographies to exchange, creating more users. With the ACS's production of dark ruralities, there is an ideology of rightful displacement of free people of color from the United States and Indigenous Africans from their lands. Rather than a developmentalist discourse of opening up, I locate an ideology of closing off, one buttressed by a "non-ethics of war," to reference cultural scholar Nelson Maldonado-Torres.[19]

Maldonado-Torres, drawing from colonial theorist and psychiatrist Frantz Fanon, states that "the coloniality of Being primarily refers to the normalization of the extraordinary events that take place in war. While in war there is murder and rape, in the hell of the colonial world murder and rape become day-to-day occurrences and menaces."[20] With the ACS infrastructure of ships, free American Negroes, liberated slaves, and Indigenous Liberians were not viewed as additional bodies to forge a connection. These Black bodies were the objects of a warlike removal. The removal was facilitated by the entanglement of dark ruralities, produced via chiaroscuro, with the idea of colonial enlightenment mediated by ACS ships. Further, the production of African rural darkness is relationally connected to the material and discursive understandings of urban America, to which I turn next.

Entangled Dark Ruralities and the Geographic Lite

The production of dark ruralities was entangled with an idea of what I call the *geographic lite*. Such a site was imagined as uninhibited and uninhabited by free people of color and/or enslaved people. This state of being lite or free of Blacks was often accomplished by force, although Blacks still settled and made livelihoods for themselves in these regions, albeit in a climate of violence. Because White supremacist power was not able to actualize the complete exclusion of Blacks from these regions, the geographic lite in the ACS archive was an imagined urban or rural space, an idyllic environmental goal. In the ACS's print and oral media, agents describe regions free of Blacks as enabling domestic tranquility, creating homogenous and perfect ecologies. The geographic lite takes the rural and the urban not as unilateral and polar sites. Rather I draw from Nigel Thrift to consider the relationality in the past between designated urban and rural cultural landscapes in which people lived, the types of spatial arrangements produced to manage those landscapes, and the images of environments inhabitants produced.[21] In the ACS archive, one finds a transatlantic dynamic in which the discursive positioning of American geographies as free of Blacks justified the disregard of their rights to US lands, as well as the settlement of free people of color and freed slaves across the ocean in West Africa.

Just as metaphors of light formed the bases of justifications of African colonization, literal light as fire was used as a means of removal to create the geographic lite. The use of the term *lite* draws on associations with the state of being free of or low in an undesired substance, as well as with the word *light*. In Western culture, *light* represents transparency, honesty, veracity, and purity. It is juxtaposed to *darkness*, the metaphor and metonym for crime, deceit, and licentiousness—all activities that supposedly take place at night. Lights such as electric lamps and candlelight are technologies that mediate and represent knowledge production, giving humans extended access to learning materials and environments. Light in the form of fire is what some scholars call an *elemental medium*—elemental in the sense that it is the foundation of technical media from lightbulbs to the cathode-ray tube.[22] Fire was also foundational to managing and transforming imagined and physical regions.

Fire was a medium used to produce the urban lite, casting off free Black bodies from cities in the United States. Organized mobs of Whites used fire to terrorize free people of color in cities like Philadelphia, Pennsylvania, burning down Pennsylvania Hall, a colored children's shelter, and an African Methodist Episcopal church in 1838.[23] The free Black inhabitants of Philadelphia were met with the normalized violence of war. Fire was used to cement social and geographic boundaries between Whites and free people of color in city spaces. The ACS's *Sixteenth Annual Report* pictured urban free Blacks and slaves existing in an irremediable state for whom separation from the environment of Americans, which is to say White Americans, was the only solution: "Find them where you may, whether in Philadelphia, Cincinnati, Richmond, or Charleston—in a free or in a slave holding state, you find them, with very few exceptions, the same degraded race. No individual effort, no system of legislation, can in this this country redeem them from this condition, nor raise them to the level of the white man, nor secure to them the privileges of freemen."[24] Media historian John Durham Peters contends that light as fire is a human-nature assemblage, an integral medium to the domestication of life on soil. Fire "is an enabling environment for ash and smoke, ink and metal, chemicals and ceramics. Teamed with technique, fire makes matter malleable, turning ores into tools, cold climates into warm ones, darkness into day."[25] Fire, Peters notes, also indiscriminately destroys both the animate and inanimate if not controlled. Regarding the imaginative and material arrangement of the city, the message to Blacks was clear: leave by fire or ship. As discussed above, the light of colonial ships comprised images utilized by the ACS and its supporters to symbolize the benefits of colonizing free people of color on the coast of West Africa. Part of its effectiveness was that this imagery appeared benign compared to the use of fire.

While burning buildings inhabited by free Black people were spectacular performances of White terror, state codes were a more tacit means of producing and managing the geographic lite. The *African Repository and Colonial Journal* reported that Delaware passed a "free negro law" that stated: "Any free negro or mulatto coming into the that State from another, subject to a fine of $60 [nearly $2,000 today], and in the case of non-payment and failure of security to leave in five days, he or she is to be sold out of the State for such amount as will cover the fine and costs."[26] Influential White statesmen used legal codes to limit the settlement of free Blacks in the region, and the ACS reported the existence of such laws to lend credence to the organization, which sought to remove these unwanted bodies. White men used media from fire to print codes to ships to form spatial arrangements. These media also helped to form arrangements of being human; White men enjoyed the protections of life, liberty, and property, while others outside this category of Man did not have access to basic human rights. Even where legal codes were enforced, violent force was still deployed. The *Thirtieth Annual Report*, published in 1847, describes how a mob of Whites in Ohio used firearms to prevent four hundred slaves emancipated by their enslaver's will from residing within the state boundaries: "When they had reached the threshold of their anticipated homes, they were met by an armed company of men and driven back, and after spending most of their money, they were scattered about in the adjacent country, here a few, and there a broken fragment of a family, in a manner most shocking even to *their* ideas of the social relations."[27] The ACS extractive project was marked by a cultural logic of geographic ordering enabled by media and technologies such as fire, guns, and printed laws. Technology and infrastructure bring to mind ideas of security, progress, communication, and connection. Infrastructure supposedly recedes from awareness. With the geographic lite and Black bodies, infrastructures are not invisible. They are highly visible indicators of spatial and ontological arrangements. Infrastructures helped to bring forth and manage the boundaries of White men and their protected rural and urban spaces that were relationally tied to the violability of Black bodies and lands.

Contraction and Dark Ruralities

With the ACS infrastructure of ships, a one-directional linear flow that enabled the desired foreclosure of possibilities for resettlement of free Blacks in the United States was the infrastructural path ideal, not the circuit. And indeed a circuit is at the heart of electronic signal traffic. Energy moves through a path and returns to its starting point, and the ocean is the medium that facilitates the transmission of signals through cable wires across the sea.[28] The conductive property of the water is valued because it enables this circular traffic. At the fourteenth annual meeting of the ACS, George Washington Parke Custis, the grandson by marriage of President George Washington, arose in the hall of the House of Representatives in Washington, DC, to articulate a different image of the ocean:

> What right I, demand, have the children of Africa to an homestead in the white man's country. . . . Let the regenerated African rise to Empire; nay, let Genius flourish, and Philosophy shed its mild beams to enlighten and instruct the posterity of Ham, returning "redeemed and disenthralled," from their long captivity in the New World. But, Sir be all these benefits enjoyed by the African race under the shade of their native palms. *Let the Atlantic billow heave its high and everlasting barrier between their country and ours.*[29]

Here, the capacity of the ship to contract space and time by transporting free Negroes and emancipated slaves across the Atlantic is valorized because the expanse of the Atlantic is imagined as forming an "everlasting barrier" between Black emigrants on the coast of West Africa and White citizens in the United States. In the history of media and communication, discourses of contraction celebrate the annihilation of space and time, as with the railroad and telegraph.[30] Often overlooked in such celebratory discourses is how loss, specifically the abjection of certain marginalized bodies and/or marginalized ways of being, is often constructed as required and ordinary. *Contraction*, as I use it here, is not simply a matter of connecting Black people to Africans and Africa, as with conventional discourse. Contraction with the ACS's infrastructure is about excommunication, the exclusion of certain bodies from human communication. The idea is to transfer Black bodies and resources to a new cultural geography with the assurance that the ocean will serve as insulation to the return of flows, not as an elemental medium conducive to such flows.

This notion of ACS traffic as a linear pathway of no return was articulated by prominent members in meeting halls and surfaced as the perceived mission and purpose of the ACS expressed by constituents and reprinted, or remediated, in the organization's public-facing communications. In *The African Repository and Colonial Journal*, volume 1, no. 8, published in October 1825, the board of directors of the African School in Parsippany, New Jersey, opined the difficulties of raising funds for the education of free Negroes to serve as teachers in Liberia. The board statement expressed the following about free people of color: "They are emphatically a separate people! They must be trained and educated by themselves; and it is this dictate of the soundest wisdom to deal with them as they are. Let them so understand us—that we are instructing them not for our society—not to form our magistrates or legislators; but preparing them to go home."[31] The Atlantic Ocean is erected as a symbolic and material boundary, a wall of water that spatially and socially separates Blacks from ever being American again. In this mode of contraction, both material and symbolic links are loss. The ACS ships are not simply objects that transmit cargoes, messages, and people. As parts of human–technical–social systems, ships comprise and shape larger geographies and economies that are political, and contraction is an aspect of dynamic practices of power, specifically social ordering and place making. ACS ships, and infrastructures more broadly, are simultaneously enabling and disabling living infrastructures. Because the positionality of people positioned as White is often at the taken-for-granted center of media historiography, contraction discourses often focus on how infrastructures enable life through connection. By attending to the positioning of Blackness in relation to notions of the body, sea, and transport, this chapter contends the ship has operated like a living membrane, a selective barrier, mediating the existence of beings in the category of the human with protected rights—that is, White men.

Given the linear path ideal, one finds the use of words like *dispose* in discussions of ACS infrastructures and Black bodies. Charles J. Kilgour, associate judge of Maryland's Montgomery County circuit court, in a letter to ACS member Francis Scott Key, states:

> In a late conversation with Mr. Mortimer M'llhany, of Loudon County, Virginia, touching the Colonization Society, its objects, prospects and usefulness, he signified his wish *to dispose* [emphasis added] of his Negroes according to the plan proposed by them; and I beg leave now, through you, in his behalf, to make to the society the proposition to receive them. They have not yet been consulted on the subject, but he has no doubt but they can be induced to settle upon the shores of their ancestors; and he is anxious to have the aid of the Society in carrying into effect that object.[32]

The idea of "settling on the shores" of the Atlantic is associated with discarding Black bodies, not with ideas of preserving relationships between free Blacks and the United States. This desire to rid Black bodies by means of an infrastructure of ships was the popular perception of the ACS. This perception was produced by the organization as it created and reprinted this picture of its mission in official communications media.

The ecologies established by a circuitous route, a model of frictionless return, must be different from a linear model grounded in contraction, the tightening to eventual closing of communication. It is important to consider what it means to create environments that are no place for White men. Appropriating sociologist W. E. B. Du Bois's notion of the color line, philosopher Sylvia Wynter similarly explains her understanding of the ordering of human being in spatial terms: "This 'space of Otherness' line of nonhomogeneity had then functioned to validate the socio-ontological line now drawn between rational, political Man (Prospero, the settler of European descent) and its irrational Human Others (the categories of Caliban [i.e., the subordinated Indians and the enslaved Negroes])."[33] Contraction when speaking of colonial technological infrastructures, specifically ACS ships, is a process that disorients, rather than reorients, the Black in the larger spatial schema of Man. The development of circular routes was the norm when it came to the traffic of enslaved Black bodies. The process of contraction of colonial slave ships was situated in the hold, where Black humans lost all human connections as they were reduced to flesh chattel.[34] The socio-ontological line remained intact, as did communication as intercourse between the White men engaged in the exchange of Black flesh. However, when the Black bodies were free and/or imagined as possible citizen-subjects, the object was to limit a pathway to return. The departure and desired nonreturn of emigrants on ACS ships across the Atlantic can be seen as delineating not simply geographical and political boundaries but also human borders, marking the White Man's land from the no-man's-land.[35] The ACS infrastructural project was marked by the high mortality of Black emigrants from the beginning. The first ship, the *Elizabeth*, departed from the port of New York in 1820 with eighty Black emigrants and three ACS agents onboard, according to the first issue of the *African Repository and Colonial Journal*. Three agents and twenty-four emigrants died for various reasons within a few weeks of arrival.[36] The ACS admitted the colony of Liberia was still no place for Man and Black settlers as late as the *Seventeenth Annual Report*, published in 1833. Attorney Gerrit Smith, a representative of the auxiliary New York State Society, states that the ACS failed to create in "those institutions in Liberia, which are necessary for the physical comfort and security, and for the intellectual and moral culture of its population."[37] Liberia was a site that was in many ways a wasteland for White men to dispose of free Negroes.

Conclusion

One could easily dismiss yet another historical inquiry into the ACS's plans as part of the usual fixation with the imperialist White supremacist capitalist patriarchal ruminations of dead White men. But these words are far from dead. They are immensely agential as mediations of extraction. White supremacy in this chapter does not connote a character flaw. The point is not to retroactively reprimand White people for being racist. White supremacy is about power—the power to make and order cultural geographies, the power to set the social and spatial arrangements determining which lives are valued and whose rights are protected in time. Often overlooked in media studies is the position that media, technologies, and infrastructures mediate these arrangements. The nineteenth-century ACS infrastructure is one instance of this power.

The production of dark ruralities in the ACS archive rationalizes the displacement of Africans, and this displacement is framed as the remediation of desolate cultural milieus by ships, bringing the civilizing lights of education, religion, and science in the early 1800s. Fanon, about a century later in *The Wretched of the Earth*, without using the word *infrastructure*, theorizes that geographical ordering produced two distinct and irreconcilable ecological zones of the settler and the native. Fanon writes, "The town belonging to the colonized people, or at least the native town, the negro village, the medina, the reservation, is a place of ill fame, peopled by men of evil repute . . . they die there, it matters not where, nor how."[38] Is it possible that Fanon's words point to an elemental, fundamental cultural logic of colonial infrastructure that helps transform Black and/or Indigenous people and places into extractable matter?

In the ACS archives, the undeveloped geographies of Black people are imagined as naturally dark, bad, and deficient, but dark ruralities are created via infrastructural processes of ordering. Relationally entangled with dark ruralities is the geographic lite, imaginative spaces where the expulsion of Black bodies, often by fire and printed laws, is the order of the day. These media productions rationalize the displacement of Black people and lands, framing it as the restoration of forsaken cultural landscapes by the light of fire or the light of the ACS's infrastructure of ships. The ACS chiaroscuro method, the play of light against dark, foregrounds an extractive logic. This discursive production and extractive logic are part of historical processes that normalize the displacement of marginalized cultures, bodies, and land by various infrastructural arrangements that traverse history, from ACS ships to segregated city buses.

Notes

1 ACS, *First Annual Report* (1818), 14.
2 ACS, *First Annual Report* (1818), 150.
3 Larkin, "Politics and Poetics," 335–36.
4 Gillen et al., "Geographies," 191.
5 Gillespie, "Politics," 349.
6 Wynter, "Unsettling."
7 ACS, *Fourth Annual Report* (1821), 26.
8 Manon, "X-Ray Visions," 2.
9 ACS, *Tenth Annual Report* (1827), 10.
10 ACS, *Tenth Annual Report* (1827), 11.
11 Fanon, *Wretched*, 4.
12 McKittrick, *Demonic Grounds.*
13 ACS, *Second Annual Report* (1819), 22.
14 ACS, *Second Annual Report* (1819), 47.
15 Yusoff, *Billion Black Anthropocenes.*
16 McKittrick, *Demonic Grounds*, 10.
17 ACS, *Twelfth Annual Report* (1829), 44–45.
18 Starosielski, "Media Under Water," 54.
19 Maldonado-Torres, "On the Coloniality of Being," 251.
20 Maldonado-Torres, "On the Coloniality of Being."

21 Thrift, "27th Letter," 192.
22 Peters, *Marvelous Clouds.*
23 Cobb, *Picture Freedom.*
24 ACS, *Sixteenth Annual Report* (1833), v.
25 Peters, *Marvelous Clouds*, 117.
26 ACS, *African Repository and Colonial Journal* 28 (1852): 157.
27 ACS, *Thirtieth Annual Report* (1847), 10.
28 Starosielski, *Undersea Network.*
29 Emphasis added. ACS, *Fourteenth Annual Report* (1831), 1.
30 Carey, "Technology and Ideology"; Schivelbusch, *Railway Journey.*
31 ACS, *African Repository and Colonial Journal* 1, no. 8 (1825–26): 277.
32 ACS, *Sixth Annual Report* (1823), 44.
33 Wynter, "Unsettling," 313–14.
34 Spillers, "Mama's Baby."
35 Tawil-Souri, "Cellular Borders" 174.
36 ACS, *African Repository and Colonial Journal* 1, no. 1 (1825–26): 3–4.
37 ACS, *Seventeenth Annual Report* (1834), ix.
38 Fanon, *Wretched*, 32.

Bibliography

ACS (American Colonization Society). *African Repository and Colonial Journal.* 1825–88. https://catalog.hathitrust.org/Record/004565311.

ACS (American Colonization Society). *Annual Report of the American Society for Colonizing the Free People of Colour of the United States.* 1833–47. https://catalog.hathitrust.org/Record/008989892/Home.

Carey, James W. "Technology and Ideology: The Case of the Telegraph." In *Communication as Culture: Essays on Media and Society.* Unwin Hyman, 1989.

Cobb, Jasmine Nichole. *Picture Freedom: Remaking Black Visuality in the Early Nineteenth Century.* NYU Press, 2015. Kindle edition.

Fanon, Frantz. *The Wretched of the Earth.* Grove, 2004.

Gillen, Jamie, Tim Bunnell, and Jonathan Rigg. "Geographies of Ruralization." *Dialogues in Human Geography* 12, no. 2 (2022): 186–203. https://doi.org/10.1177/20438206221075818.

Gillespie, Tarleton. "The Politics of 'Platforms.'" *New Media and Society* 12, no. 3 (2010): 347–64. https://doi.org/10.1177/1461444809342738.

Larkin, Brian. "The Politics and Poetics of Infrastructure." *Annual Review of Anthropology* 42, no. 1 (2013): 327–43. https://doi.org/10.1146/annurev-anthro-092412-155522.

Maldonado-Torres, Nelson. "On the Coloniality of Being." *Cultural Studies* 21, no. 2–3 (2007): 240–70. https://doi.org/10.1080/09502380601162548.

Manon, Hugh S. "X-Ray Visions: Radiography, 'Chiaroscuro,' and the Fantasy of Unsuspicion in 'Film Noir.'" *Film Criticism* 32, no. 2 (2007): 2–27. https://www.jstor.org/stable/24777345.

McKittrick, Katherine. *Demonic Grounds: Black Women and the Cartographies of Struggle*. University of Minnesota Press, 2006.

Peters, John Durham. *The Marvelous Clouds: Toward a Philosophy of Elemental Media*. University of Chicago Press, 2015.

Schivelbusch, Wolfgang. *The Railway Journey: The Industrialization of Time and Space in the Nineteenth Century*. Berg, 1986.

Spillers, Hortense J. "Mama's Baby, Papa's Maybe: An American Grammar Book." *Diacritics* 17, no. 2 (1987): 65–81. https://doi.org/10.2307/464747.

Starosielski, Nicole. "Media Under Water: Friction, Flow, and the Cultural Geographies of Undersea Cables." PhD diss., University of California, Santa Barbara, 2010.

Starosielski, Nicole. *The Undersea Network: Sign, Storage, Transmission*. Duke University Press, 2015.

Tawil-Souri, Helga. "Cellular Borders: Dis/Connecting Phone Calls in Israel–Palestine." In *Signal Traffic: Critical Studies of Media Infrastructures*, edited by Lisa Parks and Nicole Starosielski. University of Illinois Press, 2015.

Thrift, Nigel, "'The 27th Letter': An Interview with Nigel Thrift." Conducted by Paul Harrison and Ben Anderson. In *Taking-Place: Non-Representational Theories and Geography*, edited by Ben Anderson and Paul Harrison. Ashgate, 2010.

Wynter, Sylvia. "Unsettling the Coloniality of Being/Power/Truth/Freedom: Towards the Human, After Man, Its Overrepresentation—An Argument." *CR: New Centennial Review* 3, no. 3 (2003): 257–337. https://doi.org/10.1353/ncr.2004.0015.

Yusoff, Kathryn. *A Billion Black Anthropocenes or None*. University of Minnesota Press, 2018.

PART II PRACTICING RURALITY

5

Mediating the Periphery

Metabolism and Technicity on the Outskirts of Istanbul

BURÇ KÖSTEM

I am making what is now my second trip near Küçükçekmece Lake on the western peripheries of Istanbul. I am with a small group of hikers. We are waiting to cross a ditch between a series of self-built homes and a field belonging to Istanbul University's veterinary school. Ayşe, a resident from the village nearby, returns from the field bearing two plastic bags full of wild mustard she and her friends have collected from the edges of the farm for their *börek*. The ditch serves both as obstacle and reminder of the ever-looming construction efforts in the area. While Ayşe and her friends make it easily across a narrow plank placed across the ditch, many from our hiking group struggle to follow in their steps. Ayşe's friends take their phones out to document our clumsy attempts to cross, snickering at our incompetence and promising that our videos will go viral in their various WhatsApp groups. In the meantime, we have a moment to chat.

"So are you taking these people on a stroll?" In fact, she asks, "Sen bunları mı dolandırıyorsun?," a well-crafted double entendre. *Dolandırmak* literally means to take someone on a stroll, to take someone by the long route, but it could also be used to mean to indicate swindling them (that is, taking them for a ride), as tour guides might do to unsuspecting foreigners, of which there are two in our small hiking group. Ayşe's comments point to the ignorance of urban dwellers navigating this peripheral space. Indeed, walking the peripheries is a continuous lesson in the modes of knowledge, technicity, and mediation required to navigate this space—where to cross a ditch, how to navigate the landscape, how to figure out your route, where phone reception might not work as well. It takes repeated efforts to learn such things.

Subtending our interaction with Ayşe is also an acknowledgment of why anyone might be on a walking tour of the area, which will soon be subsumed by the Istanbul Canal. The Istanbul Canal is a dredging megaproject and an accompanying urban transformation plan that promises to open a new waterway between the Sea of Marmara and the Black Sea. The sheer scale of the project is spectacular, as it plans to dredge a forty-kilometer-wide, 150-meter-deep waterway that will require thousands of construction waste trucks making thousands of journeys, will require electrical lines, gas, and water supply pipes to be rerouted, and will require the uprooting of thousands of people. The dredging of such a waterway will also prove disastrous for the unique ecology of the Sea of Marmara, which will be flooded with the more saline waters of the Black Sea, not to mention the agricultural land, forests, wallows, and lakes located across the northwest of Istanbul. These environments function as sanctuaries for hundreds of thousands of migratory birds that travel between North Africa and Europe. This area has been in a strange limbo, threatened with being subsumed not only by the canal but also by the accompanying expansion of the city westward, which includes new logistics ports, waterfront housing, roads, and bridges.

In recent history, Istanbul Canal was introduced to contemporary Turkish political discourse in 2011 by then–Prime Minister Recep Tayyip Erdoğan. It soon became part of an election promise and a broader set of goals for Turkey's economic and political future by the governing Justice and Development Party (JDP). Described as Vision 2023, a nod to the centenary of the Turkish republic founded in 1923, these promises gave JDP a platform to propagate a lofty economic vision where, by 2023, Turkey would become one of the ten largest economies in the world, increasing its trade volume to $1 trillion and bringing unemployment down to 5 percent. To achieve these goals, the regime would focus on the construction industry, which has consistently represented between 12 and 17 percent of Turkey's GDP over the course of the JDP's time in power. The JDP would focus on *megaprojects*, or complex infrastructural systems, which act simultaneously as avenues of capitalist accumulation, symbols of political legitimacy, and affective nodes of organizing political subjectivity around infrastructural progress and national resurgence.[1] One geography where this construction activity concentrated was the northern and western peripheries of Istanbul, a more sparsely populated and rural neighborhood that is adjacent to the city of 20 million people. Particularly of interest were three megaprojects: a third bridge and accompanying motorway across the Bosporus, the North Marmara Motorway; a new mega-airport adjacent to the bridge, Istanbul Airport; and the infamous Istanbul Canal. While the economic promises associated with these projects have not come about, the former two of these projects are in large part complete and operational, with the canal remaining the last step in the ongoing transformation of this geography.

In response to these social and ecological changes across the peripheries of Istanbul, a growing number of artists have taken an interest in representing and exploring the city's peripheries. More than half a dozen times over the last two years, I have walked across the path that the Istanbul Canal will subsume with different hiking groups and artists. Sometimes these walks were organized by independent groups like Hiking İstanbul. Yet for the most part, they were part of an artwork, *Between Two Seas*, a four-day walking route designed by Serkan Taycan, a photographer, artist, and designer. As the map that accompanies the *Between Two Seas* walking tour explains,

> *Between Two Seas* is a four-day walking route in the near west of İstanbul, between the Black Sea and the Marmara Sea, which allows one to experience the threatening transformation of İstanbul on foot. . . . Layer by layer, the route progresses from . . . rural and forest areas, and water basins to reach the center of the city. The trajectory passes through lignite mines, the area of the new airport, the road leading to the 3rd Bosporus Bridge, excavation dump sites, industrial sites, and housing areas, . . . and inner-city vegetable gardens. *Between Two Seas* is both a proposal and an invitation.[2]

The *Between Two Seas* route covers an area that that stretches from the coast of the Black Sea down to the southwest, near the Sea of Marmara, right alongside the path of the canal. In addition to *Between Two Seas*, Taycan has documented the transformation of the peripheries of the city in several photography series. Alongside Taycan's efforts we can add the work of photographer Bekir Dindar, filmmaker İmre Azem, and cartographer Gökçen Erkılıç, all focused on studying, representing, and thinking through the shifting peripheries and coastlines of the city, especially under the past twenty years of the JDP's rule.

This notion of the "periphery" (*çeper*) is thus ubiquitous in the artistic, activist, and academic work that seeks to represent this geography. In such work, the *çeper* is construed as a geography that is on the precipice of being swallowed by the city's expansion.[3] My encounter with Ayşe captures many of the elements that make up peripheral space. As geographer Brian Rosa describes it, urban peripheries are "residual spaces of various scales without clearly defined purposes, often re-appropriated by a variety of formal and informal uses (light industry, warehousing, recreation, squatting, etc.). Though they seem separate from the urban fabric, they serve an essential function to cities; these are spaces that were conceived for flows of materials, people and of capital, tending to pay little mind to the impact they have at the ground level."[4] The peripheries of Istanbul, that I've introduced above, *feel* distinct from the city in a way that is immediately recognizable in the visual representations of this geography—open fields with the city in the background, informal housing juxtaposed to newly built apartments, farming and extraction sites crisscrossed with electricity lines, with the resources and residues of construction as the main ecological and technical activity that produces this space.

In what follows, I offer a theoretical account of the periphery that I develop in conversation with this archive of artwork that tries to represent Istanbul's outskirts. To do so, I bring into conversation two complementary accounts of the periphery that build on and trouble one another. First, focusing on the specific social and ecological relations that urban construction is premised on, I conceptualize the periphery as a space of metabolic exchange between countervailing tendencies of ruralization and urbanization through the Marxist concept of the metabolic rift. Here I draw on the *Between Two Seas* map as well as the work of photographer Bekir Dindar in his 2018 series, *Cavity* (*Oyuk*). Drawing on the work of French philosopher of technology Gilbert Simondon, I then complicate this picture by focusing on construction and agriculture as technical activities that are constitutive of Istanbul's urban metabolisms. Here, I place my walks along the *Between Two Seas* route in conversation with other works of art that study Istanbul's peripheries. As I do so, I focus on how, undergirded by capitalist expansion, the technical assemblage of urban construction in Istanbul not only separates people from the means of their social reproduction but also alienates people from the technical means with which to engage and shape their environments.[5]

Last, I end by asking what vision of radical politics can emerge from the peripheries of urban space. As recent interventions in rural studies have articulated, theoretical accounts of urbanization often risk reducing the rural into a residual or obsolescent category, waiting to be subsumed by the urban.[6] While scholars who critique this tendency recognize capitalism's valorization of the city, they also seek to define the urban–rural less as a binary and more as a relation that both revives and exceeds processes of urbanization.[7] In following their suggestions, I end by asking what focusing on the technical production of the periphery teaches us about the shifting sights and horizons of political struggle in Turkey and beyond. I propose that an eco-Marxist account of social transformation on the peripheries of urban space requires not only a scale-down of technical systems like megaprojects but equally a scale-up of technical culture, the generalization of technical knowledge, the popularization of technical tools, and the construction of technical systems that are more amenable to collective experimentation.

The Periphery and the Metabolism of Urban Construction

Part of what makes the peripheries of Istanbul and the archive of artwork that assembles around them so interesting to study is their ability to concretize and thus better understand the material forces that constitute urban space. On the one hand, the megaprojects that gather on the peripheries of Istanbul extend the reach of urban space, transforming the city's peripheries into a logistical space that sustains the movement of energy and materials toward the city. As sociologist Sezai Ozan Zeybek writes, "Istanbul neither begins nor ends in Istanbul. It conquers other places. Millions of waste objects, alongside the fruits of its wealth, spill over to other places. Istanbul hollows out other geographies, creates fallout zones."[8] The clustering of new highways, airports, and shipping ports all across the western peripheries of Istanbul is indicative of such a relation to the rest of the city. Yet on the other hand, such megaprojects also represent an attempt to transform the peripheries into new frontiers of development. When describing the construction of projects such as the Istanbul Canal or the Istanbul Airport, it is commonplace for government officials to claim that they are not merely building a new airport or a new waterway but rather constructing a completely new city, complete with things like hospitals, hotels, and congress centers.[9] This multiplying force of urban infrastructure constantly reproduces mini cities within cities.[10]

One common material structure associated with the periphery, especially in its artistic representations, is the constant movement of construction trucks (*hafriyat kamyonu*) back and forth to urban space, carrying

5.1 The Istanbul Canal project as depicted in the office of a real estate agent in Şamlar village. Istanbul Airport is at *bottom right*; it also indicates ports, logistics centers, and new residential areas around the canal area. Photo courtesy of Nick Hobbs.

stone and rock to the city and depositing construction and demolition waste back into the peripheries.[11] In this way, the *Between Two Seas* map adds, construction trucks turn the city "inside out": "There are many excavation trucks with yellow dumpers driving back and forth. The excavation dumpsites and the former mine beds and stone quarries north of the city are here. Some of the biggest construction projects in İstanbul continue in this area. The trucks carry millions of cubic meters of excavation waste from these sites and all over İstanbul to waste dumps, turning the city inside out. İstanbul meets its own inverted layers on its outskirts. And then, an entirely different İstanbul rises upon these fragments. It is truly frightening."[12] This movement of excavation trucks relates to the problem of representing the city. As Taycan notes, Istanbul's rapid expansion has made it quite difficult to *enframe* the city in a single panoramic shot. Yet the accumulation of rubble and waste dumped by different trucks traveling from different areas of the city represents a funhouse mirror of the panoramic view. One encounters the city, but not in a transcendent view from above. Rather, it is through an encounter with the multiple layers of rubble piled on its peripheries.

A characteristic feature of the concept of periphery, then, which the figure of the hafriyat truck and construction waste illustrates, is the overlaying of spaces of production, consumption, and disposal. Throughout

5.2 Serkan Taycan, photographs from the *Shell* series, 2015.

the *Between Two Seas* map, or indeed in Taycan's 2015 photography series *Shell*, it is possible to encounter sites of disposal where construction waste and garbage have historically been hidden, where abandoned mines lie, superimposed onto sites of production, where giant sand and stone quarries operate, and where new construction projects are unfolding. Implicit is the role of consumption, with foreboding horizons of new buildings seeming to emerge and surround the city from every corner, creating an unbounded feeling of urban development. For example, in *Cavity* (*Oyuk*), Istanbul-based photographer Bekir Dindar documents the giant stone quarries around the city's western peripheries, describing them as extractive cavities that form as urban buildings continues to rise. While depicting these cavities, Dindar also captures the hafriyat truck drivers who navigate these giant chasms as well as the precarity and vulnerability of these drivers as they navigate the dangerous geography. Dindar explains, "Earth, shaped by the increasing need of the growing city, is its source and soil its coat. . . . Buildings rise; cavities deepen."[13]

One helpful way to think about this aspect of the periphery is through the Marxist concept of metabolic rift. First mentioned in the writings of Karl Marx and later elaborated by contemporary authors like John Bellamy Foster, the term *metabolic rift* is used to account for the increasingly destructive relation between processes of ecological reproduction on the one hand and capitalist production on the other.[14] Marx uses the term *metabolism* to describe how labor mediates the relation between human beings and nature, arguing, "Labor is, first of all, a process between man and nature, a process by which man, through his own actions, mediates, regulates, and controls the metabolism between himself and nature."[15] Yet this general account of metabolism is overlaid, as Kohei Saito notes, with a second-order mediation of the historically specific system of labor and production.[16] Under capitalism, the labor process is subordinated to the

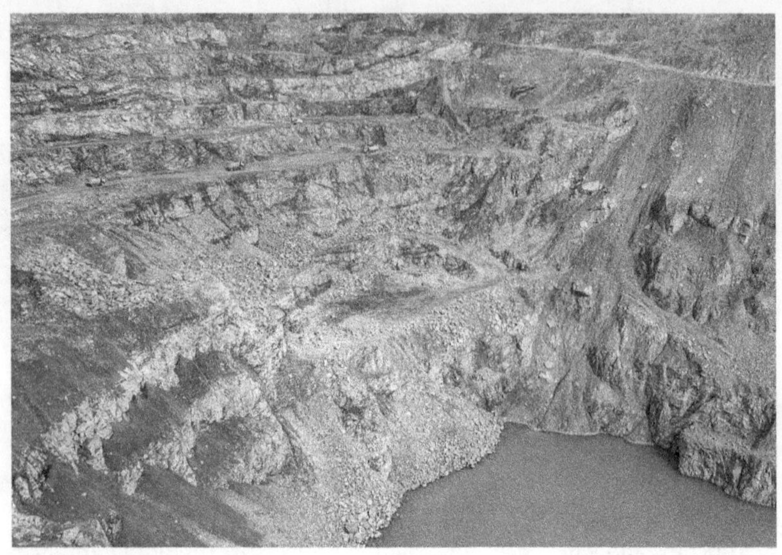

5.3 Bekir Dindar, photograph from the *Cavity* (*Oyuk*) series, 2016.

5.4 Bekir Dindar, photographs from the *Cavity* (*Oyuk*) series, 2016.

endless generation of surplus value. Therefore, capitalist metabolism entails a separation not only from labor itself but also from what we might call *ecotechnicity*, or the human ability to consciously organize and work on one's environment, which now has to be subject to a second-order mediation of generating surplus value.[17]

This separation in turn creates a rift between processes of ecological reproduction and social production. The particular rift Marx was most interested in was related to the transition from agrarian to industrial society, materialized most clearly in the spatial rift between town and country.[18] These processes of urbanization and ruralization had devastating ecological effects. In the countryside, it led to the increasing adaptation of large-scale agriculture and techniques that contributed to the displacement of nutrients from the soil such as nitrogen, phosphorus, and potassium.[19] In the city, however, the same dynamics led to waste accumulation and resultant outbreaks of disease and death.[20] At stake in the concept of the metabolic rift was how a reorganization of the relations of production under industrial capitalism, the fact that a critical mass of workers in the city were compelled to sell their labor to reproduce their lives, simultaneously prompted a reorganization of nature, creating two complementary spatial and technical arrangements, rural and urban, increasingly at odds with cycles of ecological reproduction.

During Marx's time and for much of the twentieth century, the abolition of the separation between town and country, the redistribution of agricultural production into local communities, and the disaggregation of industrial production throughout the countryside was an important horizon of revolutionary politics.[21] Yet in the peripheries of contemporary Istanbul, the distinction between town and country already appears blurred, though not through revolutionary struggle. It is important to point out the ubiquity of infrastructure and mediation in the peripheries from Ayşe's WhatsApp groups to the GPS system I used to navigate my route. In addition to being a mixture of urban and rural forms, such peripheral areas also contain the multiple interactions, mediations, and sites of struggle between them.[22] Far from being undeveloped or abandoned wastelands, the urban peripheries of Istanbul have experienced rapid development and migration creating new population centers and cities within cities that are at odds with the modes of subsistence surrounding them. Consider that right next to the field I am crossing is Kayaşehir, one of the largest housing projects constructed by the Mass Housing Development Administration in the neighborhood of

5.5 The view between Istanbul University's veterinary farm and the Faculty of Agriculture. Photo courtesy of Nick Hobbs.

Başakşehir. While villages like that of Ayşe's have been shrinking in size, places like Başakşehir have experienced rapid growth, from five thousand people inhabiting the area in 2010 to nearly thirty thousand people in 2011 after the project was constructed, and to over a hundred thousand people today.

This hybrid character of the peripheries serves to accentuate the metabolic rifts that continue to subtend the distinction between rural and urban forms, collapsing together spaces of consumption and production.[23] The hybridity of the periphery then is less evidence of the rural having already been subsumed by the urban and more a demonstration of how the rifts that constituted these distinct spatial arrangements can also be collapsed onto the same geography. To paraphrase the introduction to this volume, perhaps it makes less sense to talk about the periphery as an object of analysis than it does to talk about the peripheral as a condition and peripheralization as a hybrid process, with the metabolic rift as one of the mechanisms of this process.[24] All along the *Between Two Seas* route, I have encountered traces of this process, with the decline of agricultural work and its related migration happening away from the peripheries, in part a consequence of the disruptive effects of megaprojects. The towns along the peripheries of Istanbul had already experienced a decline in population as a result of the globalization of Istanbul's food systems and supply chains. Throughout the *Between Two Seas* route, one encounters many urban and peripheral forms of agriculture such as *bahçes* (urban gardens), *bostans* (orchards), and *bağs* (vineyards), since sacrificed to new infrastructure projects.[25] This globalization of Istanbul's food systems was a conscious result of government policies like removing agricultural subsidies, permitting wide use of synthetic fertilizers, dissolving farmers' cooperatives, and weakening agricultural labor in the face of global supply chains, all of which rendered the production processes more reliant on the technologies and standards enforced by larger companies.[26] Moreover, the sprawl of urban space and the increase of urban populations created rising land prices that further squeezed out agricultural production from the peripheries.[27]

This displacement from urban peripheries was also accompanied by a precaritization and immiseration of workers in the city, especially in the post-1990s era. This worsening of working conditions formed the material basis of Turkey's construction boom. Such poor working conditions were reinforced in the case of Istanbul, through successive waves of racialized migrant workers who often fled state violence perpetrated by the Turkish military. As of September 2018, there were around 2 million workers formally employed as part of the industry. Nearly 35.8 percent more worked informally, a considerable number of whom are refugees, migrants, and other racialized minorities. In the last fifteen years, even among insured workers, one out of every three workers has been subject to workplace accidents.[28] And the number of workers who lost their lives in 2024 was 484, as of 2024, making construction the deadliest industry in the country.[29]

Relatedly, the waste industry downstream from construction requires significant amounts of fixed capital for entry. Thus, it is incredibly dependent on the boom-and-bust cycles of construction for survival. The imperatives for competition are significant in the face of the fickle economic prospects encountered by truck drivers and smaller hafriyat companies.[30] Indeed, many of the smaller hafriyat firms that first entered the industry during the early years of Turkey's construction boom have now gone bankrupt.[31] Hafriyat truck drivers, especially those working for subcontractors, often note that they are unable to get paid for three or four months in a row during even the best of times.[32] In the face of such regulation and competition, smaller companies are able to survive through means other than carrying hafriyat to officially designated disposal sites. They either sort through the hafriyat for valuable metals and minerals or dump it illegally.

Such precarity directly contributes to the metabolic relations constituted by construction bringing about what Esra Sert describes as a "metabolic flow of iron and cement."[33] For Sert, in the post-1990s period, Istanbul's urban metabolism is constituted by the movement of iron and cement into the city and the movement of construction debris and waste out of the city, with the latter either dumped illegally onto the city's peripheries or transformed into land reclamation projects that redesign the city's waterfront. It is this interaction between the construction of buildings and megaprojects on the one hand (what Sert calls "towers") and land reclamation projects that transforms the city's coastline on the other (what she calls "voids") that constitutes the city's metabolism.[34] Construction waste is laden with materials like asbestos, radon gas, copper, and mercury, which, when dumped illegally, pose health risks for construction workers, people inhabiting the peripheries, and those living in the surrounding environment.

The sheer volume of materials produced in this construction boom has its own perhaps even more salient metabolic effects. Between 2018 and 2022, Turkey has produced on average above 80 million metric tons of cement per year—fifth in the world behind China, India, Vietnam, and the United States.[35] The production of such large volumes of cement involves the use of materials like construction sand, limestone, and water, all of which insert the city into planetary metabolic relations. A key component of cement production is fossil fuels. Fossil fuels are used to fire up the kilns in which limestone will be processed and transformed into a substance known as clinker. Once cooled, clinker can then be mixed with various other substances and ground into cement. This process emits a significant amount of carbon, which comes both from the consumption of fossil fuels in firing up the kiln and from the chemical reaction of limestone to heat.[36] Worldwide, the cement industry accounts for 8 percent of yearly global carbon emissions. In this sense, Istanbul's urban construction is implicated in the planetary metabolism of carbon. Thus, a metabolism created by a precarious labor force, cheap credit, cheap construction materials, and increasingly productive agricultural work finds itself increasingly at odds with the cycle of planetary carbon. Through the artwork that attends to it, the periphery presents itself as an interval through which one can map out this metabolic relation.

The Periphery and Ecotechnicity

In thinking through the metabolic rifts that contemporary capitalism brings about, Jasper Bernes writes that eco-Marxist debates need a theory of technology that can reckon with path dependency.[37] Rather than focusing on the merits or drawbacks of discrete technological objects, Bernes diverts our attention to the aggregate character of complex technical ensembles and how they fit together, cohering in ways that strongly determine and influence their future use, whether in urban or rural settings. The story of infrastructural development and urbanization in Istanbul's peripheries is itself one of such path dependency. Consider the three infrastructure projects I pointed out: the Istanbul Canal, the Istanbul Airport, and the North Marmara Motorway. More than individual technical systems, these infrastructure projects reinforce one another, marking a logistical, social, and ecological shift in Istanbul that, while dependent on past tendencies, also brings about significant ecological transformation. The North Marmara Motorway helped ease the movement of hafriyat trucks in and out of the city's peripheries. The grounds of Istanbul Airport, built near the village of Yeniköy, used to feature long-abandoned opencut lignite mines that had since been transformed into ecologically diverse lakelets and wallows. The construction of the North Marmara Motorway allowed the transportation of thousands of trucks carrying excavation materials to the area, which were then used to fill in such lakelets and provide the airport with a more solid foundation. The motorway also provided an alternative route to reach Istanbul Airport for people traveling in and out of the city, while the construction of the airport itself inevitably drew newcomers to the area. While the project seems to be on pause for now, if financing for it can be secured, the Istanbul Canal will create ports for additional shipping traffic along this route while also creating new residential spaces that would absorb any surplus capital directed to the region.

The technical assemblage of megaproject construction creates path dependencies that reinforce one another. The metabolic effects of ecological destruction are long term; it is difficult to simply rebuild an ecology like the northern forests. While they haven't fulfilled their economic promise, collectively, these three projects have begun to transform this geography into a logistical space that manages the flow of people and goods toward the rest of the city. They have destroyed the lakelets and grazing grounds that formerly sustained the area's small-scale animal farming. They have partially destroyed the forests north of the city, which many argue will prove disastrous for the city's air quality, further introducing rifts within the city's metabolic interactions with its surroundings. Such transformation also means that rural modes of subsistence have been rendered obsolete in many villages along the western peripheries of Istanbul. Should the canal be completed, anywhere from 50 to 70 percent of the area it subsumes will be agricultural land, a significant portion of which is still in use today.[38] In turn, this destruction will help to reinforce the city's reliance on logistical systems and supply chains to help provision for its basic needs such as food.

These series of mutually reinforcing technical systems also have an alienating effect. In her essay accompanying the *Between Two Seas* booklet, Merve Ünsal argues that through exploring the concept of the periphery, one can understand the city's "growth as a form of alienation."[39] In Taycan's work, this alienation is treated as a side effect of the spatial separation between urban and rural space. Living in the urban core, it is easy to ignore the giant stone quarries that Taycan and Dindar document—which is perhaps why it feels so horrific to confront them in artistic work. Yet the alienation inherent in confronting the peripheries is not merely a psychological dissociation or a loss of subjectivity.[40] Nor is it an effect of physical distance. In fact, this sense of dissociation between urban and rural space can be generalized into a broader phenomenon: how technical systems reconfigure one's ability to interact with and shape one's environment.

One way to understand this alienation is through the work of Gilbert Simondon, whose early writings on the philosophy of technology are especially helpful. Simondon conceives of technology as an aspect of human culture that has thus far been ignored.[41] Specifically, Simondon notes two countervailing cultural narratives through which technology is misunderstood. On the one hand, cultural narratives portray humans as tool bearers and technology as comprising mere instruments of human labor.[42] On the other hand, technical systems are portrayed as evil masters that rule over and control humans, transforming human labor into the "conscious organs" of mechanical reproduction.[43]

These two images of technicity are visible in the artwork engaging with the peripheries of Istanbul. On the one hand, construction appears as a primal, almost artisanal activity involving the mastery of elements such as sand, water, and rock. This is the image of construction we see in the paintings of artists like Mustafa Pancar, a member of the Hafriyat Collective of the late 1990s and early 2000s, whose 1996 work *Hafriyat* is reminiscent of a sandbox. Yet in more recent depictions, such as the photographs of

Dindar, construction machinery appears as foreign and frightening, part of a series of technical systems that subsume and incorporate humans into their operation. Indeed, for the drivers, technicity is most visible as an extensive system of the GPS tracking devices through which the Istanbul municipality and private companies manage and watch their movements and apportion fines in case of violations. For the inhabitants of the peripheries, the movement of hafriyat trucks throughout this space is regarded as an especially hostile and oppressive presence, often described as hafriyat terror, because the fast-moving trucks are prone to dropping construction waste onto unsuspecting pedestrians and drivers, resulting in death. It is perhaps because of this monstrous side of urban construction that İmre Azem's 2011 documentary, *Ecumenopolis: The City Without Limits*, about urban transformation in Istanbul, depicts the city being attacked by a cyborg monster that hovers above the city, sucking up construction waste in the center and dumping new buildings in the periphery.[44]

For Simondon, these countervailing cultural understandings of technology—as potential rival or as mere instrument—relate to a broader technical alienation that is pervasive in contemporary society. This mode of alienation describes not only the loss of ownership the worker experiences in the means of production or in the fruits of labor but also, and more fundamentally, an alienation from the essence of technicity itself. When Simondon published his early work, he seems to have had in mind the regime of factory work that alienated workers from the ability to shape, choose, and collectively maintain the technical ensembles that they were being asked to work with, creating a society with an impoverished technical culture. Yet this line of thought is insightful when thinking through the concept of metabolism, especially as it manifests in the peripheries of Istanbul.

Marxist theorist Søren Mau notes the pitfalls of the concept of metabolism—how it erroneously posits an original unity or harmony between ecological reproduction and social production that is subsequently troubled by the intervention of capitalism and industrialism. Instead, Mau bases his critique of capitalism less on ideals of harmony and more on those of collective freedom and subsistence. Premising social reproduction on capital's own self-valorization, capitalism deprives us of the means to organize, shape, and technically mediate our environments.[45] The problem with Istanbul's metabolism of construction is that it makes it impossible for the vast majority of the city's inhabitants to collectively shape their metabolic interactions with the city; it makes it impossible to

5.6 Mustafa Pancar, *Hafriyat*, 1996. Oil on canvas, 145 × 200 cm.

5.7 Still from İmre Azem, *Ecumenopolis* (2011).

exercise and organize collective freedom, provisioning, care, and limitation around the city's relation to its environments.

The notion of ecotechnicity adds to this critique by focusing on the technical conditions of that social reproduction. I have noted above how walking the peripheries of the city is a mediated act, requiring various forms of technics, knowledge, and expertise. Throughout my walks, I often had to rely on printed maps of the *Between Two Seas* route, the wisdom of previous walkers, spotty phone signals, and dying batteries. I have wrecked multiple pairs of shoes and finished multiple bottles of sunscreen. I have relied on the knowledge of many of the inhabitants of the peripheries—people like Ayşe—who taught me multiple things about the land, including how to look for wild mustard, where to look for blackberries, where to collect mussels, and how to get through a fence. The experience of walking offers moments of encounter with the peripheries that are invaluable. As Türeli and Al write, "Walking initiates new forms of knowledge, solidarity and resistance": "Specifically, the contribution of artistic projects such as *Between Two Seas* has been to highlight the lack of knowledge about the periphery of a city undergoing immense transformation, turn that into a collective learning opportunity, and create a community around walking in the periphery."[46] Yet the affective tenor of these encounters can also be ambivalent. When walking the peripheries, one often encountered the effects of airport, highway, and dredging projects as so many obstacles that foreclosed and subsumed one's ability to shape one's environment. Some walks were exhausting, especially in the summer. Others left walkers with a sense of distance from the inhabitants of the peripheries. Activists often note the divergent interests and forms of knowledge between their organizations and the more localized struggles of the people who inhabit the peripheries. Moreover, construction in the peripheries often operated under conditions of secrecy and arbitrary expulsion. As we walked the *Between Two Seas* route, sometimes our path was rerouted or blocked by gendarmes or private security guards who told us that the area was no longer accessible on account of ongoing construction activity along the canal's path. Nor was it uncommon for local communities to side with the gendarme, convinced that our presence would undermine the technical operations of construction activity, best left to the experts.

Such experiences can be interpreted as encounters with a sort of ecotechnical alienation, indicating not only a lack of legal ownership or physical control over the means of ecological and social reproduction but also an absence of technical means, know-how, and expertise around them.

Walking, while generating its own forms of technical know-how and encounter with the land's physical affordances, nonetheless highlighted the sheer complexity and of urban infrastructure and the extent to which infrastructural maintenance, repair, and provisioning are embedded within the bureaucratic, technical, and disciplinary apparatuses of state and capital.

Conclusion

What alternative vision of politics emerges from this peripheral account? Simondon's proposed solution for the predicament of technical alienation described above is the full abolition of labor in favor of "technical activity."[47] In contrast to the alienated account of technology, Simondon argues that technicity is a "work of organization" that continues life processes by other means.[48] The technical individual organizes and acts on its milieu and in doing so opens us up to transindividuality. That is, through technicity, one can begin to participate in collective life, shaping and interacting with our collective environments.[49] This would require a new relation to technical objects, one that includes but goes beyond collective ownership, to the ability to collectively maintain, provision, and repair our technical ensembles.

In much of the eco-Marxist work that imagines what a future society might look like, communism is conceived of as requiring a scale-down of technical ensembles—the ability to break apart industrial agriculture and create local practices of farming and provisioning, and the ability to turn off mechanical production when the sun is out and to take long afternoon naps.[50] So much of the contemporary ecological left is focused on this scale-down—no more megaprojects, technical systems that rely on a smaller throughput of materials and energy, more localized infrastructure and transport systems.[51] Yet one might add to this that an eco-Marxist horizon would equally require a scale-up of technical culture, the generalization of what Simondon calls technical activity.

In an era of increasing disasters and ecological collapse on the one hand and rapid urbanization and immiseration on the other, a more generalized condition of technical knowledge and expertise around infrastructure would be critical for radical projects that attempt to create spaces of autonomy and resistance. Despite being governed by engineers and experts, such autoconstruction is far from uncommon. Thinking through the technical assemblage of reinforced concrete (the process of embedding concrete structures with rebar to improve their durability), Adrian Forty explains, "reinforced concrete is one of the new 'technologies of poverty'—in overall quantity consumed, its use by self-builders in poor countries probably exceeds all other applications. In the shanty towns of the world, its use is characterized by ingenuity rather than innovation: . . . What matters is the way small amounts of reinforced concrete are made to go a long way."[52] Ultimately, the scale-up of technical culture I've discussed would depend on technologies that are themselves more open to technical intervention and more driven by ingenuity than solely by innovation. Such a technical culture would have to forgo the characteristic "opaqueness" of urban infrastructure—that is, its ability to operate in the background of social relations; instead, it might require a more "convivial" social relation to technology, such as the popularization of tool-lending libraries and DIY cafés along with collective construction efforts.[53] What would it mean to construct infrastructural systems that can be collectively repaired and maintained, autonomously of security officers, experts, the gendarme, or technical and repressive state apparatuses?

There are already hints of such questions being raised in urban and rural struggles. For example, in the aftermath of the historic Gezi rebellion in 2013, where millions across Turkey took to the streets in protest of the government, multiple occupation houses slowly opened in different parts of Istanbul. Such occupation houses would often take over unfinished buildings, repairing and transforming them into sites of experimentation with radical politics, albeit often in limited ways.[54] It is striking that the two most enduring such occupation houses, in the Yeldeğirmeni and Caferağa neighborhoods, both involved extensive and communal construction efforts: fixing roofs, removing bags upon bags of construction debris, creating electrical outlets, installing heating systems, and building carpentry workshops and photography darkrooms.[55] Both occupation initiatives also attempted to build their own common gardens and harvest festivals as gestures toward food sovereignty. Indeed, beyond such occupation attempts, autoconstruction remains part of the repertoire of urban resistance for marginalized and racialized peoples across Istanbul.

Today, construction, extraction, and accumulation cycles and their attendant modes of resistance and struggle have shifted from the days of Gezi, moving beyond urban centers to also include sites facing peripheralization. Under such circumstances, new horizons of resistance have become more important in Turkish politics through various struggles around megaproject construction, coal and gold mining, ship breaking, and extraction that take place outside of Istanbul yet nonetheless bear on its politics. Moving beyond Turkey's borders, in places like Rojava, the autonomous region to the south of Turkey, home to Kurdish, Arab and Assyrian peoples, collective practices of agroforestry and of autoconstruction and water management are conceived of as in direct opposition to the construction driven growth model that Istanbul's megaprojects represent.[56] Last, the concept of ecotechnicity can help us interrogate the technical and metabolic relations that subtend these struggles and to what extent they are connected to and distinct from the urban struggles that Gezi represented.

Notes

1 Tuğal, "Politicized Megaprojects."
2 Taycan, *İki Deniz Arası.*
3 Ünsal, "On the Periphery," 7.
4 Rosa, "When Is a Wasteland?"
5 Mau, *Mute Compulsion*, 129.
6 Gillen et al., "Geographies of Ruralization."
7 Gillen et al., "Geographies of Ruralization," 199.
8 Zeybek, "İstanbul'un Yuttukları ve Kustukları," my translation.
9 Baysal, "Terricide."
10 Easterling, *Extrastatecraft.*
11 The word *hafriyat*, from the Arabic *hafr* (to dig), literally means "excavation," but the term is also used to describe construction waste and debris.
12 Taycan, *İki Deniz Arası.*
13 Dindar, *Oyuk.*
14 For more on the intellectual genealogy of the concept, see Foster, *Marx's Ecology*; Kohei Saito, *Marx in the Anthropocene.*
15 Marx, *Capital*, 1:283.
16 Saito, *Marx in the Anthropocene*, 20.
17 Mau, *Mute Compulsion*, 102; Saito, *Marx in the Anthropocene*, 24. I owe the term *ecotechnicity* to an ACLA panel organized by Derek Woods and Thomas Patrick Pringle. While this specific use of the term is my own, their work has helped me immensely.
18 Marx, *Capital*, 1:637.
19 Foster, *Marx's Ecology*, 150.
20 Foster, *Marx's Ecology*, 151.
21 Bernes, "Belly."

22 Qviström, "Peri-Urban Landscapes," 427.
23 Parsons and Lawreniuk, "Geographies of Ruralisation," 204–7.
24 Introduction to this volume.
25 Taycan, *İki Deniz Arası.*
26 Keyder and Yenal, "Agrarian Change."
27 Turkkan, "Feeding Global İstanbul," 183.
28 "2024 yılında en az 1897 işçi."
29 "2024 yılında en az 1897 işci."
30 Öztürk, "Seeking the Excavation."
31 Interview with hafriyat truck drivers, April 2021.
32 Interview with hafriyat drivers, April–May 2021.
33 Sert, "Urban Metabolism."
34 Sert, "Urban Metabolism," 288–90.
35 As reported by Statistica (https://www.statista.com/statistics/267364/world-cement-production-by-country/).
36 Forty, *Concrete and Culture*, 53.
37 Bernes, "Belly," 335.
38 *Evrensel*, "Istanbul Canal Will Greatly Damage."
39 Ünsal, "On the Periphery."
40 I invoke the concept of *alienation* less as the loss of an inherent subjective essence and more as an inability to intervene in and govern the infrastructures that subtend daily life. See Read, *Politics of Transindividuality.*
41 Simondon, *On the Mode of Existence*, 16.
42 Simondon, *On the Mode of Existence*, 17.
43 Read, *Politics of Transindividuality*, 105.
44 Released originally as *Ekümenopolis.*
45 Mau, *Mute Compulsion*, 101–2.
46 Türeli and Al, "Walking in the Periphery," 332, 330.
47 Simondon, *On the Mode of Existence*, 252.
48 Simondon, *On the Mode of Existence*, 21.
49 Simondon, *On the Mode of Existence*, 248.
50 Bernes, "Belly."
51 Schmelzer et al, *Future*, 242.
52 Forty, *Concrete and Culture*, 40.
53 Schmelzer et al., *Future*, 230.
54 Eray Çaylı reminds us that such gestures of occupation can unwittingly extend practices of domination, especially when they aren't premised on confronting histories of state violence. Çaylı notes how some of the supposedly abandoned occupation houses had in fact been evacuated by non-Muslims fleeing the violence of the Turkish state in the twentieth century. When alerted to this fact, such occupation efforts were abandoned, Çaylı, *Climate Aesthetics*, 82–83.

55 *Yeniden İnşa Et*, 81–82.

56 Internationalist Commune of Rojava, *Make Rojava Green Again.*

Bibliography

Baysal, Efe. "Terricide: Poisoning the Lungs of Istanbul." *Research and Policy on Turkey* 2, no. 1 (2017): 10–24. https://doi.org/10.1080/23760818.2016.1272264.

Bernes, Jasper. "The Belly of the Revolution: Agriculture, Energy, and the Future of Communism." In *Materialism and the Critique of Energy*, edited by Brent Ryan Bellamy and Jeff Diamanti. MCM, 2018.

Çaylı, Eray. *İklimin Estetiği* [Climate aesthetics]. Everest Yayınları, 2020.

Dindar, Bekir. *Oyuk* (Cavity). Photographs, 2016. http://www.bekirdindar.com/.

Easterling, Keller. *Extrastatecraft: The Power of Infrastructure Space*. Illustrated ed. Verso, 2016.

Ekümenopolis [*Ecumenopolis: City Without Limits*]. Film. York Street Productions International, 2011.

Evrensel. "Kanal İstanbul Tarım, Hayvancılık Ve Yaban Hayatına Büyük Zarar Verecek" [The Istanbul Canal will greatly damage agriculture, livestock, and wildlife]. February 19, 2021. https://www.evrensel.net/haber/426304/kanal-istanbul-tarim-hayvancilik-ve-yaban-hayatina-buyuk-zarar-verecek.

Forty, Adrian. *Concrete and Culture: A Material History*. Reaktion, 2012.

Foster, John Bellamy. *Marx's Ecology: Materialism and Nature*. Monthly Review Press, 2000.

Gillen, Jamie, Tim Bunnell, and Jonathan Rigg. "Geographies of Ruralization." *Dialogues in Human Geography* 12, no. 2 (2022): 186–203. https://doi.org/10.1177/20438206221075818.

Internationalist Commune of Rojava. *Make Rojava Green Again*. Foreword by Debbie Bookchin. Dog Sections Press, 2018.

Keyder, Çağlar, and Zafer Yenal. "Agrarian Change Under Globalization: Markets and Insecurity in Turkish Agriculture." *Journal of Agrarian Change* 11, no. 1 (2011): 60–86. https://doi.org/10.1111/j.1471-0366.2010.00294.x.

Marx, Karl. *Capital: Volume 1: A Critique of Political Economy*. 1867. Edited by Ernest Mandel. Translated by Ben Fowkes. Penguin Classics, 1992.

Mau, Søren. *Mute Compulsion: A Marxist Theory of the Economic Power of Capital*. Verso, 2023.

Öztürk, Deniz. "Seeking the Excavation in Istanbul: Actors, Network and the Flow." Master's Thesis. Mimar Sinan Güzel Sanatlar Üniversitesi

[Mimar Sinan Fine Arts University], 2019. https://tez.yok.gov.tr/UlusalTezMerkezi/tezDetay.jsp?id=IV-JJT1WTIsat4S-fsJydA&no=Oy_ls8XtgjeKUyigwRjheQ.

Parsons, Laurie, and Sabina Lawreniuk. “Geographies of Ruralisation or Ruralities? The Death and Life of a Category.” *Dialogues in Human Geography* 12, no. 2 (2022): 204–7. https://doi.org/10.1177/20438206221102937.

Qviström, Mattias. “Peri-Urban Landscapes: From Disorder to Hybridity.” In *The Routledge Companion to Landscape Studies*, edited by Peter Howard, Ian Thompson, Emma Waterton, and Mick Atha. Routledge, 2012.

Read, Jason. *The Politics of Transindividuality*. Haymarket, 2017.

Rosa, Brian. “When Is a Wasteland? A Critical Understanding of Infrastructure and Residual Spaces.” Urb-Act (blog), January 31, 2022. https://urbact.eu/articles/when-wasteland-critical-understanding-infrastructure-and-residual-spaces-brian-rosa.

Saito, Kohei. *Marx in the Anthropocene*. Cambridge University Press, 2023.

Schmelzer, Matthias, Aaron Vansintjan, and Andrea Vetter. *The Future Is Degrowth: A Guide to a World Beyond Capitalism*. Verso, 2022.

Sert, Esra. “Urban Metabolism of İstanbul: Waterfronts as Metabolized Socio-Natures Between 1839 and 2019.” PhD diss., Middle East Technical University, December 25, 2020. https://open.metu.edu.tr/handle/11511/89696.

Simondon, Gilbert. *On the Mode of Existence of Technical Objects*. Univocal, 2016.

Taycan, Serkan. *İki Deniz Arasi: Karadeniz ile Marmara Denizleri Arasinda Dört Günlük Yürüyüş Rotasi* [Between two seas: A four-day walking route between the Black Sea and the Marmara Sea]. Film. IKSV, 2014.

Tuğal, Cihan. “Politicized Megaprojects and Public Sector Interventions: Mass Consent Under Neoliberal Statism.” *Critical Sociology* 49, no. 3 (2022): 457–73. https://doi.org/10.1177/08969205221086284.

Türeli, Ipek, and Meltem Al. “Walking in the Periphery: Activist Art and Urban Resistance to Neoliberalism in Istanbul.” *Review of Middle East Studies* 52, no. 2 (2018): 310–33. https://doi.org/10.1017/rms.2018.96.

Turkkan, Candan. “Feeding Global Istanbul.” In *Feeding Istanbul: The Political Economy of Urban Provisioning*. Brill, 2021. https://doi.org/10.1163/9789004424500_006.

“2024 yılında en az 1897 işçi, is cinayetlerinde hayatini kaybetti” [“In 2024, at least 1897 workers died in workplaces”]. ISIG Meclisi, Health and Safety Labor Watch/TURKEY, July 13, 2026. https://isigmeclisi.org/21145-ic-cepheyi-saglamlastirma-siyasetinin-ortmeye-calistigi-gerce.

Ünsal, Merve. “On the Periphery.” In *Tumulus*, by Serkan Taycan. Edited by Başak Şenova. Mas Matbaacılık, 2014.

Yeniden İnşa Et—Caferağa Ve Yeldeğirmeni Dayanışmaları Yatay Örgütlenme Deneyimi [Rebuild: Caferağa and Yeldeğirmeni solidarity horizontal organization experience]. Nota Bene, 2020.

Zeybek, Sezai Ozan. "İstanbul'un Yuttukları ve Kustukları: Köpekler ve Nesneler Üzerinden İstanbul Tahlili" [What Istanbul swallows and regurgitates: An analysis of Istanbul through dogs and objects]. In *Yeni Istanbul Calismalari* [New Istanbul studies], edited by Ayfer Bartu Candan and Cenk Özbay. Istanbul, 2014.

6

Domestic Solar Media in Rural Tanzania

Toward an Energy–Media Matrix

LISA PARKS

There is nothing natural about rurality. The concept emerges historically in relation to processes of settlement, territorial division, agriculture, and resource accumulation.[1] As James Scott explains in his account of the earliest states, "Crops allowed populations to concentrate and settle, providing a necessary condition for state formation."[2] Questioning the tendencies of civilizational narratives, Scott goes on to suggest that "the standard narrative—namely that people couldn't wait to abandon mobility altogether and 'settle down'—may also be mistaken."[3] His deep history of the state reminds us that complex dynamics of settlement and mobility help to determine (that is, to set the limits of) what rurality is and is not. These dynamics—which are both physical and imaginary—have produced ruralities in a variety of sites and ways over time.

This chapter explores media ruralities in contemporary Tanzania, focusing on the use of solar media setups to energize media devices. During the 1960s, concepts associated with rural village life played a vital role in President Julius Nyerere's 1967 vision for Tanzania's political independence and national socialist program. Specifically, Nyerere mobilized the concept of *ujamaa*, which translates to "familyhood" in Swahili, as "the basis of African socialism."[4] This term came to define a social and economic project emerging from rural Tanzanian villages where people live and work cooperatively, generate *kujitegemea* or "self-reliance," and are "uncaptured" by capitalist and state forces of change.[5] The concept of ujamaa invoked "an idealized construction of traditional African forms of kinship and extended family—one that emphasized reciprocity, collective effort, and an open version of community."[6] Although the political and economic landscape of Tanzania shifted after Nyerere left the presidency in 1985, the concept of ujamaa has persisted in the postliberalization era as a "political language shared at all levels of Tanzanian society, providing notions, ideas, images, and metaphors of power to speak and act in the present."[7] As such, the concept provides important context for understanding the energizing and usage of media technologies in rural Tanzania.

This chapter also takes inspiration from environmental media scholars who suggest media studies is at a crucial juncture. Scholars can either continue to study media systems as business as usual, or they can start to historicize, analyze, and publicize how media production, distribution, and consumption contribute to climate disruption and planetary precarity. As Nicole Starosielski and Janet Walker observe in *Sustainable Media*, "The burgeoning area of ecomedia or environmental media studies is beginning to grapple with the massive amount of energy drawn and expended to power media systems, the extraction of materials to construct media technologies, and the toxicities of use and disposal." Their coedited collection sets out to "articulate the enmeshment of media practices (both textual and technological), infrastructures (physical and social), and resources (natural and human), and reaches toward ecological alternatives."[8]

This enmeshment is precisely what Cajetan Iheka investigates in his book *African Eco-Media*. Engaging with photos, films, and video art, Iheka analyzes "the ecological footprint of media technologies in Africa" with "an attunement to labor, materiality, and toxicity from an African standpoint."[9] He rightly encourages further consideration of Africa in environmental media studies, and he notes the continent's particular relations with global supply chains and economies of extraction, waste, and conservation. As he observes, "Africa remains marginal to discussions of materiality in media studies and in the subfield of ecomedia studies, yet the fossil fuels and ore crucial to media's workings, like oil and coltan, are mined on the continent, and castoff electronic devices often end up there."[10] Attentive to energy issues, Iheka offers the term "imperfect media" to point out "low-carbon media practices and the infrastructures of finitude that are critical for ameliorating ecological precarity in the future."[11] Iheka's idea of "imperfect media" resonates with a long history of writings related to non-Western film and media, sometimes referred to as the Third and Fourth Cinemas, whether Julio Garcia Espinosa's "imperfect cinema" or the "accented" cinemas of exilic or diasporic formations studied by Hamid Naficy.[12] The conditions that inform "imperfect media" also link to Nadia Bozak's pathbreaking research on what she calls the "resource image," Jenna Burrell's book on "invisible users" in Ghana, and Ramon Lobato and Julian Thomas's work on "informal media" in African/global south contexts.[13]

While the media research mentioned above addresses important environmental and materialist questions, it often sidelines the issue of rurality. In media and communication studies, the conceptualization of media rurality has emerged from North American and European scholars who have historicized and analyzed relations of empire, the establishment of networked transportation and communication technologies, and the compression of space and time.[14] This research has generated important site-specific studies by scholars such as Darin Barney as well.[15] In the context of the global south, studies of media and rurality have emerged from fields such as anthropology, international development studies, and rural sociology. Further, scholars from the interdisciplinary field of ICT4D (Information and Communications Technology for Development) research have examined information and communication technologies, often in rural settings, in relation to economic and social so-called development, often without confronting or critiquing problematic teleological assumptions related to such development.[16] Some scholars have questioned whether such ICT4D research should continue and have suggested the need for reframing such work around issues of justice and Indigenous knowledges.[17]

Building on this research, in this chapter I explore what scholars and publics can learn about media sustainability from rural Tanzanians. More specifically, this chapter examines Tanzanians' use of solar panels to energize media technologies in rural villages in eastern Tanzania, on the outskirts of Dar es Salaam. Drawing on interviews conducted in sixteen homes in the villages of Mbezi-Singida, Chanika, and Mwasonga in 2019 as part of the Social IT Solutions workshop with the Dar es Salaam Institute of Technology (DIT), the chapter explores how rural Tanzanians talk about and use solar panels to energize media technologies such as lightbulbs, radios, mobile phones, stereo systems, and television sets. It sets out to answer the following questions: What kinds of energy-related activities do rural Tanzanians engage in to participate in the global media economy? How do rural Tanzanians who live off the grid power up and recharge their devices? How do these practices relate to or inflect the concept of sustainable media? Analysis of interview data revealed that the principle of ujamaa—and especially its relation to familyhood, rural village life, and self-reliance—persists implicitly in the ways rural Tanzanians energize and organize media technologies. The lack of state funding for rural electrification and media equipment has produced Tanzanian media ruralities that are characterized by energy-conscious, self-reliant, sociotechnical practices at the village level.[18] Rural Tanzanians use media technologies in modulated ways based on factors like weather conditions, financial flows, equipment quality, and technical knowledge, and they have formulated sustainable media practices, in part as an effect of rurality.

Selling Solar Panels

Solar panels are part of an off-grid energy economy that has taken shape in East Africa over the past twenty years.[19] The solar panel emerges from financially and environmentally fraught global and regional industries yet has become integral to rural Tanzanians' use of media devices and production of media ruralities. Between 2009 and 2019, the use of mobile phones, the internet, and social media increased dramatically in Tanzania. By 2019, 70 percent of the country's more than 57 million people used mobile phones, 40 percent used the internet, and 17 percent of homes had television.[20] Despite this use, most Tanzanians did not have electrical grid access in their homes. By 2016, only 32.8 percent of Tanzanian households had grid electricity. Of those, 24.7 percent were electrified through solar power. In the villages focused on in this study, 65.9 percent to 75.4 percent of households were not connected to electrical power.[21] The state's recent rural electrification efforts include the use of minigrids.[22] Despite this, most Tanzanians cannot afford home electrical service and adopt an opportunistic relation to grid power, endeavoring to rent or poach it to recharge devices, whether at work, at community centers, or via neighbors.[23] These conditions are not the exception but rather the norm for most people in rural communities of Tanzania and East Africa.

In their important critique of "off-grid infrastructures of inclusion" in East Africa, Jamie Cross and Tom Neumark suggest that "off-grid solar power can be an uncertain, ambivalent, and potentially contradictory mechanism of inclusion. Rather than resolving inequities, the inclusion of low-income populations in off-grid energy markets has exposed people to new forms of financial and social discipline, and risks further entrenching forms of disempowerment, inequality, and subalternity."[24] Drawing on interviews with frontline staff in solar companies, Cross and Neumark are concerned with the ways that off-grid infrastructures are "underpinned by a complex sociotechnical apparatus for lending, collecting and monitoring repayments, and, for producing creditworthy consumers."[25] This apparatus is organized to support global corporations, produce an emergent consumer class, and cheat low-income communities in East Africa and elsewhere.

While these issues are crucial, and Tanzanian media ruralities are no doubt interwoven with the political and economic forces that Cross and Neumark address, this chapter focuses on the domestic contexts of solar panel use in rural Tanzania and the ways people imagine the relationship between solar energy and mediation. I approach rural Tanzanians not only as interpolated consumers but as people who engage in sociotechnical relations and media cultures in specific ways. Given the discrepancy between the numbers of mobile phone users (70 percent) and households with electrical access (32.8 percent), many rural Tanzanians use small solar panels to energize their phones and other media devices. The solar panel has become a vital technology, supporting everything from domestic illumination to television viewing to social media participation. Consumer-grade solar panels are sold at a large public market, called Kariakoo, in the center of Dar es Salaam. Chinese companies ship solar panels and other electronic equipment to the port city, where they are imported and sold locally and regionally by Tanzanian merchants. People from rural communities either come to this urban market by bus, minivan, or car to buy solar panels or purchase them at smaller retail shops closer to rural villages.

We interviewed two solar merchants in Kariakoo. The first, ProSolar, claims to be the oldest solar company in Dar es Salaam and has been operating for more than twenty years, since 2001. The company sells individual solar panels, cables, and connectors as well as solar packages to individual consumers and communities. Solar panels of various sizes were displayed on the shop's back wall, enabling the sales staff to explain issues of size and power capacity to customers (figure 6.1). The most commonly sold kit is the medium-size one, which in 2019 cost from TZS300,000 to TZS500,000 (USD127 to USD213). The shop's shelves were jam-packed with solar-powered devices including TVs, batteries, lightbulbs, lanterns, and small refrigerators, most of which, the owner explained, were made in China and branded with the ProSolar name. The company's owner indicated that ProSolar regularly sends a truck into rural communities to sell solar equipment both to local retailers and direct to consumers.

6.1 A staff member at the ProSolar store in Dar es Salaam, Tanzania, with various sizes of solar panels displayed in the back of the store. Photograph by Lisa Parks.

The second solar merchant, RexEnergy Limited, has been in business in Tanzania since 2013. Rather than selling individual solar panels, the company specializes in packaging solar setups for homes and communities. One of its owners, a former banker, sought to develop a consistent revenue stream using a pay-as-you-go subscriber model. In this model, a basic solar package—a solar panel, cables, rechargeable battery, and lightbulbs—is delivered to and installed in a home and includes a maintenance contract. The installation integrates the site into a service network that is remotely monitored and administered by RexEnergy. The consumer is required to pay a monthly fee for a solar package, which can be done by mobile phone. The prices of solar packages from RexEnergy in 2019 were as follows: 40-watt panel, TZS575,000 (USD245); 100-watt panel, TZS1,290,000 (USD550); and 150-watt panel, TZS2,340,000 (USD997).

The RexEnergy shop includes a physical model of the equipment setup so that consumers can learn about it, try it out, and understand how it works (figure 6.2). One of the biggest challenges, according to sales staff, is that subscribers are often unfamiliar with the technical operation of a solar package. If the system fails, they often try to repair it

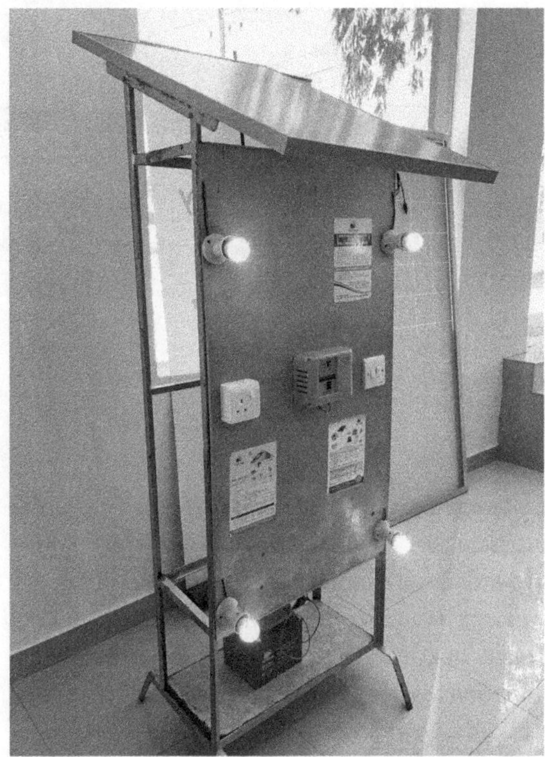

6.2 An in-store display at RexEnergy demonstrates the signal flow of a domestic solar panel setup to help educate prospective customers about its operation. Photograph by Lisa Parks.

without sufficient knowledge, which poses safety risks. Moreover, when repairs are attempted at home, customers often create more problems as they rearrange the equipment incorrectly, making it harder for RexEnergy to monitor service issues remotely. Staff also reported that there have been many problems with users hooking up multiple kinds of devices (such as TV sets) directly to the battery rather than using the solar package as an electrical network with an outlet. At the time of our interview, RexEnergy had a thousand pay-as-you-go customers throughout Tanzania and was trying to expand its business by building solar minigrids for rural communities.

Solar Media Setups

Given the high monthly costs associated with solar subscription service, most rural Tanzanians purchase and use a single small- or medium-size solar panel and battery for domestic charging of media devices. In the villages where we conducted interviews, solar panels appeared on the roof-

6.3 The solar panels on the rooftops of different types of homes in rural villages of eastern Tanzania are used to energize media devices. Photographs by Lisa Parks.

tops of multiple homes scattered throughout the area (figure 6.3). Just as media scholars have suggested that satellite dishes indicate which homes downlink TV content from elsewhere and produce "landscapes of taste,"[26] the solar panel signals a home's capacity to (re)charge and operate electronic and/or media devices in a given milieu. In rural Tanzania and beyond, the solar panel functions as both sign and capacitor of electronic mediation. While solar panels are visible on rooftops, the everyday labor of energizing media, such as electric lightbulbs, mobile phones, radios, TVs, or tablets, generally is not. To use such technologies in rural Tanzania, people climb ladders to their rooftops to mount solar panels for maximum sun exposure, read technical manuals, assemble solar media setups inside their homes, and maintain the equipment. Consistent with the principle of ujamaa, they also assist others in the village with solar panel installation and use and share technical knowledge and experience. Many people in the world, especially in the global south, labor each day to energize media devices so they can participate in a life that is electronically and digitally mediated.[27]

A series of anecdotes helps situate this discussion in rural Tanzania. In the village of Mwasonga, a father of four children uses a small solar panel and battery for mobile phone charging. Sometimes he uses it to watch football (soccer) matches on TV, but, he explained, his system does not last long enough to watch an entire game. In the doorway of another home, a young woman tells us she has one small solar panel for the sole purpose of charging her mobile phone to use Facebook, which she loves. Outside another nearby home, an elderly farmer and his grandson mention their family uses mobile phones but has no solar panel or electricity at home. To recharge their phones, they either take them to the village's charging station or a neighbor's house.

For many rural Tanzanians, acts of charging mobile phones, engaging with social media, and watching football matches (or portions thereof) are contingent on solar panel use. As a result, some rural Tanzanians are keenly aware of the relationship between the sun and their media devices. A young man named Abdul from Mbezi-Singida stated, "This solar panel, when there are good sun rays, works perfect, like it is today. But if the day is covered by clouds, it is hard to get energy from it." A man named Yasini from Chanika emphasized the limited lifespan of solar equipment, explaining, "Solar [panels] are only efficient for a few years after installation. During cloudy or rainy days, we don't get enough power even for lighting." An elderly man named Mrisho from Chanika proclaimed, "I can watch my television nicely without any problem since today the sun is good." Such comments reveal the possibility of electronic or digital media in rural Tanzania is often contingent on peoples' orientation to weather conditions such as sunshine and its capacity to help energize media devices. Interviewees demonstrated more energy consciousness and awareness than is typical in postindustrial Western contexts, where media consumers take grid access for granted and energy sources are cloaked on the back end. Interviewing rural Tanzanians introduced a differential energy consciousness within the concept of sustainable media. Indeed, many people who live in rural societies, particularly in the global south, are more acutely aware of the amount of energy required to support particular kinds of media technologies.

Interviewees' comments about solar energy prompt further thinking about mediation as a process of energy transfer. Radiation from the sun is converted by photovoltaic cells into electronic energy that powers batteries and devices used to send/receive, record, download/upload, and network audiovisual content. This energizing of media, which more palpably manifests in rural settings, is contingent on human labor and sociotechnical relations. For instance, acts of manufacturing and transporting equipment, climbing up stools and ladders onto roofs, siting and installing solar panels, (re)routing cables, and tightening connectors are all integral to electronic and/or digital mediation. The sociotechnical relations of media in rural Tanzania are organized to enable the capture and transfer of energy from one piece of equipment to another from the solar panel, through the cable, to the battery, and finally to the device. This idea of mediation as a process of energy transfer becomes clearer when considering specific users' solar media setups and comments about them.

In Chanika, seventy-five-year-old Mrisho explained that he had his solar system installed three years ago. The solar panel and battery cost him TZS400,000 (USD162) and, given his age and limited mobility, he hired a technician to install them. Mrisho kept the battery in his bedroom near his TV set so that the kids in his family would stay away from it, because, as he noted, "they are not always careful." He started using the system with a wet cell battery, but it did not last long enough, so he replaced it with a dry cell battery, which was still working. Mrisho explained that he also has a radio set that uses battery power.

Nurdin rented a home in Msongani and lived there with his family. He explained that he taught himself how to use the solar panel by reading a manual he found in the house. He learned how to connect its cables so that the battery would charge, lighting up the house and charging electrical devices like the radio and TV set. At one point, Nurdin was connecting cables to get a stereo system running, and the battery generated a spark and flame. As Nurdin explained, "With all the energy it acquires from the sun, it is very dangerous." He had to pay an electrician to come and fix it so that it would not be hazardous to operate, but he and his family remained nervous about its use.

The challenges of using solar panels were also discussed by an elderly woman known as Granny from Mbezi Singida. Her family uses a small solar panel to energize electric lights inside their home. As she explained, "We used solar only for lights. When we tried with other equipment it wasn't working well so we stopped. It lighted three bulbs but then suddenly stopped working. We are now using certain batteries from China. . . . I just want it to work well, provide light and for it to be bright! Smaller solar setups cannot even switch on radios. We even have televisions inside but we cannot turn them on. It doesn't work." Her comment resonates with scholarly attention to nonusers—that is, those who imagine being able to use devices but who may not be able to, whether because of unaffordable related costs, lack of electricity, equipment malfunction, or unfamiliarity with the technology.[28]

To assist people with energizing their devices, Sheik operated a small mobile phone charging booth in Msongani village. He used a combination of energy sources and carefully managed his costs. As Sheik explained, "I have TANESCO [the national electricity provider], but I always use it for emergency times, especially in rainy season. I prefer solar. . . . Solar is more reliable than TANESCO, but the only problem is that the solar panel system establishment cost is a little bit high compared to the standard cost of living of Msonagni people. It cost me about 700,000 TZS [USD285]to get the services operating at my booth." To cover his costs, Sheik charged people who did not have electricity at home a modest fee to charge their mobile phones.

To provide a sense of the material contexts, the photocollage in figure 6.4 features the domestic solar-powered media setups of multiple interviewees. These images are offered to generate awareness of some rural Tanzanian living environments, energy conditions, and media technologies. In rural Tanzania, media use is not a matter of simply plugging devices into a wall outlet. Rather, people exert labor to install, organize, and maintain solar media setups. Families make choices about where to mount a solar panel, which electronic devices to prioritize for domestic use, where they should be situated, when they should be used, and who, if anyone, maintains them. Given interviewees' awareness of solar energy and its variability, users also have a strong sense of energy and media limits. Rather than assume the possibility of ubiquitous connectivity or continuous streaming, rural Tanzanians anticipate signal disruptions and loss.[29] Tanzanians' energy consciousness and limited media use (and nonuse) suggests another model of sustainable media, one put into effect by limited grid power and the resulting organization and modulation of solar energy, connectivity, and mediation in rural communities.

In addition to inflecting and extending the important concept of sustainable media, rural Tanzanians who operate domestic solar media setups function as micro-level substations for the global media economy. They capture, combine, and daisy-chain human and solar energy so that it can be organized to power up media devices. Having said this, interviewees raised crucial concerns related to energizing their media devices. They emphasized the importance of weather conditions, public safety, technological knowledge, equipment quality, content duration, and service provision, to name a few. All but one of the interviewees also mentioned the importance of familial, village, or community contexts to the energizing of media in rural Tanzania, implicitly linking media technologies to the principle of ujamaa. Crucially, electronic mediation in rural Tanzania relies not only on a process of energy transfer but on village and family relations and support systems established and maintained over generations.

Solar energy is often articulated with sociocultural ideals that resist "capture." As Nadia Bozak explains, "solar power retains a certain utopian dimension in that the source itself cannot as yet be withheld from the

6.4 Domestic solar setups in rural Tanzania are used to illuminate lights, energize TV sets or stereo systems, and charge multiple mobile phones at once. Photographs by Lisa Parks and the DIT team.

world or citizen at large. Though intermittent and by most insatiable standards relatively moderate in its energy yield, the solar panel can still exist outside corporate or government-owned utilities." It "functions best on an individual or low-scale level, meeting the relatively minimal demands of the private producer/consumer or a moderately sized geographical area."[30] Indeed, there is a resonance between the capacities and scales of solar panels and the ideals and practices of ujamaa to the extent that the technology energizes media at the level of the family or village rather than the corporation or the state.

While solar power is often positioned as a preferable or more "just" energy source because it is renewable, it is important not to overlook the material effects of its supply chain, manufacturing, and logistics. China ships solar panels and electronic equipment on huge vessels into the port of Dar es Salaam, which are then transported by diesel-fueled trucks throughout the country and region. Some interviewees complained about cheap, poorly made cables used to connect solar panels to other devices, indicating the equipment often failed and had to be frequently replaced. Importing low-cost cables and electronics with shorter lifespans into Tanzania and other African countries necessitates more frequent replacement of this equipment, leading to more manufacturing and shipping and increasing the overall carbon footprint of solar panel usage. Thus, the manufacturing and transporting of solar panels may reduce their overall environmental value or effects.[31]

Beyond this, as shown in figure 6.5, solar setups in rural Tanzania combine renewable and nonrenewable sources such as wet cell (lead

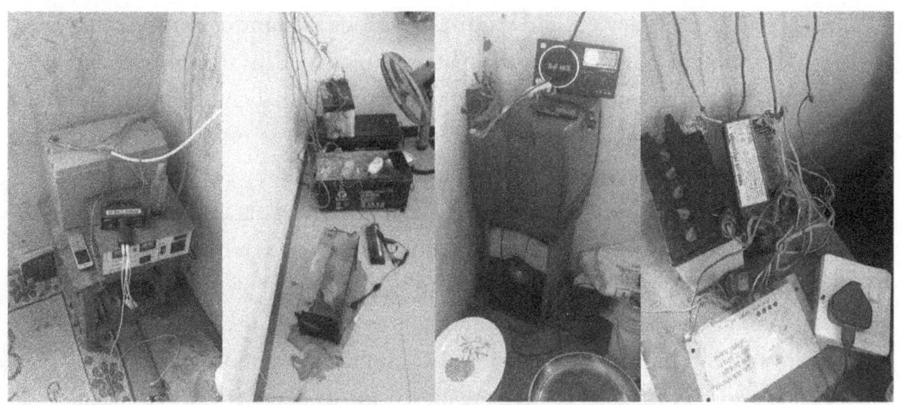

6.5 Domestic solar setups in rural Tanzania combine renewable and nonrenewable sources such as wet cell (lead acid) batteries, which contain toxic and hazardous materials. Photographs by Lisa Parks and the DIT team.

acid) batteries, which contain toxic and hazardous materials. These materials not only present risks to public health and safety but also are often not properly disposed of, thus polluting the environment and threatening humans and nonhumans alike. Thus, the renewable and nonrenewable energy mix that capacitates media in rural Tanzania is less appealing from an environmentalist perspective. Moreover, portable media devices such as mobile phones, tablets, and laptops contain lithium batteries, which emerge from another extractive industry that has far reaching detrimental effects, as Javiera Barandiarán explores in her important research on lithium mines.[32] While solar power may be preferable to fossil fuels or lithium batteries, its relation to other materials should not be ignored or unaddressed. Reviewing these relational contexts, then, complicates tendencies to imagine solar power as a sustainable energy source that is altogether isolated from more extractive or polluting energy practices.

Toward an Energy–Media Matrix

The anecdotes and images above bring forth different energy–media scenarios in rural Tanzania. The solar panel is used to illuminate a lightbulb, charge a mobile phone for Facebook use, or keep a television set running long enough to watch part of a football match. Each of these scenarios involves different devices and requires a various amount of energy, evoking the need for an energy–media matrix (table 6.1). This preliminary and speculative matrix suggests the need for a more formal and helpful schematic that would aid in generating consumer awareness and action related to energy sources and media consumption. Inspired by the energy consciousness media practices of rural Tanzanians, the diagram might function as an energy–media guide or rating system, one resonant with the MPAA's content ratings systems. Instead of determining which content is sexual or violent, or appropriate for adults or children, the energy–media ratings would indicate what types of content require more or less fossil fuel energy to consume, participate in, stream, or play back. It is a speculative diagram that identifies different media devices and provides details about the energy sources and levels needed to consume certain types of content (format, genre, and duration). This would enable media publics to formulate consciousness not only about the issue of screen time from a behavioral science or public health perspective but from an environmentalist and sustainable media perspective. A global media economy that relies on fossil fuels and hazardous materials must be questioned and altered rather than normalized and naturalized.

This energy–media matrix is intended to raise awareness about the amounts and types of energy required to consume or participate in different types of media and to generate public knowledge and deliberation regarding the carbon footprint and environmental impact of media economies. The matrix builds from research in environmental media studies with the hope of generating greater awareness of energy–media relations in different parts of the world and compelling media consumers to collectively act to reduce the global media economy's carbon footprint.[33] This can be done by adopting daily or weekly screen time limits; gravitating toward lower footprint content; inventing or using forms of media and communication that do not rely on fossil fuels; or consuming short-format media.

This analysis of solar media in rural Tanzanian media points to the environmental and social value of more modulated, limited, and sustainable media. The idea of rural Tanzanian media as sustainable resonates with film and media scholarship on African cinema, Third and Fourth Cinemas, and exilic and interstitial media. Indeed, concerns about energy and other resources in film and media production have long and complex histories, which necessarily interface with rural and environmental media studies. Suffice it to say, until recently, the critical foci in environmental

Table 6.1 Energy–Media Matrix: A Speculative Diagram

Criterion	Spectrum
Type of media content	Feature film, TV episode, cell phone video
Types of energy sources	Coal, gasoline (petrol), natural gas, lithium, hydro, wind, solar
Range of energy sources	Polysourced, monosourced
Volume of energy sources (in kilowatts)	Surge, trace
Duration of media content (in minutes)	Long, short
Resolution of media content (pixels per inch)	High, low
Environmental assessment	Unsustainable, sustainable

media studies have not looked to rural Africa for sustainable media models. Instead, Africa is often imagined or positioned as a site of "what not to do" and has been repetitively characterized as a territory of nonaccess, disconnection, or bad e-waste practices.[34] Yet as I have argued here, rural Tanzanian and other African communities are sites of important energy-conscious media practices.

Writing about science, technology, and innovation in Africa, Clapperton Mavhunga critiques narratives of diffusion that posit Africans as mere recipients of "Western" science, technology, and innovation. Mavhunga explains, "People of the South [including Africans] were already technological before colonialism happened" and should not be positioned as "the victim or subaltern of technology."[35] Implicitly echoing notions of African self-reliance embedded in the concept of ujamaa—although from a historiographic perspective—Mavhunga's scholarship prompts further thinking about the perspectives, meanings, and temporalities of technologies in rural African contexts. African people, of course, used the energy of the sun to communicate long before the invention of the solar panel or mobile phone. An elemental medium, the sun's rays illuminate, warm, and transform objects, bodies, and times and spaces. Solar radiation reflects off bodies and surfaces, enabling them to be visually perceived and made intelligible by humans and nonhumans. The sun illuminated early symbolic inscriptions and gestural forms, inaugurating a system of optical media.[36] Thus it is possible to imagine a deep history of solar mediation in rural Tanzania and elsewhere that would also engage carefully with local or regional knowledges and/or cosmologies.

In the current conjuncture, the sun's heat energizes, cultivates, and exposes as well as burns, damages, and destroys. Solar radiation is a force in what Nicole Starosielski calls "thermal media," or the "material and socially realized forms that communicate temperature, enabling heat and cold to be transmitted and received."[37] Crucially, Starosielski points out, the effects and affects of heat or solar radiation are not evenly distributed. Suffice it to say that a media history of the sun remains to be written, even though the entire planet has been reliant on its radiance. A genealogical approach, one informed by differential politics, could bring forth historical constellations of solar mediation in rural Tanzania and beyond.

Conclusion

This chapter has explored rural Tanzanians' energizing media devices with solar panels to emphasize three main points. First, by considering media ruralities in Tanzania, it is possible to recognize histories of energy consciousness and a sense of media limits that may precede environmental media studies scholars' naming of sustainable media. For decades, rural Tanzanians, and other Africans or people of the global south, have collectively energized consumer electronics with limits in mind, implicitly generating a paradigm of more modulated or sustainable use of media and information technologies. It is important that the media experiences of rural Tanzanians and other Africans and people of the global south be considered as part of the history and critique of environmental and sustainable media cultures. There is a tendency for media studies to focus on and even idly embrace energy-intensive media paradigms such as ubiquitous connectivity and streaming, which emerge from North America, Europe, and East Asia. Yet there is a crucial need for new and different paradigms and models of energy–media consumption in the context of global climate disruption. Rural Tanzanian solar media practices suggest a viable model as they mobilize a stronger sense of limits and priorities while embedding the principle of ujamaa within media ruralities.

Second, the chapter approaches media ruralities in Tanzania as sites of sociotechnical agencies and innovation rather than as they are often depicted by media scholars: as zones of cheap electronics, e-waste, or the

West's leftovers. Rural Tanzanian solar media users are much more aware about the everyday technics and materialities of media consumption than many who are socialized into conditions of ubiquitous connectivity and on-demand streaming. Furthermore, rural Tanzanians' solar media practices evoke a conceptualization of mediation as a process of energy transfer that involves users' and nonusers' awareness of solar radiation (and other energy sources) and organization of techniques to capture it, capacitate media devices, and participate via their interfaces. Much more than inserting a plug into an outlet, these conditions generate particular dispositions, mentalities, labor relations, and embodied practices relative to energy and media technologies. For instance, they require prioritizing one type of content or platform per use, normalizing limited durations and screen time, and inviting collective acts of media consumption around a single device. Sociotechnical agencies are derived from working with and within limits, not from operating within a culture of mythologized bandwidth plentitude.

Finally, it is important to recognize that there is a correlation between energy consciousness, rurality, and class. Low-income communities in rural areas have historically used less energy than those in urban centers. This is not to construct rural villages in Tanzania as innocent or neutral when it comes to energy consumption. Rather, it is to point out that there is much to learn from rural peoples and communities in Tanzania and beyond, who have contended with energy limits for decades. It is time to rethink energy media practices from the positions and perspectives of rural Tanzanians and others in the global south rather than to continue to fixate on allegedly more green postindustrial Western contexts, which are often among the most egregious offenders. The world needs energy–media models from those with a sense of limits and thresholds rather than those obsessed with abundance and surplus.

Notes

Acknowledgments: I am grateful to Joseph Matiko and Boniphace Elphace at the Dar es Salaam Institute of Technology in Tanzania as well as students from DIT and MIT who participated in the Social IT Solutions workshop and helped conduct interviews related to this project. These students include Iago Bojczuk, Daniel Bwere, Ethani Caphace, Alex Joseph, Pendo Kweka, Enock Lalusya, Beatrice Boniface

Manyaga, Napendael Msangi, Elisha Elia Nyagwaru, Baraka Sanane, Han Su, Rachel Thompson, and Gloria Wella. Thanks also to the organizers of the Solar Array symposium at UC Santa Barbara, where an early version of this chapter was presented. Research conducted for this project was supported by NSF award 1755106.

1 The term *rural* stems from the Latin *ruralis*, "of the countryside." This designation is imagined as other to a presumed site of centralized occupation, capital concentration, and/or administration. The rural, then, names a zone beyond (a *side* of the country), yet one that is internally dynamic and complex. The concept of rurality emanates from colonial mentalities and practices of spatial ordering and has built within it critiques of those same mentalities from localized (counter/alter) perspectives. Rurality is not innocent, natural, or neutral; it materializes historically in different ways in territorial and historical contexts and has various relations to media.

2 Scott, *Against the Grain*, 10.

3 Scott, *Against the Grain*, 10–11.

4 Nyerere, "Arusha Declaration."

5 Hyden, *Beyond Ujamaa.*

6 Lal, "Militants, Mothers," 2.

7 Fouéré, "Julius Nyerere," 14. Fouere sees ujamaa as a flexible concept that actors can draw on to pursue different agendas and suggests that it is "used to give intelligibility to the present and to produce alternative representations of good leadership and good governance in a demoralized political space" (13).

8 Starosielski and Walker, introduction to *Sustainable Media*, 3. For further discussion of environmental and ecomedia, see López et al., *Routledge Handbook.*

9 Iheka, *African Eco-Media*, 2, 9.

10 Iheka, *African Eco-Media*, 4.

11 Iheka, *African Eco-Media*, 10.

12 Espinosa, "For an Imperfect Cinema"; Naficy, *Home, Exile.*

13 Lobato and Thomas, *Informal Media Economy*; Bozak, *Cinematic Footprint*; Burrell, *Invisible Users.*

14 See for instance, Innis, *Bias of Communication* and *Empire and Communications*; Carey, *Communication as Culture.* For more contemporary and explicit discussions of media and rurality by media and communication scholars, see Andersson and Jansson, "Rural Media Spaces," as well as articles in that special issue. See also Hobbis et al., "Rural Media Studies."

15 Barney, "To Hear the Whistle Blow."

16 Heeks, "Information Systems"; Heeks, "ICT4D 2.0"; Avgerou, "Information Systems"; Escobar, *Encountering Development.*

17 Masiero, "Should We Still Be Doing ICT4D Research?"

18 For a historical discussion of the relationship between the national media system in Tanzania and the principle of ujamaa, see Abdalla, "Afrocentric Media Governance."

19 Cross and Neumark, "Solar Power"; also see Cross, "Capturing Crisis"; Degani, "Modal Reasoning."

20 World Resources Institute, "Mapping Energy Access: Tanzania."

21 World Resources Institute, "Mapping Energy Access: Tanzania."

22 Wen et al., "Off-Grid."

23 For a study of electricity in urban parts of Tanzania, see Degani, *City Electric*.

24 Cross and Neumark, "Solar Power," 907.

25 Cross and Neumark, "Solar Power," 906.

26 Brunsdon, "Satellite Dishes." See also Moores, "Satellite TV."

27 Markham, *Digital Life*.

28 See Oudshoorn and Pinch, introduction to *How Users Matter*; Wyche and Baumer, "Imagined Facebook."

29 For an analysis of signal disruptions in the context of rural Zambia, see Parks, "Reinventing Television."

30 Bozak, *Cinematic Footprint*, 34.

31 See Chen, "Social-Environmental Dilemmas." See also Mulvaney, *Solar Power*.

32 Barandiarán, "Lithium."

33 See, for example, Maxwell and Miller, *Greening the Media*; Iheka, *African Eco-Media*; Szeman et al., *Fueling Culture*; Starosielski and Walker, *Sustainable Media*; Marks, "Survey"; López et al., *Routledge Handbook*.

34 Maxwell and Miller, *Greening the Media*.

35 Mavhunga, *What Do Science*, 5.

36 For a discussion of historical writing techniques in Africa, see this book by a Zimbabwean farmer and visual designer: Mafundikwa, *Afrikan Alphabets*. For a European history of optical media, see Kittler, *Optical Media*.

37 Starosielski, *Media Hot and Cold*, 6.

Bibliography

Abdalla, Khamis Juma. "Afrocentric Media Governance in Tanzania's Ujamaa (Socialism) Era. "*International Journal of Academic and Applied Research* 2, no. 10 (2018): 1–14.

Andersson, Magnus, and André Jansson. "Rural Media Spaces: Communication Geography on New Terrain." In "Rural Media Spaces," special issue, *Culture Unbound* 2, no. 2 (2010): 121–30.

Avgerou, Chrisanthi. “Information Systems in Developing Countries: A Critical Research Review.” *Journal of Information Technology* 23, no. 3 (2008): 133–46.

Barandiarán, Javiera. “Lithium and Development Imaginaries in Chile, Argentina, and Bolivia.” *World Development* 113 (2019): 381–91.

Barney, Darin. “To Hear the Whistle Blow: Technology and Politics of the Battle River Branch Line. “*Topia: Canadian Journal of Cultural Studies* 25 (2011): 5–28.

Bozak, Nadia. *The Cinematic Footprint*. Rutgers University Press, 2011.

Brunsdon, Charlotte. “Satellite Dishes and the Landscapes of Taste. “*New Formations* 15 (1991): 23–40.

Burrell, Jenna. *Invisible Users: Youth in the Internet Cafes of Urban Ghana*. MIT Press, 2012.

Carey, James. *Communication as Culture: Essays on Media and Society*. Routledge, 1992.

Chen, Jia-Ching. “Social-Environmental Dilemmas of Planning an ‘Ecological Civilisation’ in China.” In *The Routledge Companion to Planning in the Global South*, edited by Gautam Bhan, Smita Srinivas, and Vanessa Watson. Routledge, 2017.

Cross, Jamie. “Capturing Crisis: Solar Power and Humanitarian Energy Markets in Africa. “*Cambridge Journal of Anthropology* 38, no. 2 (2020): 105–24.

Cross, Jamie, and Tom Neumark. “Solar Power and Its Discontents: Critiquing Off-Grid Infrastructures of Inclusion in East Africa. “*Development and Change* 52, no. 4 (2021): 902–26.

Degani, Michael. *The City Electric: Infrastructure and Ingenuity in Postsocialist Tanzania*. Duke University Press, 2022.

Degani, Michael. “Modal Reasoning in Dar es Salaam’s Power Network.” *American Ethnologist* 44, no. 2 (2017): 300–314.

Escobar, Arturo. *Encountering Development: The Making and Unmaking of the Third World*. Princeton University Press, 2011.

Espinosa, Julio Garcia. “For an Imperfect Cinema.” In *Film Manifestos and Global Cinema Cultures*, edited by Scott MacKenzie. University of California Press, 2014.

Fouéré, Marie-Aude. “Julius Nyerere, Ujamaa, and Political Morality in Contemporary Tanzania. “*African Studies Review* 57, no. 1 (2014): 1–24.

Heeks, Richard. “ICT4D 2.0: The Next Phase of Applying ICT for International Development.” *Computer* 41, no. 6 (2008): 26–33.

Heeks, Richard. “Information Systems and Developing Countries: Failure, Success, and Local Improvisations.” *Information Society* 18, no. 2 (2002): 101–12.

Hobbis, Geoffrey, Marc Esteve Del Valle, and Rashid Gabdulhakov. “Rural Media Studies: Making the Case for a New Subfield.” *Media, Culture, and Society* 45, no. 7 (2023): 1489–500. https://doi.org/10.1177/01634437231179348.

Hyden, Goran. *Beyond Ujamaa in Tanzania: Underdevelopment and an Uncaptured Peasantry*. University of California Press, 1980.

Iheka, Cajetan. *African Eco-Media: Network Forms, Planetary Politics*. Duke University Press, 2021.

Innis, Harold. *The Bias of Communication*. University of Toronto Press, 1996.

Innis, Harold. *Empire and Communications*. Dundurn, 2007.

Kittler, Friedrich. *Optical Media*. Polity, 2009.

Lal, Priya. "Militants, Mothers, and the National Family: Ujamaa, Gender, and Rural Development in Postcolonial Tanzania. *Journal of African History* 51 (2010): 1–20.

Lobato, Ramon, and Julian Thomas. *The Informal Media Economy*. Polity Press, 2015.

López, Antonio, Adrian Ivakhiv, Stephen Rust, Miriam Tola, Alenda Y. Chang, and Kiu-wai Chu, editors. *The Routledge Handbook of Ecomedia Studies*. Routledge, 2023. https://doi.org/10.4324/9781003176497.

Mafundikwa, Saki. *Afrikan Alphabets: The Story of Writing in Afrika*. Mark Batty, 2006.

Markham, Tim. *Digital Life*. Polity, 2020.

Marks, Laura. "A Survey of ICT Engineering Research Confirms Streaming Media's Carbon Footprint. "*Media + Environment*, August 23, 2021. https://mediaenviron.org/post/1116-a-survey-of-ict-engineering-research-confirms-streaming-media-s-carbon-footprint-by-laura-u-marks.

Masiero, Silvia. "Should We Still Be Doing ICT4D Research?" *Electronic Journal of Information Systems in Developing Countries* 88, no. 5 (2022): e12215. https://doi.org/10.1002/isd2.12215.

Mavhunga, Clapperton. *What Do Science, Innovation, and Innovation Mean from Africa?* MIT Press, 2017.

Maxwell, Richard, and Toby Miller. *Greening the Media*. Oxford University Press, 2012.

Moores, Shaun. "Satellite TV as a Cultural Sign: Consumption, Embedding, and Articulation. "*Media, Culture, and Society* 15, no. 4 (1993): 621–39.

Mulvaney, Dustin. *Solar Power: Innovation, Sustainability, and Environmental Justice*. University of California Press, 2019.

Naficy, Hamid, editor. *Home, Exile, Homeland: Film, Media, and the Politics of Place*. Routledge, 1999.

Nyerere, Julius. "The Arusha Declaration of 1967." February 5, 1967. Transcribed by Ayanda Madyibi. https://www.marxists.org/subject/africa/nyerere/1967/arusha-declaration.htm.

Oudshoorn, Nelly, and Trevor Pinch, editors. *How Users Matter: The Co-Construction of Users and Technology*. MIT Press, 2003.

Parks, Lisa. "Reinventing Television in Rural Zambia: Energy Scarcity, Connected Viewing, and Cross-Platform Experiences in Macha." *Convergence* 22, no. 4 (2016): 440–60.

Scott, James. *Against the Grain: A Deep History of the Earliest States.* Yale University Press, 2017.

Starosielski, Nicole. *Media Hot and Cold.* Duke University Press, 2022.

Starosielski, Nicole, and Janet Walker, editors. *Sustainable Media.* Routledge, 2016.

Szeman, Imre, Jennifer Wenzel, and Patricia Yaeger, editors. *Fueling Culture: 101 Words for Energy and Environment.* Fordham University Press, 2017.

Wen, Cheng, Jon C. Lovett, Emmanuel J. Kwayu, and Consalva Msigwa. "Off-Grid Households' Preferences for Electricity Services: Policy Implications for Mini-Grid Deployment in Rural Tanzania." *Energy Policy* 172 (2023): 113304. https://doi.org/10.1016/j.enpol.2022.113304.

World Resources Institute. "Mapping Energy Access: Tanzania." ArcGIS mapping software. https://www.arcgis.com/apps/MapJournal/index.html?appid=5e060dc63172439abae54bbed8a283fb#.

Wyche, Susan, and Eric P. S. Baumer. "Imagined Facebook: An Exploratory Study Of Non-Users' Perceptions of Social Media in Rural Zambia." *New Media and Society* 19, no. 7 (2016): 1092-108. https://doi.org/10.1177/1461444815625948.

7

Hong Kong in Siliguri/ Dhulabari

Exploring Media Objects and Border Towns

ISHITA TIWARY

In mid-June 2020, Indian and Chinese troops engaged in a brawl regarding a border dispute at Galwan Valley, close to Aksai Chin. The territory is controlled by China but claimed by both countries. Accusations of overstepping the line of actual control were hurled by both sides. The reasons for this brawl seem to be geostrategic and economic in nature and have roots in India's decision to split the state of Jammu and Kashmir into distinct territories.[1] This worried Chinese authorities because the Ladakhi corridor connecting China to Pakistan constitutes a crucial element of China's Belt and Road Initiative.[2] Previously, in May 2020, India was involved in another border dispute with Nepal. On May 8, India's defense minister inaugurated a new eighty-kilometer-long road in the Himalayas, which connected India's border with China at the Lipulekh pass. The Nepali government protested, claiming that the road crossed its territory. It subsequently changed its map to reflect that Lipulekh, situated in Kalapani, belonged to it cartographically. The region is of strategic importance because it connects to the Tibetan plateau. Unsubstantiated rumors abounded that Beijing encouraged Kathmandu to take a more aggressive stand with New Delhi regarding the territorial dispute.[3]

Nepal, a small country, lies between India and China, and as such, it has long been a site of struggle for both countries to assert their geostrategic and diplomatic prowess. When I visited Kathmandu in summer

2019, the signboards I saw all around me illustrated these tensions. The signboards are written in a mix of English, Hindi, and Mandarin. My parents, who accompanied me on this trip, were stunned by the preponderance of Mandarin-language signs across the city, as they assumed the closeness between India and Nepal supersedes that of Nepal and China because of the shared religious background of the two countries (both are Hindu-majority states) and the high visibility of Nepali migrants in India in everyday life. In contrast, my Nepali friends informed me that China is aiding Nepal in several infrastructure projects and Mandarin is now being taught in their school curriculum. The border bazaar discussed in this chapter is thus a site where Nepal's status as a conduit for complex flows of people, goods, and media surfaces.

I begin by referencing the border disputes as well as my own fieldwork experience in 2019 to draw attention to different ways of defining the border. The border disputes highlight the geopolitical, cartographic, and formal relationships and tensions between the three states. The border here is a site where formal negotiations over power, ownership, and territorial claims take place. It becomes the space where geopolitical and diplomatic tensions materialize. However, border cultures demand to be read beyond security concerns. In this chapter, I turn our attention toward a different inhabitation of the border and thus to different patterns and epistemologies of migration. Specifically, I focus on the border bazaar of Dhulabari, a small town in Nepal. I place it in relation to the larger border town Siliguri in India, arguing that Dhulabari's mediated rurality can only be understood in terms of its location in infrastructural networks for the circulation of people, media, and other commodities between Siliguri, Dhulabari, and China. Thus, the form of rurality emerging in Dhulabari might have less to do with its agrarian surroundings and more to do with the conditions and experiences of migration, borders, and commodity circulation in which it is now implicated. In this framing, the pirate bazaars, where media commodities are exchanged, are doubly sites of mediation of these emerging social forms—sites where Dhulabari's rurality is being pulled in multiple directions, defined by something other than growing the rice, wheat, and tea associated with its more traditionally agrarian rurality. I highlight how pirate bazaars in these border zones are reflexive of the migrant experience.

In this chapter, I focus on transit zones of the border pirate bazaars of Dhulabari in Nepal and the Hong Kong Market of Siliguri in India. By combining ethnographic and archival approaches, I explicate complex everyday transactions that take place between the three countries by tracking the migration of media equipment, media content, and personnel from China to India via the Nepal border. I concentrate on the movement of pirated videocassettes, VCRs, and other video technologies and map them onto these bazaar spaces. With a migrant workforce consisting of Indians, Nepalis, and Bangladeshis, there is a complex web of class, religious, caste, and linguistic dynamics that influence the way these trade networks are run. The demand for these goods, primarily media hardware and software, creates infrastructures such as video theaters/parlors and fosters demand for additional commodities, popularized by films, thereby indicating an aspirational desire to be connected to the larger world via commodity culture. I do this work by tracing how the moving media commodity mediates the relationship between the workers, the bazaar spaces, and the consumer desires unleashed by the arrival of liberalization.

As set out in the introduction to this volume, media rurality conceptualizes rurality as "dynamic, . . . situated, and emergent" and ruralization as a "set of multiple ongoing relational processes by which rural places, people, and experiences are continuously formed, deformed, and reformed." Tying these processes together is the idea of mediation. I trace the relationship between Siliguri and Dhulabari as an illustration of media rurality. These markets are in a continuous process of making and unmaking themselves in response to the processes of economic liberalization. This chapter maps how the media objects that circulate in these markets mediate such dynamic and changing relationships across space and over time.

The Advent of Liberalization

The 1980s marked a turning point in China and India's patterns of media consumption. Processes of liberalization in both countries opened their economies to larger global markets, heralding a market for electronic goods and consumer culture in both India and China. In India, analog video technology became contraband and started circulating via gray markets. During the 1980s, China acted as a conduit for video technology and realized the business potential that lay in the production of cheap technologies. This was a period where a horizontal network of traders and couriers from China proliferated. Thus, after the end of the video era in India in the early to mid-1990s, media equipment such as DV camcorders and later smartphones produced in China flooded the market.[4] This chapter is situated in the crucial moment of transformation (1980s–1990s) when both countries entered the global market economy.

Methodologically, I followed the media object as I tried to trace the routes of people that populate the pirate bazaar spaces. The question I asked myself in the field was, how does one go about finding these routes, the kind of people engaged in this job, the forms of media technologies, and the content that were traveling across the borders, especially when no formal archive documents this? I did this by taking an ethnoarchival approach.

For example, while perusing the Delhi state archives, I came across multiple maps—land, air, and sea—that illustrated travel routes from India to other parts of the world. The sea routes, Silk Road routes, imperial air routes, and national carrier Air India's air routes gave me clues to the history of routes that connected India, China, and Nepal. The traditional image of the border is still inscribed onto maps in which discrete sovereign territories are separated by lines and marked by different colors. The image has been produced by the modern history of the state. Borders are not merely geographical margins or territorial edges. They are also complex social institutions marked by tensions of border crossing. This image is evocative and helpful to map out a history of these transactions. But what were these routes?

During the monsoon season, on a cloudy Sunday in July 2019, my colleagues who were helping me with translation work and I made a trip from Siliguri, India, to Dhulabari, Nepal, via the Asian Highway. Our approach to the border was accompanied by a diverse set of expectations as to what the border would look like and what crossing it might entail. However, when we reached the border, chaos greeted us. It was an open border, with the Mechi River on both sides and people ferrying easily to and fro. There were cars, rickshaws, scooters, buffalo-pulled carts, and pedestrians, with only a loose sense of the divide between the paths to and fro. The only requirement needed to pass the border area was having a national identification card, which was checked by the border police on arrival.

Siliguri is connected to Dhulabari via the Asian Highway. The Asian Highway is a cooperative project between countries in Asia and Europe to improve highway systems in Asia. Dhulabari lies next to the Mechi River, east of the Himalayas, at the international border of Nepal and India. According to one of my field respondents, Ujjawal Prasai, after a bridge was built near the Mechi River, the routing of goods started happening near

Kakarvitta, a border point between the two countries. Siliguri, located in the state of West Bengal, is connected to four international borders—China, Nepal, Bangladesh, and Bhutan—making it a strategically important city for India.[5] When it comes to the geography of Nepal, Ludwig Stiller notes that its geographical and ecological pockets create conditions by which contact and communication are difficult, if not impossible.[6] Stiller maintains that these brutal facts of geography have been major factors influencing the history of Nepal. The geographic patterns have an effect on the communication systems in Nepal. The basis of almost all transport in most hill and mountain areas is portering people by mule. The topography of the hills makes construction expensive, and chronic poverty means that labor for portering is cheap and plentiful. During my fieldwork, I found that the words *carrier* and *courier* were used to describe this practice, which plays an important role in the way goods run between the two countries.

The goods transiting this communication circuit were produced in China. Here I focus on video technologies, but popular items produced during the 1980s and 1990s included portable audio players like the Sony Walkman, stereo systems, still cameras, video cameras, VCRs, pirated videocassettes, and watches. The video goods traveled to Nepal through Tatopani. This place has an open-border policy within a hundred kilometers to allow for trading to take place between the Chinese and Nepali traders. From Tatopani, goods would find their way to Kathmandu via road and from Kathmandu via road to Dhulabari, and then from Dhulabari to Siliguri via a network of couriers. This has been the most popular route for goods to travel through since the 1990s. Before that, these goods also came via Hong Kong–Bangkok–Singapore, through shipping ports to Calcutta, and via Calcutta to Nepal. One of the customs officials I interviewed in Delhi noted that Nepal was dependent on India for its supply of salt.[7] This fact was exploited by pirates. During the sea route from Calcutta to Nepal, pirates would drill holes in cartons carrying salt and let them empty out. Once empty, they would fill them up with pirated goods. The carton was cleared for customs in Calcutta. The exchange was done after the check and during travel, so the swap would go unnoticed by officials on both sides of the border. This thus became one of the ways that pirated goods traveled.

These paths followed traces of the historical silk, opium, and imperial/colonial routes in the region (figure 7.1). The opium route connected the ports of Calcutta and Bombay to the Canton region in China. The Silk Road

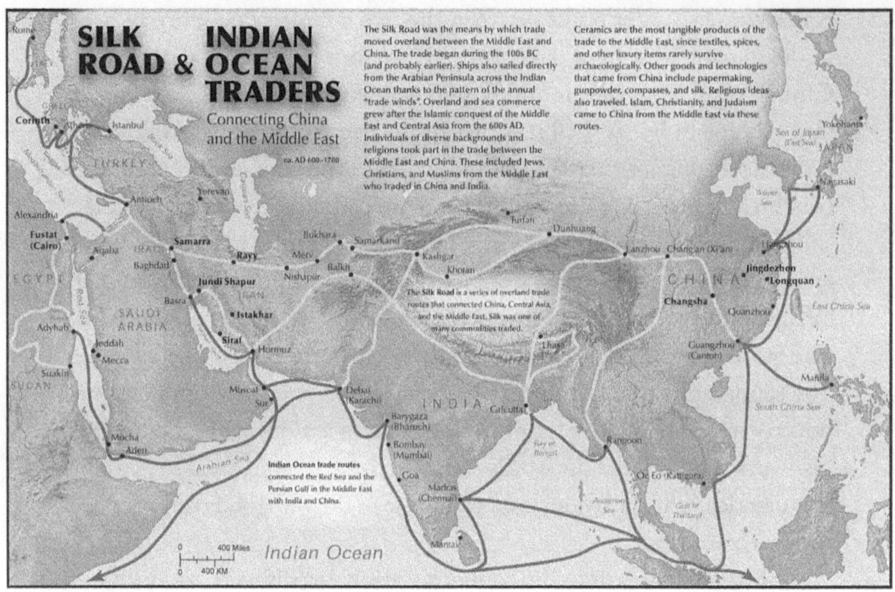

7.1 The Silk Road. The Port of Calcutta is connected to China via Hong Kong. Source: University of Chicago, Institute for the Study of Ancient Cultures, https://isac.uchicago.edu/museum-exhibits/special-exhibits/silk-road-and-indian-ocean-traders-connecting-china-and-middle-east.

connected the ports of Calcutta to China via the Nepal region. The imperial airway route does not show a direct route to China, but it does include Hong Kong as part of its routes. It is also important to note that the air space between India and China only opened up in the early 2000s. These cartographic histories laid the foundation for everyday use by traders and pirates as well as the economic and political connections between the three countries.

This topography is essential to understanding how bazaar spaces came to flourish at these places. Because no direct route linked India to China, port cities like Calcutta became important in ferrying goods. Ports like Hong Kong, as well as the land route of Nepal, become important midpoints for these goods to travel. Present-day contraband routes follow these historical paths, which set the stage for bazaars trading in pirated goods. These routes and their bazaars are key sites where desires and aspirations regarding being connected to modern global culture are mediated, as evidenced by the flows of media commodities between Dhulabari Market in Nepal and Hong Kong Market in Siliguri from the mid-1980s onward. Markets in South Asia have been an essential feature of the historical landscape, central places of exchange at which peasants, townspeople, landholders, and rulers have historically converged "to conduct wholesale and retail trade, to gather news and information, and to engage in various social, cultural, religious, and political activities. . . . Markets provide a compressed display of an area's economy, technology, and society—in brief, of the local way of life."[8] Both the Dhulabari and Hong Kong markets trace their origins to the late 1960s to early 1970s. However, it was only after China and India started opening up their economies in the 1980s that these markets reached their peak, thanks to the trade of electronic media goods. The flow of these electronics into India was dependent on transit via Dhulabari. Such goods would normally be subjected to high tariff duties under the strict import regime imposed by India. However, such import restrictions were not present in Nepal, and the porous border allowed goods to pass without restriction into India for sale at lower prices.

The destination for these goods is Siliguri's main commercial nodal point, Hong Kong Market, whose name evokes Hong Kong's status as "not so much a place as a space of transit."[9] This recalls Ujjawal Prasai's observation to me: "Back in the day, people would sell these goods as though they were from Hong Kong, electronics to slippers, everything would sell under the name. Hong Kong was like a brand name. People started calling it Hong Kong Market and it stayed on" (figure 7.2).[10] As one of the more powerful shopkeepers and traders in the market, Daman was more circumspect, perhaps even a little dismissive, when he told me, "Every country has a Chinese market, selling the cheapest stuff China makes."[11] This notion of Hong Kong goods having a brand value challenges the established idea of the East always desiring Western modernity and suggests a south–south fascination and desire. Moreover, in the case of Hong Kong Market, we observe a topographic desire, where Hong Kong becomes a literal and metaphorical place of transit on which aspirations of modernity and consumption are displaced. This imagination of Hong Kong emerged through a combination of the liberalism markets and of border regimes and is being mediated by the moving media object, thus subverting the traditional image of the border as a fixed geopolitical space.

The Mediated Media Object, Identity, and the Woman Carrier

One of the common refrains I heard in both these border towns was, "We all are migrants here." In these places, practices of mediation directly shape immigrant experience. Mediated practices have always been part of immigrant experience and their tenuous transnational lives. Immigrants

7.2 Entrance to Hong Kong Market, Siliguri. Source: Photo by Ishita Tiwary.

use media in innovative ways to negotiate cultural space and forge transplanted communities. These practices themselves interact with earlier historical patterns. Moreover, as in the formation of Hong Kong Market, Siliguri suggests, these border pirate markets emerged because of the demand for electronic media goods, creating migration patterns, topographical imaginaries, and complex flows and negotiations of identities in the bazaar spaces. This is consistent with AbdouMaliq Simone's work on urban spaces, where ready-made formats specifying forms of movement, forms of living, and forms of work are absent and people themselves become the infrastructures.[12] When people are infrastructures, "their selves, situations, and bodies bear responsibility for articulating different locations, resources and stories into viable opportunities for everyday survival."[13] One's family or ethnic background, one's personal style and location in residential arrangements—these become the "stuff of shifting circuitries of connections."[14] By taking on the infrastructural form of intersections, people bear the "full brunt of absorbing . . . contradictions, of cushioning their impact, and of deflecting them onto someone else."[15]

The concept of people as infrastructures can be a useful way to understand the relationship between media technology, migration patterns, and identity in these border towns. As noted earlier in this chapter, these markets arose as people saw an opportunity for trade in electronic goods. It is noted by almost all vendors that from the late 1970s to the mid-1990s was the golden era for these markets. The cross-border flow of media technologies and content also facilitated a cross-border flow of migrant traders. This is perhaps best manifested in the figure of Lalit Agarwal, a Nepali trader of Indian origin whom I met in Dhulabari Market, Nepal.

A self-professed Hindutva socialist, Lalit now runs a ready-made garment shop, but his family initially started the business by trading video and still cameras.[16] He is a card-carrying member of the Socialist Party, which seems to be the dominant political platform in Dhulabari, even as he professes his belief in the ideology of Hindutva and the current Indian prime minister, Narendra Modi (who belongs to the Bhartiya Janta Party), because of his allegiance to India, the homeland of his family: "I am loyal to Nepal, but India is in my heart. It is my motherland."[17] With vermilion on his forehead, he seriously informs me that socialism is inscribed in our shastras (sacred Hindu texts); he is concerned whether I, as a recent migrant to Canada, have betrayed my caste allegiance and given up vegetarianism. After further probing and finally becoming convinced that I am not a journalist, he turns self-reflexive and observes that as a result of Dhulabari's location (on the border), there seems to be free movement of people to conduct trade, unlike landlocked trading cities such as Kathmandu, Delhi, or Bihar.

Lalit's story illustrates that borders constitute the space where the nation is experienced most intimately.[18] The movement of people to and fro in these border spaces makes them infrastructures crucial to the formation of these bazaar spaces. Yet in this transnational, liminal space, the question of identity is even more firmly tied to the idea of nation. The borderland is thus a site of chauvinism on a daily basis, where citizenship and cultural identity are monitored, even as it is also a dynamic, liminal, and interdependent locale.[19] This sentiment was echoed by almost all the traders I spoke to in Dhulabari and Hong Kong Market. Regarding Siliguri, Daman said, "It's a business city, it's a migrant city, it's a mix, multicultural people. For instance, the Bengalis here are different from the ones you find in Kolkata . . . different types of Bengalis, different types of Nepalis. As for me, I am a Punjabi. My hometown is Jalandhar. I feel confused too at times, but ultimately this is where I live and work."[20] But people like Lalit and Daman, who are the elite traders of the town, belong to the status quo because of their upper caste and class privilege despite their migrant status.

The majority of the traders who run the smaller shops usually either belong to the lower caste or are migrants from Bangladesh who don't speak the Indian dialect of Bangla, which linguistically signifies their difference. Complex negotiations of identity in relation to class, gender, religion, and linguistics are articulated through the way trade gets conducted at the border, which determines who becomes the most precarious in this chain. These traders found their way into these markets in about the late 1970s to the early 1980s; they described it as a golden age, a time where Indians were obsessed with imported electronic media goods—so much so that even movie stars would come and visit these bazaars.

The business started to flounder in Dhulabari Market thanks to India opening up its custom regime after the liberalization measures of the 1990s.[21] The upper-caste Hindu traders deserted the market to go back to Siliguri, India. Today they are found at a commercial marketplace called Golden Plaza, where premium imported Chinese goods are sold. As Lalit observed, "Those who have legal permits come over to Siliguri. Those who deal in banned products stay over in Dhulabari."[22] He then described the trade process: "That this is a black market; only insiders know. It is because foreign goods used to be legal in Nepal, but when they would travel to India, they would become black. Goods would come legally but were illegal. We used to have a setting with custom officials. We would say it is from Japan, but it was actually made in China. The Chinese can write anything on the label."[23] This conflation of media, migration, capital, and identity illustrates Mezzadra and Neilson's claim that "borders are becoming finely tuned instruments for managing, calibrating and governing global passages of people, money, and things."[24] Yet the border can also function as a metaphor for various kinds of marginalization that are materialized by architectural and structural measures that also lead to the social regulation of gender.[25] The multiplication of labor in border spaces can also result in further fragmentation of levels of precarity and risk taking, thereby creating its own hierarchies. In these border towns, if the men do the legal or white work of trading, then it is the women who run the illicit goods across the border and carry the risk literally, thus finding themselves at the bottom of these hierarchies.

Crossing the India–Nepal border, I spot a caravan of humpbacked saree-clad women, slowly making a perilous journey across the Panitanki border crossing, which lies between the border trading towns of Siliguri in India and Dhulabari in Nepal (figure 7.3). I suspected that these humps were decoys for the illicit goods that these women were ferrying across

7.3 A woman ferries goods across the border in a fake humpback. Source: Photo by Ishita Tiwary.

the border. My suspicions were confirmed when I spoke to shopkeepers at Hong Kong Market, Siliguri, and my colleagues in India who grew up in Siliguri. They told me they heard stories of how women were often used as couriers or carriers, sometimes using the ruse of pregnancy to carry the goods on their bellies. The goods being carried across the border included things like audio- and videocassettes, Walkmans, and VHS players. The women's bodies thus become the medium of transportation for smuggled equipment, underlining the precarious nature of their work. In border zones, everybody is transformed into a transnational subject, with "bodies that allow themselves to be marked, to be exchanged, to be turned into a commodity, and to be recycled will be granted the visa for mobility in the transnational space. This is the most vulnerable, penetrable site, the place where anxieties about national identity concentrate."[26] In these sites, power relationships rearticulate the border through the practice of crossing, which is not necessarily a happy one.

These power relationships were made visible to me as a researcher in my lack of access to these women. They were a silent, spectral presence. I could not interview them because I did not know who they were or where they lived. It was suggested to me, as I tried to gain access and interview these women, that this work of ferrying goods across the border is not constant, and because of the perilous nature of this work, women don't do this regularly to avoid getting caught by the authorities. The asymmetry of the gender power relationships was materialized in the form of my access as a researcher. The shopkeepers and traders were men, whom I could speak to and cite in this essay even if I don't name some of them because of their power status in the market space. But the unnamed women did not even have that voice. What I infer about gender and labor relations is culled from gaps, silences, and the rumor mill. In her study of La Salada, Latin America's largest illicit market, social scientist Veronica Gago examines the labor and entrepreneurial practices of migrant communities in this border section of Buenos Aires. Gago's argument is a consideration of how migrant communities created an "informal and subterranean" popular economy in La Salada.[27] This was built on heterogeneous labor and entrepreneurial practices. Such transformation of the urban space, she argues, operates spatially and temporally. The spatial transformation occurred as an outcome of the collapse of traditional masculinized practices of employment, family support, and the welfare state in the wake of crisis. In response to such crisis, "feminized figures (the unemployed, women, youth, and migrants)" began to produce informal economies through their occupation of "the street both as an everyday public space and as a domestic space."[28] Can one extend this notion to these women, who belong to rural backgrounds, are economically marginalized and ferry these goods from one urban market to another across borders? Earlier I referred to the practice of portering and carrying while describing the topography of these border bazaars; thus, can one conjecture that word *carrier* has its roots in the topography of space?

According to Daman, these carriers worked at night, helped by the open border regime, and "it was women who did 90 percent of the work."[29] These carriers would get anywhere between 5 and 20 percent of the commission. As security, they had to pay the entire consignment value to the shopkeeper before bringing goods across the border. One can conjecture, then, that these women must be borrowing money at interest from moneylenders in order to enter this profession. Women were valued couriers because they could hide the goods in their bodies in the form of pregnant bellies or hunchbacks. Moreover, they were less likely to be bodily checked by the border security forces. It seems that the traders and customs officials at the border are doing some moral outsourcing, where the belief persists—at least nominally—that a woman's body is inviolable. I also learned from my conversations with custom officials that these women would cultivate a network of informants at the border. For instance, if they got to know a routine check was going to take place, they would change their routes and plans and wait for it to be over before continuing their journey. These couriers would carry several people's orders. The goods were legally imported for Nepal; they were transited from India to Nepal and Nepal to India via couriers, as Mamraj told me, adding, "No loss for us—we got security for that money. Once the product is delivered, the money is released."[30] It seems that the desire for modernity and consumption via media technologies and content that gave rise to traders and consumers was built on the bodies of vulnerable women. What emerges, then, is a pirate modernity based on exploiting women physically and financially through a whole economy of debt and insecurity.

Desiring Consumption

These desires for consumer goods manifested themselves through infrastructures—video theaters/parlors, commodity cultures, and shadow bazaars, or the secret markets housed within the gray market in the 1980s and 1990s.[31] In this context, video theaters/parlors comprise a media rurality that is both distinctive and contrary to stereotypes of rural spaces (especially in the global south) as spaces where media and media cultures are simply absent. In this section, as the example of Chunniji's theaters will illustrate, one observes how the circuits described throughout the chapter demonstrate the opposite: that transnational circuits through which media commodities and devices are circulated are generating rural media practices and experience that are not well described by conventional urban–rural binaries.

The most popular types of films in these bazaar spaces were the latest Bollywood and Hollywood releases, kung fu films, and Western pornographic films. These were ripped and then copied onto blank cassettes and later onto CDs that came from China and were dubbed in Kathmandu.[32] The demand for these films led to the cropping up of video theaters/parlors, a semiformal space that mimicked the structure of cinema. As Daman points out, "These were hall prints. We never had the option of going to a PVR (a monopolistic multiplex chain in India)—then how could we watch the latest English movies?" This signals a temporal desire to be connected to global culture.[33] This desire to be connected to global culture can be thought to be manifested through the framework of "visuality as infrastructure."[34] The framework situates film within a paradigm of representational politics and lays the groundwork for projects that spread and multiply registers of visibility. For instance, the production of new films can lead to building new roads and forms of access.[35] In the case of border pirate bazaars, we see visuality as infrastructure manifesting itself in the creation of media consumption spaces. This is illustrated by the example of Chunniji and his erstwhile video parlor.

Chunniji set up his own video parlor at Dhulabari (figure 7.4). Collecting parts from the gray market, he would assemble radio sets, TV sets, and VCRs. Building on this demand, he opened his own video parlor that could accommodate up to a hundred people. The video parlor had a morning show, a matinee, and a prime-time show, mimicking the structure of cinema theater. Reflecting on that time, he stated, "First all foreign goods used to come here. Now everything is finished. People from as far as Kolkata used to come here. Things you couldn't get in India, you could get here. . . . I started my video business that time. The blank cassettes and CDs used to come from China. Back then it used to cost 1,500 rupees [USD10], and I would charge 5 to 15 rupees [USD4 to USD10] for a person per show, depending on the film. I would make posters for the three shows—12/3/7, aka the morning/matinee/night show."[36] I further pressed Chunniji about the kind of films he showed. He told me he showed all kinds of films—kung fu, Bollywood, and English films. When I ask him about what kind of English film, he refuses to meet my eyes, turns sideways, and keeps muttering, "English films." One of my male colleagues comes to his rescue and names a big Bollywood film of the 1990s. Relief falls across Chunniji's face, and with the change of subject, we keep chatting about his parlor. The unnamed English film here refers to pornographic films, an experience I repeatedly encountered in the field. It seemed that the men wanted to protect me morally by not naming the genre. At best they would refer to it as "those types of films" for me to get the meaning. First, this example shows that access to world cinema was dependent on circuits of piracy. Second, although women were allowed to watch Hindi- or other Indian-language films, the space was designed primarily for men, as evinced by the pornography anecdote; moreover, these spaces were also run by men. Third, Chunniji's video parlor is now extinct; it is now a kiosk for him to make ends meet. Chunniji's story and those

7.4 Chunniji's video parlor has been converted into a mobile phone kiosk. Source: Photo by Ishita Tiwary.

of others like him are a response to the unequal flow of capital wrought by globalization, which leads to a dynamic where media is constantly in a process of breakdown, repair, and piratical consumption.

In the video era, new spaces emerged for the consumption of cinema alongside a series of negotiations trying to define and create the means for the regulation of the new medium. These new practices often moved between legal and illegal zones and were subjected to a wave of regulations that were stubbornly resisted. As Mayur Suresh has reflected, these networks and practices constantly disaggregated and recycled the modernity that video was supposed to usher in.[37] Moreover, with video and leisure culture, a new imagination of the spectator was possible. In the case of the border town, the migrant becomes the new spectator and the new video entrepreneur.

The consumption of these films led to a desire for media commodities. The most popular commodities were VCRs, video cameras, film cameras, videocassettes, Walkmans, radio sets, and stereo players. The popular electronic brands were Canon, Yashica, Casio, Sony, and Aiwa. Owing to the popularity of kung fu films, demand for white canvas shoes was at an all-time high. These goods were first copy. *First copy* is the best duplicate of a brand, which often comes out before the brand releases it. This process of faking globalization is one in which fakes appear when cities are "just about to enter the world economy and become exposed to media representations of global commodities."[38] Thus, a discussion on counterfeit goods should focus on the struggle over legality and legitimacy.[39] Indeed, in discussions with traders, they would often invoke either the moral angle or try to justify that their goods are as good as the original. As one of the big importers told me, "We sell quality products even from China. We have premium products."[40] A small-scale trader told me, "Nowadays many people don't like Made in China brand, so we sell it as PRC. China is seen as *sasta samaan* [cheap goods]."[41] Dharmendra, an Indian origin trader at Dhulabari, said indignantly, "Original media equipment used to come before. Very high quality from Japan. Now it's Chinese [implying lower quality]. Chinese is all fake brands. Even then, what we are doing is totally legal. How can you call gray market as piracy—blank CD main jo daalna hai [what is put on the blank CD]. How is it our headache?"[42]

This is expressed in the bazaar space through infrastructures of secret markets, or markets within a larger market. These bazaar spaces have parallel showrooms that house the first copies in plain sight. Only those with local knowledge can access these spaces. One needs to know the right people or code words to get into these shadow pirate bazaars. I was challenged by all the big shopkeepers in both Siliguri and Dhulabari to find these shadow bazaars. They wagered I would fail. They were right. Although I was privy to these shadow bazaars in Delhi, my hometown, I could not find them in these border bazaars. Perhaps one needs insider knowledge to access these spaces. Today, Dhulabari is now a shadow of itself, struggling to survive because all the vendors have left for India (figure 7.5). Hong Kong Market is still thriving, however. Almost every shopkeeper has gone through various product cycles. If the late 1970s to the mid-1990s was the era of electronic media goods and content, then in the present, it is all about mobile phones, athleisure, and Korean beauty products.

Conclusion

In studies of migration and media, scholarship focuses on two dominant approaches: *media about migration*, where migration is the topic of media content; and *migrant uses of media*, as in the ways that people on the move

7.5 Present-day Dhulabari Market, Nepal. Source: Photo by Ishita Tiwary.

mobilize media texts and technologies to facilitate such movement or to advocate for themselves. The framework of ruralization allows me to center the mediating aspects of *media objects* and to follow their journey as I traced the relationship between the markets of Siliguri and Dhulabari. Thus, by methodologically centering the media object via a south–south pattern, I want to illustrate different ontologies of globalization in this chapter. If in the 1980s and 1990s it was audiovisual equipment that was the dominant and most popular commodity in bazaar spaces, in contemporary times it is the mobile phone. The media object thus represents the desire to be connected to broader global culture. The media object is a specific kind of commodity, one dependent on circulation and subject to various trade agreements and copyright regimes. As questions of topography, identity, gender, labor, modernity, and consumer culture collectively illustrate, there seems to be a highly intimate relationship between the moving media object and the people who move (with) it. In the border bazaars, this intimate relationship is manifested by the way pirate bazaars reflect the migrant experience. By tracking the movement of the media object, we see that on the one hand, it engenders the mobility of people and things, the formation of bazaar spaces, and the creation of infrastructures of video theaters/parlors. On the other hand, it multiplies and reinforces existing caste, class, and gender hierarchies embedded in society for running the trade in these border spaces. Placing the media object at the center of inquiry in border spaces and following its routes calls for a transinterdisciplinary approach to the field that includes archival, ethnographic, and formal methods capable of surfacing these distinctive sites of media rurality.

Notes

1 On October 31, India formally revoked the constitutional status of the disputed state of Jammu and Kashmir. The states were split into two and were stripped of their constitutional autonomy so as to fully integrate them into India. The state's constitution, flag, and penal code were nullified, and the region is now subject to the same central laws as the rest of the country. For more, see Ellis-Petersen, "India Strips Kashmir"; NDTV, "Jammu and Kashmir."

2 For more, see Griffiths, "Why Are China and India Fighting"; Al-Jazeera, "Five Things."

3 For more, see Xavier, "Interpreting"; Saran, "As Nepal Paints Itself"; Shrestha, "Nepal–India–China Tri-Junction."

4 The phenomenon of China producing cheap goods for global markets has its roots in Cold War spatial constructs that saw the manufacture of cheap products from socialist economies. This in turn led to the implementation of antidumping laws in the 1970s, which were meant to keep commodities produced in socialist economies away from capitalist markets in Europe; see Chari and Verdery, "Thinking Between the Posts." It is these legacies, coupled with the experience of liberalization in both countries, that I seek to examine through media circulation practices. At this point, it is crucial to note that the term *Chindia* gains currency after the 1990s, and it is important to emphasize that China and India have been the main focus of economic growth narratives during this period. Significantly, both countries have had historical overlaps. For example, India gained independence in 1947, the People's Republic of China was established in 1949, both countries adopted a socialist model of economy, they went to war with each other in 1962, and they began the process of liberalization roughly around the same period (1978–80).

5 Interview with Ujjwal Prasai, July 5, 2019.

6 Stiller, *Nepal.*

7 Interview with Mr. Khanna, August 2, 2019.

8 Yang, *Bazaar India*, 1–2.

9 Abbas, *Hong Kong*, 4.

10 Interview with Ujjawal Prasai, July 26, 2019.

11 Interview with Daman, July 27, 2019.
12 Simone, "People as Infrastructure"; Simone, *City Life.*
13 Simone, *City Life*, 124.
14 Simone, *City Life*, 125.
15 Simone, *City Life*, 124–25.
16 Hindutva is the ideology of the Hindu right represented by the Bharatiya Janta Party in India. Hindutva is the political ideology of Hindu nationalism. It tries to define Indian culture in terms of Hindu values and is critical of ideologies of pluralism and secularism.
17 Interview with Lalit Agarwal, July 29, 2019.
18 Aggarwal, *Beyond Lines of Control.*
19 Aggarwal, *Beyond Lines of Control.*
20 Interview with Daman, July 27, 2019.
21 In 1990, India faced a serious economic crisis, where the government was close to default and had foreign reserves that could only finance two weeks' worth of imports. This situation led to sweeping economic reforms under then–prime minister P. V. Narsimha Rao and his finance minister, Manmohan Singh. The new trade policy focused on disinvestment in public sector companies, reducing excise duties, and abolishing export subsidies.
22 Interview with Lalit Agarwal, July 29, 2019.
23 Agarwal interview.
24 Mezzadra and Nielsen, *Border as Method*, 3.
25 Aggarwal, *Beyond Lines of Control.*
26 Biemann, "Performing the Border," 3.
27 Gago, *Neoliberalism from Below*, 17
28 Gago, *Neoliberalism from Below*, 7.
29 Interview with Daman, July 28, 2019.
30 Interview with Mamraj, July 29, 2019.
31 To think about these infrastructures, I draw on Meghan Morris's work on Hong Kong action cinema and Neikolie Kuotsu's work on piracy and the northeast in India: Morris, "Transnational Imagination"; Kuotsu, "Architecture." Morris's work underlines the transnational histories of the direct-to-tape Hong Kong action film, which is useful to situate the popularity of these films in Southeast Asia. In contrast, Kuotsu's work problematizes the notion of the nation-state and identity by looking at the consumption and circulation of South Korean new wave films in the northeast region of India. His work explicates not only the tensions between region and nation but also the various modernities that are present in the Asian region. Morris's and Kuotsu's work gives me a crucial frame to explicate the desires of modernity and consumer cultures present in these transit zones, exemplified by the pirate bazaar spaces of Hong Kong Market, Siliguri, and Dhulabari Market, Nepal.

32 Later, this network of ripping shifted toward Bangladesh, Pakistan, and the Gulf region.

33 Interview with Daman, July 27, 2019.

34 Campos Johnson, "Visuality."

35 Thinking of visuality as infrastructure is thus a useful way of bypassing representational approaches; it enables a historical narrative about the changing historical conditions of visuality, including the multiplication, expansion, and thickening of visual forms as well as their increasing centrality to social, economic, and political processes so that we might want to talk about visual forms as a critical infrastructure of global capitalism. See Campos Johnson, "Visuality."

36 Interview with Chunniji, July 29, 2019.

37 Suresh, "Video Nights."

38 Abbas 2013, *Faking Globalization*, 289.

39 Neves, *Underglobalization.*

40 Interview with Anonymous, July 28, 2019.

41 Interview with Anonymous No. 2, July 28, 2019.

42 Interview with Dharmendra, July 29, 2020.

Bibliography

Abbas, Ackbar. "Faking Globalization." In *The Visual Culture Reader*, edited by Nicholas Mirzoeff. Third ed. Routledge, 2013.

Abbas, Ackbar. *Hong Kong: Culture and the Politics of Disappearance.* University of Minnesota Press, 1997.

Aggarwal, Ravina. *Beyond Lines of Control.* Duke University Press, 2004.

Al-Jazeera. "Five Things to Know About the India-China Border Standoff." June 22, 2020. https://www.aljazeera.com/news/2020/6/22/five-things-to-know-about-the-india-china-border-standoff

Biemann, Ursula. "Performing the Border." In *Globalization on the Line: Culture, Capital, and Citizenship at the U.S. Border*, edited by Claudia Sadowski-Smith. Palgrave Macmillan, 2002.

Campos Johnson, Adriana Michele. "Visuality as Infrastructure." *Social Text* 36, no. 3 (2018): 71–91.

Chari, Sharad, and Katherine Verdery. "Thinking Between the Posts: Postcolonialism, Postsocialism, and Ethnography After the Cold War." *Comparative Studies in Society and History* 51, no. 1 (2009): 6–34.

Ellis-Petersen, Hannah, and a reporter in Srinagar. "India Strips Kashmir of Special Status and Divides It in Two." *Guardian*, October 31, 2019. https://www.theguardian.com/world/2019/oct/31/india-strips-kashmir-of-special-status-and-divides-it-in-two.

Gago, Veronica. *Neoliberalism from Below: Popular Pragmatics and Baroque Economies*. Duke University Press, 2017.

Griffiths, James. "Why Are China and India Fighting over an Inhospitable Strip of the Himalayas?" CNN, June 18, 2020. https://www.cnn.com/2020/06/17/asia/india-china-aksai-chin-himalayas-intl-hnk/index.html.

Kuotsu, Neikolie. "Architecture of Pirate Film Cultures: Encounters with the Korean Wave in North East India." *Inter-Asia Cultural Studies* 14, no. 4 (2013): 579–99.

Mezzadra, Sandro, and Brett Nielsen. *Border as Method, or The Multiplication of Labor*. Duke University Press, 2013.

Morris, Meghan. "Transnational Imagination in Action Cinema: Hong Kong and the Making of Global Popular Culture." *Inter-Asia Cultural Studies* 5, no. 2 (2004): 181–99.

NDTV. "Jammu and Kashmir Not a State from Today, Officially Split into 2 Union Territories." October 31, 2019. https://www.ndtv.com/india-news/jammu-and-kashmir-not-a-state-from-midnight-officially-split-into-2-union-territories-2124866.

Neves, Joshua. *Underglobalization: Beijing's Media Urbanism and the Chimera of Legitimacy*. Duke University Press, 2020.

Saran, Shyam. "As Nepal Paints Itself into a Corner on Kalapani Issue, India Must Tread Carefully." *Indian Express*, June 12, 2020. https://indianexpress.com/article/opinion/columns/india-nepal-new-map-kalapani-6454647/.

Shrestha, Buddhi Narayan. "Nepal–India–China Tri-Junction." Kathmandu Post, July 26, 2020. https://kathmandupost.com/columns/2020/07/26/nepal-india-china-tri-junction.

Simone, AbdouMaliq. *City Life from Jakarta to Dakar: Movements at the Crossroads*. Routledge, 2010.

Simone, AbdouMaliq. "People as Infrastructure: Intersecting Fragments in Johannesburg." *Public Culture* 16, no. 3 (2004): 407–29.

Stiller, Ludwig. *Nepal: Growth of a Nation*. Educational Publishing House, 1993.

Suresh, Mayur. "Video Nights and Dispersed Pleasures." In *The Public Is Watching: Sex, Laws, and Videotape*, edited by Mayur Suresh and Namita Malhotra. Public Service Broadcasting Trust, 2007.

Xavier, Constantino. "Interpreting the India–Nepal Border Dispute." Brookings Institution, June 11, 2020. https://www.brookings.edu/articles/interpreting-the-india-nepal-border-dispute/.

Yang, Anand. *Bazaar India: Markets, Society, and the Colonial State in Bihar*. University of California Press, 1998.

8

The Preservation of Embodied Masculinity in Rural Tech-Altered Workplaces

JENNA BURRELL

The occupational values of traditional rural industries in the western United States center on the laboring body. Rural work is celebrated and even mythologized as physical, perilous, embodied, and done in the outdoors. So what happens when high tech (perceived as an urban, white-collar, office-based industry) makes inroads into such rural places? How do communities respond when the livestock auction becomes internet connected, drones replace cowboys for monitoring cattle, and a Facebook data center becomes the newest employer in town?

This chapter explores how rural workers reconcile new work cultures that intrude into rural sites. I investigate two key examples: the rise of the data center industry, which filled the void left by the declining timber industry, and computing technologies, which became incorporated into traditional rural workplaces, specifically cattle ranching. Who accepts computing technologies in male-dominated rural workplaces relates to how closely they identify with masculine occupational values, particularly the value placed on physical strength and endurance.[1] The high-tech industry is likewise male dominated. Yet computing technologies are aligned with contrasting values associated with masculinity, specifically the ideal of a dispassionate rationality that privileges mind over body.[2] As computing enters rural worksites, this sets the stage for a potential clash between quite different notions of workplace masculinity.

In the worksites I examine in this chapter, new computing technologies and the expansion of the high-tech industry to new rural workplaces prompt rural workers to reconsider their self-identity as workers. Other kinds of materialities and material constraints, such as regional and global economic shifts and the aging human body, also factor into work's revaluing. These various challenges bend, but do not entirely displace, traditional gendered norms around work. The traditional rural workplace and its ideal of masculine physical labor proves surprisingly durable. In fact, local leaders and workers in rural communities often see in the arrival of high tech workplaces an opportunity to protect or restore an embodied workplace masculinity that once seemed to be irrevocably in decline.

To form this argument, I draw from data collected during extended periods of fieldwork in central Oregon and the Northern California coast beginning in 2015. These sites were selected not for their representativeness but because they are places where digital and network technologies were especially present and visible. Both sites hosted key infrastructure of the global internet. In coastal California, I looked at the community along the Mendocino County coast, where, notably, the tiny town of Manchester hosts the landing station for a transpacific suboceanic fiber-optic cable. Yet locals there struggled with inadequate and unreliable internet connectivity. In central Oregon, I spent time in Prineville, a town that hosts a sprawling Facebook data center.[3] I talked to linemen, construction workers, and other trade union members who built and maintain this infrastructure. I also explored the lifestyles of fishermen, farmers, and cattle ranchers working the surrounding area. I grew interested especially in the way the bodies of rural workers figure into imaginaries of continuity and change.

Current scholarship on gender, and specifically on masculinity and manhood, challenges the idea of a stark binary that distinguishes feminine and masculine. Instead, scholars recognize the ongoing constitution and reconstitution of masculinity and the coexistence of multiple masculinities.[4] While a masculinity rooted in the body and its capacity for labor prevails as the regionally hegemonic masculinity in my field sites, the range of ways of being masculine is more varied and nuanced when viewed up close. I highlight, for example, how the self-identities of cowboys versus cattlemen shape differences in values-based reasoning about which technologies to accept and use in cattle ranching. Selective resistance to new technology is a reflection of this reasoning, not a bluntly conservative reaction against the threat of change.[5] Even as economic pressures make it sensible or necessary to utilize new technologies in work processes, these technologies do not exert a deterministic force on workplace cultures.

In my analysis, I resist operating with an essentialized notion of the rural; I challenge, as do many chapters in this volume, the notion of rurality as either homogenous or as maximally different from the urban. Within the rural regions of the American West where this fieldwork took place, a familiar gender binary often surfaced in interviews, but at the same time, masculine-identified roles in the workplace were varied. In previous interviews with women in ranching, I also found many who were rethinking and reworking the gendering of agricultural work that generally put them at a cultural and/or economic disadvantage.[6] New sources of influence arriving from the computing industry did not merely sweep away old regimes of value.

It is also worth noting certain silences in the conversations I had with my interviewees about work. Nonnormative gender identities and the existence of LGBTQ+ identities went unremarked on, save for one offhand reference to lesbians in the skilled trades. Those I interviewed presented as more or less cisgender. A few brought up their family history in the region, generally pointing to European heritage and ancestors who were homesteaders—in other words, settler colonizers in the region. A number of interviewees claimed they were entitled to shape what happened in the region because of the number of generations of their family who had lived there. At the same time, they did not comment at all on Indigenous populations, whose longevity in the area would greatly surpass that of European settlers. In this way, interviewees reinforced a colonial fiction of the American West as uninhabited before its colonization and as having historical origins that start with homesteading and ranch life. I provide this context as a way to reflect on how the following discussion of cowboy masculinity takes place where their "Indian" counterpart is absent from the conversation.

A Brief History of How Computing Became Gendered

A foundational definition of masculinity is rooted in the body and established by demonstrations of physical strength. It is, in other words, an *embodied masculinity*. Connell terms it "true masculinity."[7] Blue-collar jobs that rely on physical labor and that disproportionately employ men emphasize and value the bodily capacities of workers. Broader economies of advanced capitalist countries like the United States have shifted, however, toward service and white-collar work sectors, while agriculture and manufacturing have consolidated and automated (reducing their workforce), moved offshore, or simply declined. One result is that other occupational values, ones aligned with masculinity but that *de-emphasize* the body, have gained power and influence in workplaces, popular culture, and everyday life. However, as I will show, the influence of this new form of masculinity is uneven.

Raewyn Connell addresses the competition between different notions of masculinity and how they may change position in a hierarchy. Masculinity as physical and bodily strength is not everywhere and is not always the dominant one. "Hegemonic" masculinity is what currently "sustains a leading position in social life" as the "culturally accepted answer to the problem of the legitimacy of patriarchy, which guarantees . . . the dominant position of men."[8] Yet this definition is unstable. Alternative masculinities emerge within particular regions, subcultures, and industries, positioned in different ways in relation to or against a hegemonic one. For example, the stigmatization of male homosexuality—or, pertinent to this discussion, of the computer nerd—places it as *subordinate* in a hierarchy of masculinities. Such a hierarchy is often enforced through domination, including physical violence.[9] There are also relationships of complicity that uphold this hierarchy. While many men don't fully meet normative standards of masculinity, they still benefit from a patriarchal dividend by endorsing and enforcing those standards. The relationship between masculinities may shift, however; ultimately, alternative masculinities can be taken up by groups to construct a new hegemony.

Among these alternatives, a masculinity defined by intellect, rationality, dispassion, and absence of emotion—a decidedly *disembodied* masculinity—is closely linked to male dominated white-collar industries. It has come to be strongly associated with computer programming and work in high tech.[10] Eglash argues that this transcendence or rejection of the body has been core to the masculine nerd archetype.[11] As knowledge work, including computer programming, has come to centrally drive the economy, commanding high salaries and evident power, this archetype has challenged, if not entirely overtaken, the hegemonic position.[12] In some contexts, masculinity centered on the body and physical performance is displaced from work activities and shifted into other pursuits, such as sports. In other contexts—including rural locations excluded from developing high-tech industries—masculinities defined by technical reason have been slower to develop.

The computer nerd as a new kind of masculine archetype did not emerge with the advent of the computer. In the early history of jobs that centered on computers and computer programming, such work was neither blue nor white collar. In the midcentury, neither bodily strength nor dispassionate reason were defining occupational values of computer work. It fell instead into the category of pink-collar jobs, which were held by women. The mainframe computer was at first viewed as just another type of new office machinery, like a typewriter or switchboard, both machines operated principally by women.[13] For middle-class women, these jobs were assumed to provide only supplemental income in male-headed households or were positions they expected to hold only until marriage.[14] According to Ensmenger, programming became a male profession in the United States only after it was symbolically recast as a kind of "black art."[15] Changes in hiring practices also served a gatekeeping function. Recruiters began to use aptitude and personality tests whose designs "embody and privilege masculine characteristics."[16] Historians point out that the work itself did not radically change; it still comprised the same tasks women had been doing a few years before. Yet the focus shifted to hiring workers with advanced math skills, even though programming requires little more than high school–level math proficiency. This shift in qualifications privileged men, who were more likely to have been encouraged to develop these skills and/or were more likely to be seen as having the requisite natural ability.

The computer nerd was initially a "liminal masculine identity," one associated with men but still cast in negative terms.[17] The term was long used as a pejorative. As the significance of mastery over machines became a visible pathway to wealth and power in the market economy, the nerd identity began to challenge the embodied masculinity of physical strength by preserving the notion of masculine control.[18] In turn, this made space for those who were not perceived to be physical specimens of masculinity—think of Bill Gates, Jeff Bezos, and more recently Elon Musk—to occupy not only the highest ranks of business but status hierarchies as well.[19] Processes of globalization have also facilitated opportunities for those in geographic margins to challenge associations between Euro-American men and hegemonic masculinity. Poster argues that the economic opportunity to seek one's livelihood or fortune in computing tech in India had an equalizing effect and yielded new forms of worker agency from the margins.[20] This led to the formation of a technomasculinity rooted in individual technical prowess rather than the masculine business cultures that embed power in firms and other institutions in the global north.

The most recent refiguration of masculinity in high tech, the *brogrammer*, combines (1) the impression of physical strength and (2) the frat-guy sociability of the masculine managerial role, together with (3) the skills of the technically capable.[21] This emerging trope reworks multiple idealized dimensions of competing masculinities into a single figure. It illustrates not one singular masculinity but multiple masculinities that exist as dynamic and competing ideals that different individual workers align themselves with and that vie for the hegemonic position. News accounts in October 2022 about a planned mixed-martial arts cage match between two tech CEOs, Elon Musk and Mark Zuckerberg, show this realignment as actively underway.[22] In this case, it is enacted by a literal and public masculinity contest.[23] In this way, masculinities that were once hegemonic resist becoming entirely sidelined or made archaic. New formations of masculinity reassert and recenter old masculine values while simultaneously bolstering new ones through alignment.

This literature on the cultural history of computing sets some context for the way careers associated with computing technology might disrupt masculine occupational values in rural sites. It also demonstrates that workplace masculinity continues to be unsettled, and that a hegemonic masculinity of Euro-American workplaces has been, from various directions, challenged and reframed. In the following sections, I link the evolving occupational values and gendering of work in rural sites after computing tech is introduced to show what happens when two distinctly masculine value–aligned work worlds meet. Following Poster, I demonstrate the regional variations in how workplace masculinity comes to be defined and shaped.[24]

High Tech Comes to the "Cowboy Capital of Oregon"

When a Silicon Valley behemoth like Facebook sets up a new outpost in a rural town like Prineville, Oregon (a place previously known mostly for cattle ranching and the timber industry), it is a situation ripe for gendered occupational values to clash. Facebook sought this area for its environmental and other conditions, notably cheap electricity fueled by a New Deal–era federal hydroelectric project, a cool climate, and a favorable tax abatement scheme.[25] The skill set of the local workforce was of less interest to senior executives and decision-makers at Facebook than these other factors. The relationship between the Prineville area workforce and the new local employer was not predestined. It was partly directed and shaped by local leaders and powerful figures representing the interests of the tech industry. One major figure shaping the narrative of this cultural transition was the incoming director of Facebook's Prineville data center, Ken Patchett.

For Patchett, establishing the suitability of a data center for this region, given its existing blue-collar labor force, was part of his job representing the Facebook corporation to the local community. In an article for *Outside* magazine, Patchett is quoted as saying, "The internet is not a fake thing . . . it's a very, very real thing made with hammers and concrete. And nails. And blood and sweat and tears and people."[26] Patchett is describing the working conditions at Facebook (within the data center, at least) in a language ordinarily associated with construction, skilled trades, or agricultural and natural resources jobs. Patchett, who in 2010 was hired away from the Google data center in nearby The Dalles, Oregon, also used his own physical embodiment as a part of the message he communicated. He looked and dressed like a member of a particular class of local men. In a Wired.com article covering the data center's opening, Prineville's city engineer, Eric Klann, is quoted stating approvingly, "Ken has a little bit of cowboy in him . . . and this is a cowboy community."[27] The article goes on to detail Patchett's boots and matching belt buckle. Indeed, in video and photos from the Facebook page for the Prineville data center, Patchett is shown wearing a short-sleeved button-up shirt, his large rodeo-style belt buckle visible as he describes Facebook's efforts to make the plant maximally energy efficient and more environmentally sustainable.

Jason Carr, a former Prineville city council member, speaks further about the symbolism of Patchett's embodied presence. According to Carr, Patchett had a way about him that demanded he be taken seriously. He "knew how to be with rural type folks and mingle with them and talk to them in a way that they would understand." Carr noted that Patchett would make statements along the lines of, "I grew up on my grandparent's dairy farm in Northern California, population smaller than Prineville" and "I could get on a horse just like the rest of you" in response to locals who tried to dismiss him as an outsider. Patchett's role at Facebook was partly conceived of as a public relations role, a stark difference from his previous job at Google. In the case of Google's nearby data center, the launch was done with extreme secrecy, which consequently raised some amount of suspicion in the local community. In Prineville, by contrast, Patchett served as a public ambassador of the company to the community. When he positioned himself both as leader of this new facility and as someone raised in agriculture—with horsemanship skills, and multigenerational roots in rural life—he lightly aligned the data center with local masculine occupational values. While horsemanship certainly was not highly relevant to the data center's operations, it demonstrated the data center leader's familiarity with that work world.

In my interviews with local leaders, they framed the arrival of the data center in Prineville in terms of what became of the old labor force with this shift to a new industry. Their narrative emphasized continuity. A story often retold in Prineville was about Jason Beebe, a lifelong Prineville resident and city council member who in 2020 was elected mayor. A fellow member of the city council, Jason Carr, described him as someone who "graduated from Crook County High School, been a military career guy, worked at Les Schwab. He's about as blue collar, rural as you get. Here's this guy in his early forties who was like, 'My body, I can't do this throwing tires all the time,'" so he found work and retrained at the Facebook data center. City manager Steve Forrester, who also related Beebe's story to me, added, "Now he's in an air-conditioned environment, he doesn't get cold in the winter or hot in the summer, and he's fixing these little machines." Zuboff similarly finds that workers reevaluated their role after computerization swept through manufacturing work, shifting their day-to-day activities from hands-on work with machines to command and control from a remote computer room.[28] Some saw a subtle status gain as they moved into this more sedentary work, one with the appearance of white-collar labor, even as the work itself became more routinized, less skillful, and less interesting.

Multiple accounts of work within data centers, however, depicted it not as a relief from labor but rather as physically strenuous. The role of the data center technician, for example, entails responsibility for managing, repairing, and maintaining the facility's enormous banks of servers. In the early days, much of the work in data centers focused on managing the heat thrown off by servers that, if not kept in check, could lead them to malfunction. Gonzalez Monserrate's ethnography of data center workers casts the struggle to fight heat to avoid the dreaded downtime in masculine embodied terms.[29] One worker he interviewed described losing control over the machines and their going into a heat spiral as "emasculating." While advancements in the design of these facilities, particularly at large corporate data centers, means heat can be monitored and managed remotely, there is still a great deal of physical work. Albert G., who was employed for a time at a large corporate data center in Oregon, said of the work, "It's pretty exhausting. . . . You're on your feet all day. . . . You're lifting servers that weigh twenty, thirty pounds. Taking out bad parts, putting in good parts. Putting the heavy server back on the rack. . . . Moving entire racks. Sometimes you're up on a ladder, running cable." It was "lots of repetitive physical work." Additionally, the environment of the data center designed to manage server heat was "not the most friendly." It was very dry and incredibly loud. Albert G. said, "More fans are spinning than you could possibly imagine. The white noise is just crazy." Both accounts offer observations that radically break from popular and media depictions of tech jobs as cognitive work in pristine and quiet environments, showing an underlying truth behind data center director Ken Patchett's remarks about how the internet was built with "blood and sweat and tears."

Local leaders in Prineville made an effort to show that data center jobs were blue-collar jobs that entailed physical exertion, but they drew a link to specific local blue-collar men transitioning from traditional declining industries into these new data center jobs. In public forums, Prineville city manager Steve Forrester often referred to the people he encountered when he visited the Facebook facility. At a public event in February 2017 that was scheduled to address the housing shortage in Crook County that had resulted from the influx of out-of-towners arriving for data center work, he took the opportunity to stave off criticism that the data center project was importing outsiders to the community and had no local benefits. He said, "I'll remind folks that when I come up there to tour that facility, every single time I've been up there, without exception, I see people who used to work with me in forest products industry or people that I knew, that I grew up with, that are from Les Schwab that have transitioned into this high-tech world at a pretty good rate in terms of income. . . . We have to be able to bring some of the folks that came out of the forest products industry, that haven't been working because of what's happened up at Woodgrain or just the shrinkage of that industry and get them engaged in some of these other opportunities."[30] His obvious point was that this new industry could be viewed as a direct and immediate solution to the decline of the old one. He offered evidence, which he had seen with his own eyes, of the preservation of the local community through new employment opportunities for longtime locals.

Outside of direct employment at Facebook, the broader workforce also gained a boost from the data center, particularly in the skilled trades and specifically in construction work (building data centers) and utilities work (laying fiber-optic cables). This further reinforced particular occupational values aligned with masculinity, as such work was also rooted in physical labor. In fact, when I spoke to workers or their union reps, they insisted that it be understood as labor. Richard S., a union organizer for the International Brotherhood of Electrical Workers, challenged the perception that some skilled trade jobs were physically easy, noting that "being an electrician is labor. It's not. . . . You may see a picture of an electrician standing in front of a breaker panel and he's landing the wires. That's relatively easy work, right? [But] sometimes we're digging ditches and putting pipe in them, and pulling wire, and it can be going up and down a ladder all day." Talk with union leaders about welcoming women into these traditionally male-dominated fields was framed by existing standards of physical capability: "If you can dig a ditch just as good, as well, I don't care if you're Black, white, Hispanic, female, lesbian, doesn't matter. . . . If you can do the work, get the work done. We don't care." Who filled the roles could change, but the masculinity-reinforcing norms and standards of the work itself were inflexible.

In rural Oregon, the value of the data centers was also linked by local promoters to the restoration of male providers as primary or sole income earners in their households. This must be understood in the context of the financial crisis that predated Facebook's arrival. In the great recession starting around 2007, unemployment in Prineville exploded, approaching 20 percent, as the housing market crashed and a previously booming local construction industry dried up. The population of Prineville declined by 10 percent as residents sought stable jobs elsewhere. Through a combination of regulation, offshoring, and economic decline, the sorts of traditional rural jobs defined by an embodied masculinity—in manufacturing, natural resources, and even construction—became jobs that no longer provided a living wage or reliable income. Data centers, however, were promised to bring back "family wage jobs," a phrase used by Carr, Forrester, and others in interviews. The notion of a family wage job is gendered, alluding to the sort of job that can provide for one's dependents, namely one's (female) spouse and children. It references another kind of symbolic emasculation, one that followed from the economic changes that swept through places like Prineville, leading men to lose financial primacy in their households.[31] In sum, it was these three points centered around reinforcement of masculine workplace identity—data center jobs as physical jobs, as jobs that local workers transitioned into from other blue-collar industries, and as jobs that restored a family wage—that facilitated the arrival of Facebook by minimizing cultural disruption.

Reckoning with Change in Other Rural Worksites

Long before the arrival of the Facebook data center in Prineville, however, local leaders had been forced to contemplate how high tech might threaten existing occupational values. Computerization and automation were already factors in the declining numbers of jobs in some industries, particularly the timber industry. Challenges of hiring and labor supply forced existing ways of valuing and compensating work to be reconsidered.

Steve Forrester, the city manager for Prineville, had worked for many years as a sawmill manager. He described his thinking when he went through the process of hiring a new type of worker as his industry became more automated. He noted that today, a sawmill that once had eighty employees would now have only ten or fifteen. At his sawmill, he had hired three "programmers" already, but the system had become so sophisticated that he now needed an IT director. When he learned the pay range for such a person, his first reaction was disbelief. He recounted a conversation with his boss that went like this:

> STEVE: Well, that's more than my production guy that comes in at five in the morning and works twelve hours a day, oversees two shifts, and working his guts out . . . and I'm not going to pay this fricking IT guy that much. . . .
> BOSS: Do you need an IT director?
> STEVE: Well, yeah, I've got to have one.
> BOSS: Then you need to pay them.
> STEVE: Oh crap. OK. Well, yeah.

Forrester characterized the "production guy" as his model worker in visceral terms, as "working his guts out." He describes the job in terms of the endurance and stamina required, including long work hours that limited sleep.[32] This job reflected traditional occupational values associated with embodied masculinity. The new role of the IT director, by comparison, was more cerebral. It seemed to demand intellect but not much else. Steve described the process of hiring the IT manager initially as an adjustment to a new way of valuing work. He concluded by noting the IT manager's pay in definitive terms: "And so our IT director became one of the most highly compensated people on our team, because that's what they are, *that's what their value is.*"

Steve Forrester's narrative of two workplaces—the Facebook data center and the sawmill—were starkly different. He and other local leaders shaped the narrative of Facebook's arrival in Prineville to reassure the community and minimize resistance. Through this narrative, they also encouraged a particular group of locals—underemployed men previously working in declining rural industries—to recognize it as an opportunity for them specifically. There's an irony here. The arrival of a Silicon Valley–based company was presented as a return to well-paid blue-collar jobs while it was the local sawmill that forced rural communities to confront a new workplace reality, one that de-emphasized the laboring body. The revaluation of work Forrester faced at the sawmill followed from a seemingly inescapable material reality: the cost of hiring an IT manager and the necessity of having one to keep the sawmill economically viable. Another material reality, the limits imposed by the aging body, that likewise forced a reconsideration of embodied masculinity, will be taken up in the next section. Workers contending with their own aging bodies, or witnessing it in others, had to make accommodations. This often had implications for whether and how they used technologies.

Gendered Occupational Values Shape How the Work Is Done

The second part of my discussion further elaborates on the different masculinities in play in rural workplaces and how they shape which technologies are adopted, by whom, and for what uses. Above I described how local leaders and other rural residents grappled with the computing industry and how they reconciled new industries and new work roles with established masculine occupational values. However, new computing technologies also shaped work practices in a more fine-grained way, including decisions about how to do the work and which tools to use. I encountered varied attitudes toward digital technology among those working in traditional rural industries. The technologies mentioned in conversations included communication technologies (mobile phones, the internet) as well as automation technologies, such as drones, four-wheelers, and GPS-embedded tractors. The same technologies were embraced or accommodated by some workers and eschewed and avoided by others. Cattle ranching in particular was a fascinating space of both tech abstention and accommodation related to two contradictory masculine identities: cowboy versus cattleman.

Cowboyism is defined by a kind of symbolism and style as well as a kind of mentality and work practice. In either sense, it is not specific to raising cattle. Levy talks about long-haul truckers and their particular cowboy-isms, including "open scorn for legal and managerial authority, general resistance to rules, and proud self-reliance."[33] As noted already, the Face-book data center director, Ken Patchett, was flattered as having "a little bit of cowboy in him" by Prineville's city engineer and always attempted to look the part. However, locals used vivid and memorable terms to alert me to pretenders. Trent G., an itinerant ranch hand, described "wannabes" and "wish-they-weres" who were "all hat and no cowboy." He also derided people who bought a ranch just because they had money and a fantasy. He drew a very particular portrait of people who had maybe made a fortune on an internet company and then decided to raise Wagyu beef but did not know what they were doing.[34] Delilah B., who'd raised two sons on a family cattle ranch before moving into town, described people who come to the rodeo to play "dress-up." Jackson B., who had inherited the large cattle ranch he grew up on, took a principled stance rejecting this label for himself, noting, "I've known too many good cowboys to throw myself in with that deal. I can't throw a rope to save my life." The point made in these comments was that faux cowboys buying ranches with urban wealth or playing dress-up only superficially represented what was, in its authentic realization, a deeper practice, art, and way of life.

Whether tech was or was not "for me" was talked about in particular ways and was understood by men in relation to their gender identity and the occupational values of their industries. One way of expressing this was to talk about what tech innovations (and their local adoption) do in relation to the body. For some, digital tech was seen as a weakening force. Giorgi S., a cattle rancher, groused, "Kids today can't pick up much more than a phone." Often these men condemned social media while in the same breath disclaiming any ability or competence with computers. Trent G. mentioned that his daughter managed his résumé for him (he was often looking for jobs), noting, "I try to keep a woman around that knows that thing." Two older men I interviewed both excused their lack of knowledge about computers, joking, "That's above my pay grade" and mentioning how at school the girls learned to type while the boys learned other skills, like welding. There was a consistent association of computers with secretarial skills, typing, clerical work, and ranch bookkeeping, or in relation to jobs done part time in town by ranch wives to supplement farm income.[35] At a small local livestock auction in central Oregon, it was exclusively women who operated the computers, handling the bids submitted online or called in by phone. The auctioneer and the owner were men, as were the cowboys herding cattle through the arena, and I observed mostly men bidding on the cattle. One farmer I interviewed talked about how his wife would listen to him and use her typing skills to help put his thoughts down on paper. While she was described with affection and appreciation, her role was clearly presented as helpmeet—a useful conduit to a thinking male mind rather than as a master of the machine.

Another factor that forced a reconsideration of occupational values in rural blue-collar work was the limits set by the worker's aging body. This reckoning happened at the individual level as men in labor-intensive professions in my central Oregon and Northern California field sites anticipated their future, looking to what fate befell older members of their occupation. Jasper R., a second-generation fisherman still in his twenties, whom I met in coastal Northern California, described how grueling the work was and what it did to the bodies of fishermen over time: "If you want to talk about your back breaking . . . lifting up 150-pound crab pots . . . you don't see old fishermen all jolly, like life was good and they're all jolly. Their hands are all fucked up, they're all hunchback of Notre Dame, barely walking along to their boats. It's one of those things that puts a toll on you." In central Oregon, Trent G. described a period where he switched from working as a ranch hand to running a saddle shop. It started when a "good friend" of his family's who "had run a ranch, been on the ranch for fifty years. He got old, sick, had a heart attack . . . didn't even get a gold watch. [Just] adios!" For Trent, however, this switch to saddle making didn't stick. He said he couldn't stand being indoors so much.

When rural workers were forced to reckon with their own physical limits, tech (broadly including things like farm machinery) came to be viewed in a more positive light—as labor alleviating. Just as men reckoning with aging bodies moved into new occupations, like working at the Facebook data center, the promise of making work easier nudged them to reconsider physical effort as an occupational value. This reconsideration was linked to a competing dimension of their gender role: the priority in heteronormative households for men to be the principal income earners and providers for their families, as noted above in the way local leaders promoted data centers' provision of family wage jobs. In this context, new tech brought into traditional industries could remove physical strength as the primary determinant of employment and survival. Giorgi S., who lamented that "kids today can't pick up much more than a phone," nonetheless acknowledged how farm equipment meant that the "moms" and older men can share in more of the farm work. In his case, he had taken over ranch operations from his father and hoped to pass the ranch on to his son one day. He mentioned that his father had a degenerative disease that led to muscle wasting. The harvester they used had an automatic clutch, allowing his father to continue doing the work he had done all his life. New tech was able to compensate for his disability.

There was a division in opinions about automation, however, across men in the industry. By closely studying the lines of disagreement, two character types within cattle ranching emerged: the pragmatism of the business-minded cattleman, who embraces useful tech out of practicality, versus the cowboy, whose occupational values and relationship with tech are rooted in a more romantic embodiment and are not simply about shows of strength but the art and beauty of the work.

Those who self-identified as cowboys took a profoundly antimaterialistic stance to their work. Accumulating wealth, possessions, and property was not the way to express their self-sufficiency; it was by realizing autonomy in their day-to-day work. It was by heading out alone on a vast ranch and by being free of personal possessions. As Trent G. noted, "Cowboys don't have a nickel, but other people do." Recalling his earlier failed attempt to switch to saddle making, he described, with a note of resignation, a life defined by endless work up until the moment of death: "I'm gonna die somewhere and the next guy's gonna step over my body to do whatever I was doing when I died." This obligation or compulsion to work did not hamper Trent's sense of being self-directed. He stated, perhaps counterintuitively, that such a life was nonetheless characterized by "just being free."

A related tension surfaced in distinctions between real cowboy work and rodeo. A few of the cattle ranchers I met had, as young men, done stints touring rodeos. As a sport, rodeo seemed to take embodied masculinity to its extreme as a public ritual of strength with all of its attendant risks. Connell notes the function of sport as "symbolic proof of men's superiority and right to rule."[36] The master of ceremonies at the Crooked River Roundup, a rodeo I attended in Prineville in 2017, lauded the competitors, who "put their life on the line and that's what makes them heroes." Rodeo served a cultural role, although ranchers I spoke with clarified that rodeo did not reflect in any realistic way the work of cattle ranching. As evidence of its role in reaffirming masculine identity, as not only embodied but also heteronormative, cattle ranchers I spoke with who competed in rodeo in their youth said the big draw of going on the rodeo circuit was to "meet girls." At the rodeo itself, women's participation was strictly limited to only one of seven events (barrel racing) and to the ambassadorial role of rodeo queen. I was struck, however, by the emphasis placed on horsemanship for the rodeo queen and her court, who charged into the arena on horseback at full gallop carrying flags as they were introduced to the crowd.

The spectacle of bull riders, rodeo queens, and the huge prize money of rodeoing at the elite level was in contrast to ranch rodeos, a special subclass of less publicly known competition that Trent G. said "catered to what real cowboys do." This work was nuanced. It called for precision, not speed and bold domination. Horses were emotional, sensitive animals and responded to subtle direction. Startling your cattle, I was informed, could negatively affect beef quality or milk production, so they needed to be handled gently and with care. Yet actual ranch work was also seen in some ways as more risky than rodeo. As Trent G. noted, if you crash during a rodeo event, there's an ambulance waiting to whisk you away to the nearest hospital. On remote ranch properties, however, "you crash out here, you're liable the flies will get you before they find your body."

Lucas T. was a twentysomething self-described cowboy who worked as a ranch manager alongside his father on a large corporate-owned ranch. When asked what he liked most about cattle ranching, he said, "You get to work outside. You get to work with animals. The more knowledge you have, the more efficient you can be. . . . It's kind of an art in that aspect, I guess. It's untapped potential to learn." When I asked him about the new trend of using drones to observe the far reaches of the property, he replied with a disarming grin, "It takes me out of my cowboying job a little bit. . . . I'm a little jealous of the drone then." He fully recognized the value of using drones and the efficiency gains of doing so, but he saw the drone as an automation *threat*, one that reduced the time he spent on a part of his job that he liked.

I had similar conversations about the use of all-terrain vehicles (ATVs) instead of horses for ranch work. Mary Anne, a widow who now single-handedly ran a small calving operation, scoffed at ranchers who only want to use horses and never ATVs, saying they "think they are cowboys." Trent G., by contrast, described his aversion to contact with machinery on the ranch in terms that sounded like religious purification: "I don't want no tractor blood on me!" He added, "I spend my days in the saddle, that's it." Giorgi S. approached the four-wheeler with equanimity. He lauded the kinds of ranches that only use horses, noting, "That's old school—trying to keep the tradition alive. And I agree. We're a dying breed. There's not many of us left." For his own ranch, he noted four-wheelers were "extremely time saving. I can get a lot done. I can get places fast. Horses are good, you can only do certain things on horses, but when we're gathering cattle, I can cover the country. On a horse, it's hot this time of year. I don't want to go to that back corner and look for those few cows that might be there, but with the four-wheeler, the buggies, I'll go. . . . Yes, we use the horses. We use the mechanical machines. When the job needs to get done, I get it done regardless of which way I do it." Cattlemen like Giorgi S. were about the bottom line—what the ranch needed in order to be economically sound, so it could continue to be a profitable venture.

As I've begun to show, the cattleman, and the occupational values associated with that archetype, was distinct from the cowboy. Cody G., who had only begun ranching in his seventies after a successful career running a factory, was a unique case. He was a cattleman. His ranch was the retirement project of a work-obsessed overachiever, and he had established himself quickly in his area, becoming cattleman of the year in 2010. He approached cattle ranching with a relentless efficiency similar to the work he did at his factory, and he approached managing his ranch with technological gusto. He wrote his own code to remotely manage an irrigation system, maintained a database of cow genetics, and tagged the herd with RFID tags. Cody G. was an exception among the older set, but younger men were also very likely to use and embrace certain digital technologies, although they again worked to distinguish tech uses in a way that ultimately aligned with gendered values. Jackson B. described his love of the first-person shooter video game *Call of Duty* and explained the necessity of maintaining two internet connections so he would have adequate bandwidth and minimize latency for gaming—a challenge on his remote ranch. Lucas T., who eschewed drones, talked about how it would be extremely helpful to have mobile phone service at the far reaches of the ranch. He described how his father had recently used an online livestock auction for the first time and found the internet extremely useful for doing research and purchasing equipment. Delilah B., who raised two sons on a remote cattle ranch outside of Prineville, told me one of her sons seemed inclined toward computers in high school and that he'd even built one from parts. Ultimately, however, she says he hated all the sitting around that this kind of work entailed. He came back home after dropping out of college to manage the family cattle ranch with his father and brother. Following from this passion for computers, she said he brought a distinctively "logical" approach to ranch work.

Cattlemen like Cody G., Giorgi S., Delilah B.'s son, and Jackson B. emphasized a kind of occupational masculinity associated with corporate executives. They took a reasoned approach to farming, built robust businesses and took a command-and-control approach over herds, staff, land, and equipment. By contrast, cowboys like Trent G. and Lucas B. sought a kind of solo frontiersman's independence. But both cowboys and cattlemen firmly rejected a role that required yielding to or negotiating with others. Giorgi S. spoke about why his ranch didn't do direct sales to consumers. Imagining his father promoting their beef in an urban supermarket, he said, "Dad could not. He'd be in jail. He'd kill somebody, with the stupid questions people ask." Jackson B. wanted to set up an Airbnb on his property, but he asked a neighbor's wife to manage it. He did so because he wanted to avoid "people problems." Cattlemen avoided people problems by having their say over business decisions and ranch management. Cowboys did the same by working solo and only with animals. Trent G., the itinerant ranch hand, was committed to no particular place and to no particular boss but instead moved constantly from place to place, through Texas, Nevada, and California, and then to Oregon. On a return trip to the area a year after I had interviewed him, he was nowhere to be found, having apparently moved on once again.

These differentiations of masculine identity in traditional industries in the rural West meant there was neither a wholesale acceptance and accommodation nor an absolute rejection of tech. Presumptions that traditionally minded workers are luddites, that they avoid tech out of fear or in an absolute sense, are inapt in this context. Workers also did not hold completely unique or idiosyncratic orientations to tech. Instead, distinct patterns of tech adoption and use aligned with certain gendered values. Such values offered a way of discerning which technologies and applications to embrace, and for what uses.

Conclusion

Foundational work in gender theory argues that there is not a singular definition of masculinity (set in opposition to femininity) but rather *multiple* masculinities that exist in relation to one another. Raewyn Connell posits that the relationships of complicity, subordination, or marginalization serve to reinforce a dominant (or hegemonic) form: the idealized way of being a man.[37] My purpose in this chapter was to show and describe the multiple masculinities in rural workplaces in the American West, including within industries, but also to show how they shaped the adoption and use of computing technologies in particular. I have also sought to show the dynamics that led an established hegemonic masculinity to remain intact even with the arrival of high tech, which one might imagine seems primed to import a new disembodied masculinity of rationalism. Understanding occupational values and how they guide thinking about new tech helps to explain how the community in Prineville received and accommodated the data centers that came to their town.

In central Oregon, an embodied masculinity that was regionally hegemonic was shaped through workplace practices. For cattlemen, the physicality of work with two-ton animals out on the land every day reinforced this value even as the cattleman's pragmatism led him to embrace automation technology. For cowboys, the autonomy of working on remote ranchlands with only animals for companions defined their sense of what made their work meaningful. I have shown how prevailing masculine values within work cultures are confronted by new technology. Sometimes the response is to abstain. Sometimes abstaining simply means delegating that necessary work to others, as was the case with men who left computer work to the women (wives, daughters) in their lives. Sometimes the response was to accommodate, as when the labor savings of farm equipment (like ATVs and drones) overcame the desire to privilege ranching by horseback as more rugged, artful, or authentic. Some tech remade the industry entirely, as automation in sawmills did, thrusting many workers into unemployment while promoting a new class of IT experts to manage sawmill operations. Some tech presented new appealing options, as did data center jobs at the new Facebook data center facility, for men seeking to recapture the family wage job and to reverse the decline in blue-collar jobs that uphold a particular kind of embodied masculinity.

The link between masculinity and work relates also to a very human desire or need for purpose, meaning, and dignity in work.[38] Workers seem to seek not just a sense of general skill and accomplishment (though this is part of what composes a sense of dignity) but also a sense of alignment with their identity. This is where gender can come into play. While austere and abstracted economic notions of work suggest that rural economic transitions hinge on labor force readiness (linked to skills) and adequate pay, my interviewees, including rural leaders, workers, union representatives, and employers, indicate there is a more affective and social element to these transitions. It was also evident in central Oregon that traditional rural industries not only shaped values within workplace settings but also generally shaped the regional culture. Though both the education and health care fields employ far more workers in Prineville, it was historical industries (the timber industry, cattle ranching) that were promoted as giving the town and area its distinct identity, branded by the local rodeo as "the cowboy capital of Oregon." Elected officials, many of whom had pursued their own livelihoods in these industries, gave special consideration and priority to agricultural, timber industry, and manufacturing workers in planning for the region's economic future.

The discourse of the digital divide has been rooted in two general explanations, access and skill. Access narratives suggest that inadequate equipment or infrastructure is at the root of the so-called digital divide. Skills narratives suggest that a lack of training, education, or exposure is at the root of a so-called digital divide. This has also been shown for some time now as a conceptually limiting way of understanding why people do or do not adopt or incorporate digital technologies into their lives.[39] Such arid economic assessments miss how critical meaningful work is for workers. Meaningful work is established within cultural contexts. This chapter has shown how rural communities can find meaning in new industries by bridging traditional ways of life and new economic opportunities in a way that symbolically smooths this break with the past.

Notes

Acknowledgment: This research was supported by NSF grant 1734431 and a noncontractual research gift from Intel Corporation.

1 By *male-dominated industry*, I mean an industry that disproportionately and predominantly employs those who present and identify as men, where power in these industries is even more disproportionately concentrated in male-identified workers (as managers, capitalist owners, investors, etc.), and where masculine occupational values prevail as workplace ideals.

2 See Connell, *Masculinities*, chap. 7, on the particular notion of masculine authority as rooted in disembodied reason.

3 Since the time of my fieldwork, the company (and its Prineville data center) has been rebranded Meta. I refer to the company throughout this chapter by the name in use at the time of my fieldwork.

4 Connell, *Masculinities*; Howson and Hearn, "Hegemony"; Poster, "Subversions."

5 See Ems, *Virtually Amish,* for examples of value-driven selection of tech in the rural, religious order of the Amish, which is often misunderstood as a default rejection of technology.
6 Kahn and Burrell, "Sociocultural Explanation."
7 Connell, *Masculinities.*
8 Connell, *Masculinities,* 77. In making sense of the high-tech industry as a space dominated especially by male engineers, a number of scholars have turned to the notion of hegemonic masculinity and its alternatives; see Eglash, "Race, Sex"; Kendall, "Nerd Nation."
9 Pascoe, *Dude, You're a Fag.*
10 Ensmenger, *Computer Boys*; Hicks, *Programmed Inequality.*
11 Eglash, "Race, Sex."
12 Kendall, "Nerd Nation."
13 Ensmenger, *Computer Boys.* Engsmenger notes that ENIAC had a certain visual similarity to a switchboard.
14 Ensmenger, *Computer Boys*; Hicks, *Programmed Inequality.*
15 Ensmenger, *Computer Boys,* 35.
16 Ensmenger, *Computer Boys,* 77.
17 Kendall, "Nerd Nation," 261.
18 Kendall, "Nerd Nation."
19 Crandall et al., "Magicians."
20 Poster, "Subversions."
21 On the emerging trope of the brogrammer, there is little academic literature, but see Hicks, "De-Brogramming," for a provocative think piece. On the manager's role as "embodied," see Shoshana Zuboff, *In the Age of the Smart Machine,* in which she talks about the "salience of bodily presence in shared action contexts as the core of the manager's world" and describes how management culture focused on oral communication above other forms.
22 Kay, "We've Reached Peak Tech Bro."
23 Berdahl et al., "Work as a Masculinity Contest."
24 Poster, "Subversions."
25 Burrell, "On Half-Built Assemblages."
26 Streep, "How Big Data Saved the Mountain Town."
27 Metz, "Welcome to Prineville."
28 Zuboff, *In the Age of the Smart Machine.*
29 Gonzalez Monserrate, "Thermotemporalities."
30 Woodgrain was a wood products manufacturing facility in Prineville that shut down and laid off all of its workers in 2015.
31 See similar accounts of the loss of the family wage and the related disruption to masculine identity with the decline of the timber industry in Oregon (Stein, *Stranger Next Door*) and Northern California (Sherman, *Those Who Work*). See May, "Historical Problem," on the gendered

history of this term in labor and unionization movements, which goes back to at least the 1820s.

32 On overcoming sleep deprivation as a masculine trait in work settings, see Karen Levy's description of long-haul truckers: Levy, "Digital Surveillance." See also Cooper, "Being the 'Go-To Guy,'" on how extreme working hours among high-tech knowledge workers is seen as a demonstration of commitment.

33 Levy, "Digital Surveillance."

34 An interviewee in Sherman, *Those Who Work*, also noted internet money bringing outsiders into their community.

35 Kahn and Burrell, "Sociocultural Explanation."

36 Connell, *Masculinities*, 54.

37 Connell, *Masculinities.*

38 Lamont, *Dignity of Working Men*; Hodson, *Dignity at Work*; Ticona, *Left to Our Own Devices.*

39 Kvasny, "Cultural (Re)production"; Eubanks, *Digital Dead End*; Greene, *Promise*; Kahn and Burrell, "Sociocultural Explanation."

Bibliography

Berdahl, Jennifer L., Marianne Cooper, Peter Glick, Robert W. Livingston, and Joan C. Williams. "Work as a Masculinity Contest." *Journal of Social Issues* 74, no. 3 (2018): 422–48. doi.org/10.1111/josi.12289.

Burrell, Jenna. "On Half-Built Assemblages: Waiting for a Data Center in Prineville, Oregon." *Engaging Science, Technology, and Society* 6 (2020): 283–305. doi.org/10.17351/ests2020.447.

Connell, R. W. *Masculinities*. 2nd ed. University of California Press, 2005.

Cooper, Marianne. "Being the 'Go-To Guy': Fatherhood, Masculinity, and the Organization of Work in Silicon Valley." *Qualitative Sociology* 23, no. 4 (2000): 379–405. doi.org/10.2307/j.ctv16f6j36.

Crandall, Emily K., Rachel H. Brown, and John McMahon. "Magicians of the Twenty-First Century: Enchantment, Domination, and the Politics of Work in Silicon Valley." *Theory and Event* 24, no. 3 (2021): 841–73. doi.org/10.1353/tae.2021.0045.

Eglash, Ron. "Race, Sex, and Nerds." *Social Text* 20, no. 2 (2002): 49–64. doi.org/10.1215/01642472-20-2_71-49.

Ems, Lindsay. *Virtually Amish: Preserving Community at the Internet's Margins*. MIT Press, 2022.

Ensmenger, Nathan. *The Computer Boys Take Over: Computers, Programmers, and the Politics of Technical Expertise*. MIT Press, 2010.

Eubanks, Virginia. *Digital Dead End: Fighting for Social Justice in the Information Age.* MIT Press, 2012.

Gonzalez Monserrate, Steven. "Thermotemporalities, Thermomasculinities: Uptime, Downtime, and Server Heat in the Digital Anthropocene." *New Media and Society* 25, no. 2 (2023): 324–44. doi.org/10.1177/14614448221149938.

Greene, Daniel. *The Promise of Access: Technology, Inequality, and the Political Economy of Hope.* MIT Press, 2021.

Hicks, Mar. "De-Brogramming the History of Computing." *IEEE Annals of the History of Computing* 35, no. 1 (2013): 86–87.

Hicks, Mar. *Programmed Inequality: How Britain Discarded Women Technologists and Lost Its Edge in Computing.* MIT Press, 2018.

Hodson, Randy. *Dignity at Work.* Cambridge University Press, 2001.

Howson, Richard, and Jeff Hearn. "Hegemony, Hegemonic Masculinity, and Beyond." In *Routledge International Handbook of Masculinity Studies,* edited by Lucas Gottzén, Ulf Mellström, and Tamara Shefer. Routledge, 2019.

Kahn, Zoe, and Jenna Burrell. "A Sociocultural Explanation of Internet-Enabled Work in Rural Regions." *ACM Transactions on Computer–Human Interaction* 28, no. 3 (2021): 1–22. doi.org/10.1145/3443705.

Kay, Grace. 2023. "We've Reached Peak Tech Bro, with Billionaires Suggesting Cage Fights and Bench-Press Contests." *Business Insider.* Accessed August 18, 2023. https://www.businessinsider.com/tech-bro-culture-peaks-with-elon-musk-mark-zuckerberg-fight-2023-6.

Kendall, Lori. "Nerd Nation: Images of Nerds in US Popular Culture." *International Journal of Cultural Studies* 2, no. 2 (1999): 260–83. doi.org/10.1177/136787799900200206.

Kvasny, Lynette. "Cultural (Re)production of Digital Inequality in a US Community Technology Initiative." *Information, Communication, and Society* 9, no. 2 (2006): 160–81. doi.org/10.1080/13691180600630740.

Lamont, Michèle. *The Dignity of Working Men: Morality and the Boundaries of Race, Class, and Immigration.* Harvard University Press, 2000. https://doi.org/10.2307/j.ctvk12rpt.

Levy, Karen. "Digital Surveillance in the Hypermasculine Workplace." *Feminist Media Studies* 16, no. 2 (2016): 361–65. doi.org/10.1080/14680777.2016.1138607.

May, Martha. "The Historical Problem of the Family Wage: The Ford Motor Company and the Five Dollar Day." In *The Intersection of Work and Family Life,* edited by Nancy F. Cott. De Gruyter, 1992. doi.org/10.1515/9783110969467.371.

Metz, Cade. "Welcome to Prineville, Oregon: Population, 800 Million. "*Wired,* December 1, 2011. https://www.wired.com/2011/12/facebook-data-center/.

Pascoe, Cheri J. *Dude, You're a Fag: Masculinity and Sexuality in High School.* With a new preface. University of California Press, 2012.

Poster, Winifred R. "Subversions of Techno-Masculinity: Indian ICT Professionals in the Global Economy." In *Rethinking Transnational Men: Beyond, Between, and Within Nations*, edited by Jeff Hearn, Marina Blagojevic, and Katherine Harrison. Routledge, 2013.

Sherman, Jennifer. *Those Who Work, Those Who Don't: Poverty, Morality, and Family in Rural America.* University of Minnesota Press, 2009.

Stein, Arlene. *The Stranger Next Door: The Story of a Small Community's Battle over Sex, Faith, and Civil Rights.* Beacon Press, 2005.

Streep, Abe. "How Big Data Saved the Mountain Town." *Outside*, August 4, 2017. https://www.outsideonline.com/adventure-travel/destinations/north-america/living-cloud/.

Ticona, Julia. *Left to Our Own Devices: Coping with Insecure Work in a Digital Age.* Oxford University Press, 2022.

Zuboff, Shoshana. *In the Age of the Smart Machine: The Future of Work and Power.* Basic Books, 1988.

PART III POLITICAL RURALITIES

9

Gas Can Imaginaries

On the Politics of Combustion, Anti-Urban Resentment, and Playing Indian at the 2022 Freedom Convoy

JORDAN B. KINDER

Some two weeks into the 2022 Freedom Convoy's occupation of Canada's capital city of Ottawa, the Ottawa Police Service announced that those transporting fuel into the marked-off red zone could now be charged and arrested. Recognizing petroleum and diesel as the occupation's lifeblood, representatives of the repressive state apparatuses sought to halt the flow of fuel as a determinant move to dismantle the blockade. Protesters responded by carrying empty gas cans or ones filled with water in an effort to overwhelm police capabilities in enforcing the injunction. A sea of red and yellow converged into the cordoned-off red zone of Ottawa's Centretown, resulting in at least seven arrests and over a hundred tickets (figure 9.1).

Fuel, of course, was an essential ingredient in the recipe of combustion that was the Freedom Convoy, just as it was in a previous convoy in 2019. Back then, an oil industry–adjacent small business owner and Canadian yellow vester, Glen Carritt, staged a dress rehearsal of the Freedom Convoy. After raising over CAD100,000 (USD72,571) on GoFundMe, United We Roll made two demands as it made its way from Red Deer, Alberta, to Ottawa: intensified resource extraction through the support of Canada's oil and gas industry, and strengthened borders by ending a United

9.1 Drawing by Jordan B. Kinder.

Nations migration pact. Pipelines and borders were two pillars of the Canadian yellow vest movement, to which United We Roll once owed its namesake before revision (after word got out about just how far right the yellow vesters on this side of the Atlantic were). Many figures associated with United We Roll would appear in the 2022 Freedom Convoy, including Chris Barber, James Bauder, Pat King, and Tamara Lich. With a pandemic veil garnering millions of dollars in donations that subsumed and contained these originary aims of more extraction and stronger borders, the fossil-fueled foundations of the 2022 Freedom Convoy were in clear view during the gas can swarm.

As far as official narratives can be determined, the 2022 Freedom Convoy was born from frustration with border restrictions placed on truckers traveling between the United States and Canada that required them to be vaccinated against COVID-19. Previously truckers had an exemption as essential workers, which was upended by this new requirement coming into place in January 2022. In the setting of a demonstration that mobilized the truck as a tool of resistance, truckers as a primary political subject, and the communal road trip as political praxis, the gas can carried significant symbolic weight. As much as the convoy was about what participants perceived as tyrannical government overreach into their daily lives through pandemic restrictions and mandates, it was equally about maintenance of the current fossil-fueled order. That the gas can took on such material-symbolic significance reveals as much.

This chapter centers the gas can as a medium of expression for the 2022 Freedom Convoy. After providing a brief narrative timeline of the convoy, I outline the contours of a politics of combustion that follows a growing fossil-fascist creep in Canada, where I understand fossil fascism alongside Cara Daggett, who theorizes the concept as an authoritarian desire expressed through attachments to the burning of fossil fuels.[1] I then propose that anti-urban resentment mediated through a broadly rural political identification underwrites these politics of combustion on display at the convoy. Finally, I detail how nativism inflected the convoy's political form and content. The convoy itself is a quintessentially petrocultural form. The combustion of diesel and gasoline enabled demonstrators to travel to and occupy the city that most sharply represents the divide between a perceived urban eastern elite and a rural western working class that has long occupied the western Canadian political imagination. In addressing this divide as it influenced the convoy, my chapter puts pressure on what the category of the rural obscures and reveals in the era of planetary warming on the one hand and in a nation constituted by historical and ongoing settler colonialism on the other in a specifically Canadian context. A gas can is a containerized contingency plan that ensures continued travel fueled by combustion without interruption. In other words, it is a *medium* of containment, contingency, and combustion that, in the context of the convoy, mediated a series of imaginaries in these terms from which this chapter derives its title. In the current conjuncture defined in large part by the fossil economy, whose burdens, including those to resolve its contradictions through policy measures like increased taxation at the point of consumption, are shouldered by the many for the benefit of the few, the gas can imagination and the politics of combustion expressed by this imagination are last resorts in maintaining business as usual on a warming planet.

A Brief Narrative Timeline of the Freedom Convoy and Its Afterlife

Lasting over three weeks, the 2022 Freedom Convoy wasn't even a convoy—it was *many*. Driving out of Delta in southern British Columbia on January 23, the western contingent traveled eastward, picking up more along the way as it traversed Alberta, Saskatchewan, and Manitoba before settling into Ontario. As this flank gained momentum, so too did a set of internal Ontario convoys as they traveled to Ottawa, as did eastern contingents, including a Quebecois one associated with Les Farfadaas. Les Farfadaas is a Quebec-based group formed to protest COVID-19 public health measures that has its roots in La Meute, a far-right anti-Islam and anti-immigration group—roots that parallel those of the yellow vests and United We Roll.[2] These convoys, which in form and content belied the presentation as a singular one by and for truckers, convened on Ottawa and quickly transformed into an occupation as, among other things, vehicles were disabled to create barricades while encampments were set up with running vehicles constantly fed with a supply of fuel made possible by gas cans. These efforts were bankrolled by crowdfunding organized by Lich and others. On GoFundMe, donations reached over CAD10 million (USD7.25 million) before being frozen and ultimately refunded because it violated the platform's terms surrounding the support of violence and harassment. Fundraising efforts shifted to GiveSendGo, a Christian crowdfunding platform, which also received several million dollars.[3] E-transfers to Lich, cash donations, and other methods like cryptocurrency donation were also used.

As the occupation took hold in Ottawa, related demonstrations raised the temperature. Most prominently, these included blockades at border crossings between the United States and Canada: the Pacific Highway crossing in Surrey, British Columbia, Coutts in Alberta, Emerson in Manitoba, and the Ambassador Bridge in Windsor, Ontario. At Coutts, a cache of weapons including long guns, handguns, ammo, and body armor were discovered and intercepted by the Royal Canadian Mounted Police (RCMP). Thirteen people would be arrested as a result. Compounding the threat of violence represented by these seizures and arrests were the economic tolls of the border blockades. Windsor is an automobile manufacturing hub and corridor where just-in-time-manufacturing is in full force, which limits holding stock in production, with Detroit serving as a sort of production sibling that allows for precise manufacturing temporalities. A single car part, for instance, might travel across the Ambassador Bridge several times over the course of a day for machining before finally being installed into a vehicle on the production line. In response to these deepening economic impacts and the escalation of the occupation in Ottawa represented by, for instance, the installation of an inflatable hot tub, barbecue stations, and a blow-up bouncy castle for children to play in, Prime Minister Justin Trudeau invoked the Emergencies Act. This invocation established a state of exception in which police at all levels were afforded extra powers. Three days later, the convoy was dismantled, key actors arrested, and the act repealed.

The Emergencies Act requires that a public inquiry into its use be convened within a year of its invocation. Taking place over thirty-six days from October 13 to December 2, 2022, and commissioned by Justice Paul Rouleau, the Public Order Emergency Commission saw witnesses ranging from Ottawa residents and convoy organizers to the chief of Ottawa Police Services, the deputy director of operations at the Canadian Security and Intelligence Service, and Justin Trudeau himself give evidence to the commission and later being cross-examined by lawyers representing the convoy organizers, the Union of British Columbia Indian Chiefs, and others. I listened to all of these sessions, which amount to over three hundred hours and reveal the complex dynamics of the convoy and its afterlives, including the role of social and alternative media in shaping the convoy and disseminating its messages; the realpolitik of police bureaucracy that resulted in lackluster enforcement by municipal, provincial, and federal forces; the economic impacts and threats to trade relations with the United States, particularly in the automobile sector; and the ways that idling vehicles, horns, and exhaust were mobilized as media of disruption to Ottawa residents. It is this weaponization of vehicles and fuel through combustion enabled by the gas cans on which this chapter focuses, arguing that as much as the convoy was occasioned by pandemic restrictions, it was equally about the right to burn fossil fuels. As witness testimony from the inquiry outlined, the most pronounced economic anxiety generated from the United States was related to future capacity surrounding the production of electric vehicles (EVs) and perceived threats about how this could impact Canada's place in EV incentives as part of the United States' Build Back Better legislation.[4] Inadvertently, this episode drew attention to the kinds of frictions that emerge out of confrontations between federal policy and the realities of everyday life outside Canada's urban centers. Canada's federal government announced in 2021 that "all new light-duty cars and passenger trucks sales [will] be zero-emission by 2035."[5] Yet well into 2023, as an article by Doug Johnson, a climate reporter for the Weather Network, details, widespread electric vehicle adoption remains technologically challenging outside urban areas thanks to a lack of charging stations, grid capacity, and more.[6] If it carries on this way, such infrastructural inadequacy will arguably foment existing cultural resistance to electric vehicles and further stoke anti-urban resentment, which I discuss later in this chapter, ultimately culminating in deepening a political identification with the right to burn that was on display at the 2022 Freedom Convoy.

A Politics of Combustion

At her bail hearing on February 19, 2022, after being arrested and charged with counseling to commit mischief, organizer Tamara Lich appeared in clothing familiar to anyone with knowledge of the pro-oil movement in Canada: a hoodie with a design reading "I ♡ [Canadian] Oil & Gas." These hoodies are produced by the pro-oil group Canada Action, started by Calgarian real estate agent Cody Battershill, who was once described by the *National Post* as a "one-man oil sands advocate."[7] The logo appears in many places across western Canada primarily—as bumper stickers on pickup trucks and even under the ice at the Saddledome arena, home to the Calgary Flames. Lich's wardrobe choice revealed further links between United We Roll and the Freedom Convoy by mobilizing the worldwide stage set by her arrest to show support for Canada's oil and gas industry. A courtroom sketch of Lich in this hoodie would make the rounds across legacy and social media channels nationally and internationally.

Gas and diesel were not mere material ingredients for the convoy; their combustion was not a passive necessity but actively celebrated. Gas cans, of course, played a determinant role in the convoy, as they enabled the practices of occupation and blockade. Whereas fuel more generally provided the materials for travel, which was as much part of the event as the occupation itself, the gas can was the medium for fueling the occupation materially and ideologically. The primary mechanism through which occupation was staged was the act of idling, a combustive relation through and through. In other words, the combustion of gasoline and diesel made the convoy possible, mediated as it was by the gas can at the site of occupation, which fueled an increasingly tangible *politics of combustion*. Materially speaking and in contemporary usage, combustion refers to "the action or process of burning." But it is the common usage of combustion in the seventeenth and eighteenth centuries that clarifies the ideological companion to this material act through which a politics of combustion forms. As the *Oxford English Dictionary* puts it, combustion in this setting refers to "violent excitement or commotion, disorder, confusion, tumult, hubbub." In the case of the convoy, the source of this wellspring of combustion comes from that same "white and masculine backlash animated by climate change denial and hyper-extractivism" that Winona LaDuke and Deborah Cowen suggest influenced United We Roll. A politics of combustion, then, is precisely the kinds of politics LaDuke and Cowen argue "has taken hold across the continent."[8]

The individual act of combusting fossil fuels is often invoked as a pivot point to address questions of culpability and responsibility in the Anthropocene. Turning to this act, environmental philosopher and literary theorist Timothy Morton describes the individual act of ignition in relation to the scalar problem of global warming, the latter of which Morton sees as a "hyperobject," or "things that are massively distributed in time and space relative to humans." "The reason why I am turning my key—the reason why the key turn sends a signal to the fuel injection system, which starts the motor—is one result of a series of decisions about objects, motion, space, and time," Morton speculates.[9] By turning to the ignition of an internal combustion engine as the result of "a series of decisions" whose totality of effects reshapes planetary relations through global warming, Morton provides an entry point in unpacking the political ecology of the convoy. In Morton's meditation, ignition is an impulsive, unconscious action that inadvertently sets into motion an event with consequences that outweigh the severity of the action; for the convoy, ignition is a conscious act that expresses a politics of combustion.[10]

A politics of combustion pivots on two central practices whose means meet their ends: the wanton combustion of fossil fuels, and the disruption or disturbance of those perceived as liberal or progressive subjects. The practice of *rolling coal* perhaps best exemplifies this politics: Diesel engines are modified to emit black, sooty fumes, often at specific targets, like electric vehicles, cyclists, or pedestrians.[11] In the case of the Freedom Convoy, a politics of combustion took shape as the disruption of day-to-day lives of residents of Ottawa. Tactics against residents included physically barricading streets with big rigs, trucks like the one pictured in figure 9.2, and sedans, as well as endless aural-assault honking with train and vehicle horns. The vehicle, in other words, was weaponized spatially and aurally as a medium of combustive politics.

Generally understood to be passive or accidental violence, motor vehicles have become a prominent medium of violence in recent years. At the 2017 Unite the Right rally in Charlottesville, Virginia, for instance, a self-identified white supremacist intentionally drove into a crowd of counterprotestors, injuring thirty-five people and killing one. In Texas in fall 2021, several cyclists were put into the hospital by a teenager who

9.2 Drawing by Jordan B. Kinder.

was rolling coal onto them and lost control of his truck.[12] More recently, during a memorial march for residential school survivors, a white man in a Chevrolet Silverado drove into the walking crowd of demonstrators. Although witnesses say that the man shouted threats and racial slurs, the RCMP initially said that the driver was simply "impatient" and no charges were immediately laid.[13] In 2024, however, the driver was ultimately convicted of dangerous driving and sentenced to a nine-month conditional sentence to be served in the community.[14]

Linked with the weaponization of vehicles in the Freedom Convoy, combustion serves a necropolitical impulse in microscopic scales of immediate physical violence and macroscopic scales of slow ecological violence. Testimony throughout the commission articulated the far-reaching impacts of idling vehicles and the use of horns on physical and mental well-being. Some of the first questions of witnesses Victoria De La Ronde and Zexi Li, two Ottawa residents willing to represent those living downtown, revolved around this impact. De La Ronde, for instance, detailed how she was affected by "the fumes and other smells," which led to persistent, lasting effects like "a physical trigger" when she "gets a smell of gas."[15] While under cross-examination from the lawyer representing the convoy participants, Li described an experience of "intimidation with a truck."[16] And Councillor Mathieu Fleury directly invoked weaponization, stating, "Well, for us, you know, having the physical truck on the street created a big weapon in the spirit of the noise, the pollution of the fumes, the ability for folks to operate their businesses."[17] During testimony from convoy organizers and participants, these effects were diminished as if they were overly sensitive urbanites. When asked by Paul Champ, the lawyer representing the residents and businesses of Ottawa, about the fumes, Lich responded: "Well, Mr. Champ, my ex-husband was a tool push on a drilling rig, and I have spent many days on the site of a drilling rig, and there is a lot of diesel fumes and there is a lot of noise."[18] Throughout the inquiry, combustion emerged as a central tactic not only to keep the occupation going but also to disrupt the mental and physical health of residents.

Anti-Urban Resentment and the Convoy's Petit Bourgeois Foundations

The Freedom Convoy nurtured the existing divide between the interests of a western, more rural working class as embodied in the figure of the trucker and an eastern urban elite. This tension has a long history, both real and exaggerated, captured as it is by feelings of "Western alienation"—feelings that have sparked a western separatist movement akin to Brexit whose interests were represented at the convoy.[19] Certainly the population distribution of Canada points to a material condition from which this narrative of alienation and abandonment takes its cues. Almost half of the Canadian population lives in the Quebec City–Windsor corridor of 1,150 kilometers (710 miles), and more than 90 percent live within 241 kilometers (150 miles), of the US border. This population distribution, concentrated in the geographic south and east, which shapes distribution of resources and political representation, often gets reduced by right-wing populists to a divide between rural and urban that simultaneously invokes those between the working class and elites.

I've avoided invoking the rural as a category in my work on the cultural politics of energy, media, infrastructure, and environment in Canada. In some settings, including the ones in which I live and work, the category of the rural obscures more than it reveals. Historically speaking, this hasn't always been the case; the Canadian prairies, for instance, were once a hotbed for trade union organizing and revolutionary socialism primarily in the agrarian sector. And beyond the borders of Canada or the United States, where rurality runs parallel to indigeneity rather than in tension with it, rural spaces remain key sites for what we might understand as land-based, life-affirming activism against colonial infrastructures, as Ayesha Vemuri's chapter in this volume, which provides an account of Indigenous resistance to flood management infrastructure and wildlife conservation in India, shows. However, today, as in the past, the categories of urban and rural in Canada efface the settler colonial histories, relations, and economies on which urban and rural spaces have been built and reproduced. Focusing on rural and urban as politically determinative categories can serve to erase the fact that urban and rural formations alike are grafted onto Indigenous territories, whether treatied, ceded, or unceded. Additionally, as a category of political identification, the rural obscures questions of class. A mining company owner could just as easily live in Prince George, British Columbia, and a fast-food worker in downtown Ottawa, Ontario. This raises the question of the role played by rurality and its invocation in the context of the convoy. The ambiguous constellation of signification generated by *western*, *working class*, and especially *rural* helped garner support for the convoy in ways that obscured other, more salient elements of the political forces and facts underlying it. This branding was effective, leading to, for instance, Marxian economist Richard Wolff's seeing promise in the convoy because regardless of ideological underpinnings, the convoy saw workers band together to demand change of working conditions akin to a strike.[20]

We know in hindsight, however, that while working conditions related to vaccine mandates and bodily autonomy may have sparked the demonstration, there were more grievances animating the protest that overdetermined its impetus. But we also knew this in the convoy's early days. Interviewing one of the Ontario-based organizers, Jason LaFace, for *BayToday* on January 26, 2022, Casey Taylor details how LaFace compared mask mandates to Jews living under the Nazi regime while making goals clear as his "team" had been working with constitutional lawyers to "compel the government to dissolve government."[21] Under the auspices of Canada Unity, Bauder authored a six-page memorandum of understanding circulated online and in person that contained demands including the usual suspects, such as abolishing all pandemic restrictions and mandates as well as a call to overthrow and dissolve the federal government.[22] Such easy Holocaust comparisons and an attempted coup d'état expose the fascist elements of the convoy.

As the dress rehearsal organized in 2019 showed, having the resources to drive big rigs across the country and prop them up for weeks on end reveals levels of access arguably unavailable to working classes. In an early piece critiquing representations of the convoy as a working-class endeavor, union researcher Adam D. K. King describes the findings of his and his coauthors' work, which reveals the uniquely exploitative conditions of the Canadian trucking industry today, including how the industry is responsible for almost 80 percent of labor standard violations reported to the federal government and that, more generally, small businesses of under a hundred employees are responsible for 89 percent of all labor violations reported.[23] There are, in other words, classed stratifications within the industry based on questions of ownership. Those on the bottom rung experience some of the most violations firsthand, whereas a class of small business owners and owner-operators have a much different experience and, in some cases, even run the companies responsible for such violations. This leads King to a speculative point that explains why the Canadian Trucking Alliance spoke out against the convoy, given that most of its members are vaccinated. "Rather than exploited workers in a deregulated industry," King writes, "my guess is that the 'truckers' actually present in Ottawa were by and large self-employed owner-operators." "It was," he continues, "a 'revolt' of the petit-bourgeoisie, financially backed by wealthy right-wing grifters." The commission would all but confirm these suspicions as those of the main organizers who actually are truckers—Brigitte Belton and Chris Barber in particular—are owner-operators, falling outside of conventional Marxist definitions of the working class and squarely into the category of petit bourgeoisie, given their relationship to means of production. It remains unclear where Bauder, another trucker-organizer, falls. In this context, rurality and its classed implications were instrumentalized affectively to provide ideological cover for a political mobilization that had very little to do with actual rural or working-class interests. This is a key mediating function played by rurality in the present conjuncture.

It is thus urgent, given these political-economic conditions, to draw attention to what becomes obscured by invocations of rurality in Canada, particularly when allegedly rural, working-class energies are tapped into for increasingly fossil-fascist ends. Rural resentment has some explanatory power here. For political scientist Katherine J. Cramer, rural resentment describes a political affect that emerges from perceived divides between rural and urban sensibilities, as well as a sense that the urban, often understood as elite, gets priority in political and economic settings. This resentment takes the form of antagonism toward the resented others and the people, institutions, and discourses that advantage them.[24] Such resentment is often invoked to explain what kinds of geographic political-economic conditions and affective relations led to Donald J. Trump's winning the 2016 US presidential election and the subsequent emboldening of the far right in the United States and elsewhere, and thus has garnered increased attention in recent years by academics and journalists. With Trump's second win and early actions in office, this emboldening has approached a zenith by becoming fully institutionalized.

The degree to which the conditions breeding such resentment are empirically valid remains difficult to determine. Nevertheless, the affective responses to these conditions—emerging out of what Cramer calls "rural consciousness"[25]—certainly are real, stoked as they are by perceived limits imposed on their freedom in the form of a laundry list of grievances: masks during the COVID-19 pandemic, taxation, rising gas prices, electric vehicles, limits on offensive speech through political correctness, and more. Through these restrictions on freedoms, the unfair share is redistributed to an array of others in uneven ways, like minorities (visible and otherwise), environmentalists, 2SLGBTQIA+ peoples and, crucially, Indigenous peoples. This resentment, then, is enacted against these others and the cultural and political elites who serve them. By mobilizing these existing conditions of resentment, the architects of the convoy targeted urban spaces, which served as a site where this unequal redistribution occurs.

The convoy itself—and further the abstraction of *freedom* around which the convoy organized its political desires—in other words emerged from a distinct form of resentment that positions itself primarily against the urban as if it were performing class struggle. The first prototypical suburb of Levittown, Pennsylvania, anticipated such a tethering of freedom to fossil fuels and nonurban spaces; it would become the stage for a riot sparked by truckers during the 1970s energy crisis over high fuel costs and big government. Geographer Matt Huber turns to this episode in his analysis of how neoliberalism influenced American conceptualizations of an apolitical freedom in the later twentieth century, with social and economic relations materially fueled by petroleum and centralized in the suburb.[26] As the motivating grievances scaled out, the freedom construed as under threat by the Freedom Convoy worked in ways that track with Huber's analysis. This anti-urban sentiment materialized in how space was occupied, including the bouncy castles (figure 9.3) or inflatable hot tubs (figure 9.4), rendering downtown Ottawa into a kind of lawn that requires spatial affordances more common to suburban and rural environments that claimed the land as their own.

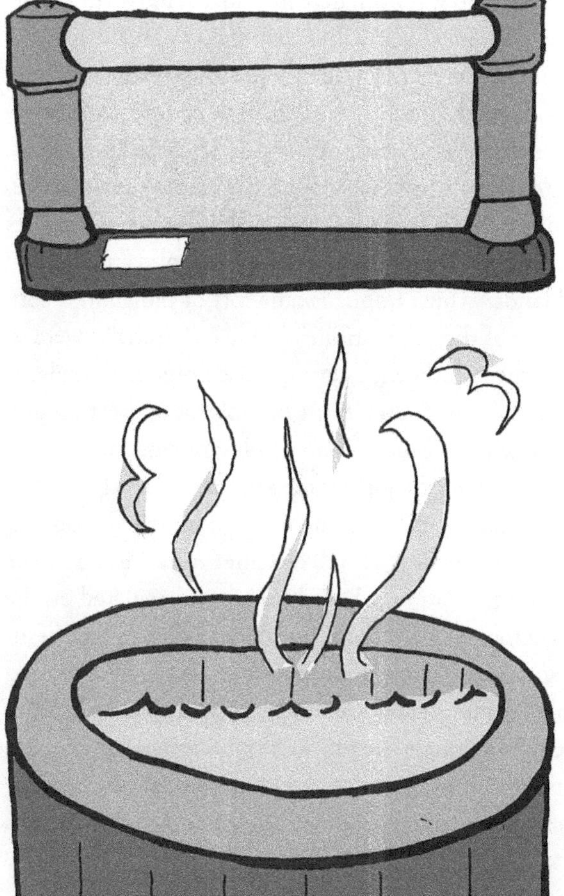

9.3 and 9.4 Drawings by Jordan B. Kinder.

Playing Indian: Blockades, Convoys, Occupations

Nativist claims to Canada as land and nation also inflected the convoy's ideological project. During the early stages of Pat King's testimony, a lawyer for the commission played a compilation video of statements made on King's podcast that, among other racist sentiments, offered a definition of Indigenous peoples as anyone born in Canada. King stated: "It is

100 percent every person who was born here in Canada, in North America, you are Indigenous. People don't realize that. If you were born of the land, you are Indigenous of the land."[27] King's statements extend a long history of nativism in Canada, and indeed North America, while offering a modulated echo of the well-worn notion of old-stock Canadians. Popular commentary online and off drew connections between the strategies and tactics of Indigenous land and water defenders, namely blockades and occupation as well as the "use of Indigenous ceremonies and cultural items on traditional lands," which implied some sort of Indigenous contingent and sanctioning of the occupation.[28] These connections were so heavily leaned on that many Indigenous groups made official statements against the premises of the convoy and its use of Indigenous practices and symbols. Reporting for *Indian Country Today*, Miles Morrisseau gathered these statements, which included a joint statement from the Algonquins of Pikwakanagan, the Algonquin Anishinabeg Nation Tribal Council, and the Kitigan Zibi Anishinabeg, as well as Chief Allan Adam of the Athabasca Chipewyan First Nation, whose comments speculated on the discrepancy between enforcement of the Coutts blockade and those of Indigenous blockades and occupations. "If this blockade had been organized by Indigenous people," Adam states, "we have no doubt that authorities would respond quickly to remove the blockade and utilize the law that has been created to do so," citing Alberta's Critical Infrastructure Defense Act.[29] The use of an upside-down Canadian flag as a protest symbol was also once common at anticolonial protests, but it is now arguably more associated with the intersecting interests on display at the convoy.

Appropriation of indigeneity by settlers is nothing new. Dakota historian Phil Deloria's influential 1999 book on the subject, *Playing Indian*, begins at a quintessential episode of US history: the Boston Tea Party, which saw the Sons of Liberty dump tea at Griffin's Wharf on December 16, 1773, in protest against the British Tea Act that exempted the British East India Company from taxes on the sale of tea in the American colonies. This episode is a constitutive one in establishing features commonly understood to be American because it would ultimately lead to the American Revolution. Lesser known, perhaps, is that during this act of dissent, these Sons of Liberty played Indian during their act of defiance, presumably for disguise. As Deloria describes the event, "a chorus of war woops sounded outside the hall, and a party of what looked like Indian men sprinted down the street to the wharves."[30] While the United States has a different relationship with Indigenous peoples and Deloria's thesis centers around how white Americans played Indian to carve out a new identity in contrast to older European ones, elements of Deloria's concept travel across borders. Playing Indian was in full force during the convoy, particularly in terms of its formal elements of occupation and blockade and how it was policed.

For those of us who have spent time at marches or demonstrations, a discernable uncanniness inflects the Freedom Convoy in the shadow of increasingly visible Indigenous resistance—call it defamiliarization, call it strategizing, or call it simulacral. It's difficult to call it anything but uncanny in the wake of the 2020 blockades, wherein railway lines were blocked in support of the Wet'suwet'en, who were experiencing violence from the RCMP enforcing an injunction from the supreme court of British Columbia in support of TC Energy's Coastal GasLink pipeline. The most prominent of these blockades was between Montreal and Toronto on Tyendinaga Mohawk territory, running for weeks as tensions in British Columbia heightened. It was in response to these actions that Alberta's Critical Infrastructure Defense Act was enshrined in law as Bill 1, a designation that symbolically highlights how significant the protection of critical infrastructure, which includes railways, borders, and pipelines, is viewed by the province.

These blockades in support of the Wet'suwet'en resistance to Coastal GasLink were part of a much longer tradition of Indigenous resistance that targets the circulation of capital and commodities through infrastructural choke points in spaces outside of urban environments and closer to points of extraction and circulation of resources. Michii Saagig Nishnaabeg theorist Leanne Betasamosake Simpson offers an extended meditation on the significance of blockades in struggles against settler colonialism for this reason, which she does with reference to beavers and their practices of building such as damming (figure 9.5). "Blockades are both a refusal and an affirmation," she writes. "An affirmation of a different political economy. A world built upon a different set of relationships and ethics. An affirmation of life."[31] In this capacity of negation and affirmation, blockades often have two consequences that are its means and ends: disruption of the circulation of capital and affirmation of other modes of living.

But in the stage set through Facebook feeds, Twitch streams, Telegram channels, and Zello channels, the enemy for the convoy wasn't capital or its various manifestations; it was the allegedly elite city dwellers of downtown Ottawa, imagined by participants to be part of an elite political class responsible for restricted freedoms. Literary and cultural theorist Joshua

9.5 Drawing by Jordan B. Kinder.

Clover, whose most recent work examines what we might understand as the politics and aesthetics of strikes and riots, weighed in on the Freedom Convoy as an uncanny sight. First highlighting how the convoy "recalled the history of Indigenous border blockades that began in earnest with the Cornwall Bridge blockade in 1968 by the Kanien'kéhaka of Akwesasne," Clover describes how the power of blockades resides in the truism that they are a tactic available to all.[32] Blockades "require[e] no privileged access to the production process, unfolding in ambiguously public space policed by the state, often interfering with the circulation of commodities." As he puts it, however, the convoy carried with it a tweak on the undercurrents to circulation: "the circulation struggle," he writes, "but make it nationalist." Moreover, the convoy required "privileged access to the production process"—big rigs like those pictured in figure 9.6. To clarify, Clover suggests that we turn to the sources of action rather than the tactics themselves—to the content rather than the form. "We are not on the side of the riot but of the George Floyd Uprising; not in solidarity with blockades but with the history of Indigenous land and water protection," he writes. "As we think about how to fight, as we must, swiftly, it will be helpful to reflect on sites of vulnerability, how they may have changed, and what this makes possible." Clover's comments are in friction with Wolff's in important ways, which isn't to endorse an ideological project of purity politics that so often disrupt possibilities for coalition building but rather

9.6 Drawing by Jordan B. Kinder.

to identify where those coalitions may not be worth pursuing. Clover reminds us that the aims and targets of dissent are important components of movements worth supporting for more equitable futures.

The convoy's adoption of tactics typically associated with mobilizations defending Indigenous territory and asserting Indigenous jurisdiction was not sufficient to legitimize its political project as a freedom struggle. However, during the Public Order Emergency Commission hearings, questions of indigeneity were also invoked in other ways. First, indigeneity was often commented on or invoked by some of the convoy organizers. At least one of the convoy organizers is indeed Indigenous: Lich, whose Métis identity was often mentioned after her arrest on charges related to mischief because she cited it at her bail hearing. Questions about Lich's identity circulated across media, including through the news wing of the Aboriginal Peoples Television Network (APTN). Lich would later confirm her affiliation with the Métis Nation of Alberta to APTN by supplying a copy of her citizenship card.[33] Proving identity was important here, as

Lich's lawyer would confirm that a Gladue report will be raised in her case. As APTN details, "Gladue reports assist the court in delivering appropriate sentences if a person is convicted of a crime, taking into account the unique experiences of Indigenous offenders."[34]

But elsewhere in testimony, indigeneity was raised to claim some sort of movement authenticity. Bauder, for instance, cited his wife's Algonquin identity when being questioned about an event at Confederation Park during the occupation that displayed the kinds of cultural appropriation discussed earlier. An allegedly fake chief, "Chief J. D. Anderson," started a "sacred" fire at Confederation Park in concert with King and held a pipe ceremony, which wasn't consented to by the Algonquin Nation and ultimately, as Bauder claims, led to sour relations between Bauder and some sovereign clan mothers who were going to bring teepees to the park.[35] Coverage of this event detailed the complex territorial politics in settings like these, but it reiterated that the Algonquin Nation and other Indigenous organizations around Canada did not support the convoy.[36]

Second, indigeneity appeared in the framework of enforcement that the Ontario Provincial Police (OPP) followed throughout the convoy and occupation, "A Framework for Police Preparedness for Indigenous Critical Incidents." The framework was developed as a response to a three-year provincial inquiry that followed eight years after the Ipperwash Crisis in 1995. This crisis came as the culmination of an occupation of Ipperwash Provincial Park in Ontario by some members of Stoney Point Ojibway band, who were claiming nearby land of theirs that had been expropriated to make a military camp during World War II.[37] During this occupation, an OPP officer shot and killed Dudley George, one of the protesters. The document describes the conditions under which the framework was developed: "The OPP recognizes that conflicts may arise as Indigenous communities and the various levels of government work to resolve outstanding issues associated with matters such as land claims, self-determination and Indigenous or treaty rights, which may include activities such as hunting, fishing and extraction on ancestral or traditional territories."[38] The purpose of the framework is likewise clearly explained: "The Framework for Police Preparedness for Indigenous Critical Incidents (the Framework) provides a guideline for police response to conflict and has applicability to both Indigenous and non-Indigenous issue-related conflict." If the framework is for all critical incidents, what does the "Indigenous" in "Indigenous Critical Incidents" mean? Its origins in the Ipperwash Crisis partly explain this qualification, but the qualifier also reveals the ways in which

Indigenous movements are always already criminalized by the repressive state apparatus. Such criminalization manifests in, among other things, widespread surveillance from institutions like the Canadian Security and Intelligence Service, as Andrew Crosby and Jeffrey Monaghan meticulously detail.[39]

It is revealing that many of the convoy participants at the inquiry expressed surprise when, after weeks of nonengagement, police enforcement began with crowd-control tactics and arrests. Christopher Deering—a participant and Afghan war veteran who was wounded in combat and now suffers from PTSD—described being beaten by police during arrest and later denied medication.[40] Speaking more generally of police enforcement of COVID-19 restrictions, Tom Marazzo, a twenty-five-year veteran holding the rank of captain who used his expertise in logistics to help plan the convoy, revealed, "For the first time in my life, I was . . . actually afraid of police."[41] Later, he spoke of enforcement after the act was invoked, explaining that he "had seen some things that I never thought that I would see in this country."[42] Despite the suggestion from this testimony that police violence is unusual, police violence is a familiar sight and experience at protests across Canada. Listening to or reading Deering's testimony is difficult for these reasons. But unlike the kinds of protests I tend to attend and support, convoy participants saw the police as allies and considered enforcement a betrayal. These responses to unfortunately common, arguably constitutive police behavior show how sheltered many convoy participants were from the kinds of criminalization that Crosby and Monaghan outline. The appropriation of Indigenous identity and tactics seen at the convoy is, then, a settler colonial move that highlights the discrepancies of treatment between Indigenous land and water defenders and those of the Freedom Convoy. That Marazzo saw "things that [he] never thought that [he] would see in this country," in other words, speaks to privileged spaces in which many convoy participants inhabit, since Indigenous land and water defenders have intimately known and continue to intimately know this violence since as early as police presence was first established in 1873 with the North-West Mounted Police, a paramilitary precursor to the RCMP. Routine exposure to police violence is, of course, a widespread aspect of Indigenous experience across Canada, but it is particularly acute in rural settings where racist tactics like starlight tours, a practice of driving victims outside of town during winter, forcing them to walk back to town. The 1990 freezing death of seventeen-year-old Neil Stonechild (Saulteaux First Nations) outside Saskatoon brought attention to this practice, thanks to a provincial public inquiry.[43]

Conclusion: Combustion in the Age of Petronostalgia

Nostalgic desire for a return to a golden age of the production and consumption of oil that runs deep in the present has been theorized by those working in the energy humanities as *petronostalgia*.[44] Canadian communications scholar Douglas Tewksbury sees this tendency in the Canadian yellow vest movement. Tracking communication in the official Yellow Vests Facebook group, Tewskbury describes its politics "as a grab bag of conservative values representing a petro-nostalgic formulation of political group identity, but one where a mythologized narrative of the past is tied to fossil fuel production and extraction"—to which I would add combustion as well.[45] Nostalgia emerging as a political mode in these ways is unsurprising because settlement was animated by a desire to create such conditions. It was achieved first by violently staking claims to Indigenous territory, then by naturalizing settlement through the creation of infrastructure like the Canadian Pacific Railway as nativizing, nationalist origin stories alongside these material, territorial interventions. As I've emphasized throughout this chapter, the bonds among the Canadian yellow vests, 2019's United We Roll convoy, and the 2022 Freedom Convoy are strong, and Tewksbury's arguments apply equally to all three settings.

The 2022 Freedom Convoy was, in other words, a demonstration championing business as usual in all its connotations, including the abolishment of all pandemic public health measures against a "new normal"—phrasing used by journalists and politicians to describe postpandemic social relations that has become a target of critique from antivaxxers and others—to continued and increasing production and consumption of fossil fuels. This impetus to maintain business as usual, or an "old normal," is mediated by the material and symbolic tendencies of the gas can, which serves as a synecdoche of the aims and aspirations of the convoy and the imaginaries that fueled it. Disrupting circuits of capital was arguably an unintended aftereffect of the event rather than its prime motivation. Further, there is a marked historical and material difference between the Freedom Convoy and the reoccupation of traditional and ancestral territories that sometimes take shape as blockades. Ultimately, the convoy itself was a container of disparate yet interconnected aims, affects, and aspirations underwritten by the combustion of fossil fuels. As firmer measures to address the climate crisis, like a 2035 ban on dealerships selling new gas-powered vehicles in Canada, roll out, it seems as though a nostalgic desire for a fossil-fueled order through centering a right to combust will increasingly inspire right-wing political action. While Canada may have narrowly avoided institutionalization of fossil fascism at the federal level with the recent election of Mark Carney as Prime Minister over Pierre Poilievre, residues of the fossil fascist creep remain in commitments by the Carney government through Bill C-5 to intensify natural resource development, energy production, and construction infrastructure under rationalizations of sovereignty and national interest and security. Those of us committed to a more socially and ecologically just future must remain attentive to how the relationship between political forms and political commitments is obscured by plumes of exhaust.

Notes

1 Daggett, "Petro-Masculinity," 27.
2 CBC News, "Convoy Organizer."
3 Commission Council, "Overview Report," 51p142 and 34p90.
4 Deputy Prime Minister Chrystia Freeland quoted in POEC, *Report*, 30:5–6.
5 Transport Canada, "Building a Green Economy."
6 Johnson, "What's Needed."
7 Gerson, "One-Man Oil Sands Advocate."
8 LaDuke and Cowen, "Beyond Wiindigo Infrastructure," 246.
9 Morton, *Hyperobjects*, 1, 20.
10 Here I develop the same line of argument to discuss 2019's United We Roll convoy as I do in Kinder, *Petroturfing*, 194.
11 Weigel, "Rolling Coal."
12 Edwards, "Teenager Allegedly Hit."
13 Watson, "Driver Told Memorial Marchers."
14 Little and Robinson, "No Jail Time."
15 Victoria De La Ronde quoted in POEC, *Report*, 2:5.
16 Zexi Li quoted in POEC, *Report*, 2:38.
17 Mathieu Fleury quoted in POEC, *Report*, 2:160.
18 Tamara Lich quoted in POEC, *Report*, 17:7.
19 On the reach and political content of the Wexit movement, see Ruiz-Soler and Chun, "Regionally Alt-Right?"
20 Democracy at Work, "Ask Prof. Wolff."
21 Taylor, "Truck Convoy's Message."
22 Canada Unity, "Memorandum of Understanding."
23 King, "Trucker Convoy."

24 Cramer, *Politics of Resentment.*
25 Cramer, *Politics of Resentment.*
26 Huber, *Lifeblood*, 97–100.
27 POEC, *Report*, 15:241–43.
28 Morrisseau, "First Nations Speak Out."
29 Morrisseau, "First Nations Speak Out."
30 Deloria, *Playing Indian*, 2.
31 Simpson, *Short History*, 56.
32 Clover, "Political Economy."
33 Pimentel, "Tamara Lich."
34 APTN National News, "Lawyer for Convoy Organizer."
35 James Bauder quoted in POEC, *Report*, 16:206–8.
36 Lapierre, "Three Indigenous Chiefs."
37 Salomons, "Ipperwash Crisis."
38 Indigenous Policing Bureau, "Framework," 2.
39 Crosby and Monaghan, *Policing Indigenous Movements.*
40 Christopher Deering quoted in POEC, *Report*, 17:96–97.
41 Tom Marazzo quoted in POEC, *Report*, 15:124–25.
42 Tom Marazzo quoted in POEC, *Report*, 15:159.
43 Windspeaker Staff, "Saskatoon Police Chief."
44 Boyer et al., "Ep. #57—Timothy Mitchell"; Daggett, "Petro-Masculinity."
45 Tewksbury, "Petro-Nostalgia," 950.

Bibliography

APTN National News. "Lawyer for Convoy Organizer Tamara Lich Says a Gladue Report Will Be Part of Her Case." March 25, 2022. https://www.aptnnews.ca/videos/lawyer-for-convoy-organizer-tamara-lich-says-a-gladue-report-will-be-part-of-her-case/.

Boyer, Dominic, Cymene Howe, and Timothy Mitchell. "Ep. #57—Timothy Mitchell." *Cultures of Energy*, podcast, February 16, 2017. https://cenhs.libsyn.com/ep-57-timothy-mitchell.

Canada Unity. "Memorandum of Understanding." December 3, 2021. https://web.archive.org/web/20220201001209if_/https://canada-unity.com/wp-content/plugins/wonderplugin-pdf-embed/pdfjs/web/viewer.html?file=https%3A%2F%2Fcanada-unity.com%2Fwp-content%2Fuploads%2F2022%2F01%2FCombined-MOU-Dec03.pdf.

CBC News. "Convoy Organizer, Les Farfadaas Member from Quebec Appears in Court." March 1, 2022. https://www.cbc.ca/news/canada/ottawa/steeve-charland-freedom-convoy-court-1.6367565.

Clover, Joshua. "The Political Economy of Tactics." Verso Books blog, March 16, 2022. https://www.versobooks.com/blogs/5303-the-political-economy-of-tactics.

Commission Council. "Overview Report: Fundraising in Support of Protesters." Public Order Emergency Commission, October 2022.

Cramer, Katherine J. *The Politics of Resentment: Rural Consciousness in Wisconsin and the Rise of Scott Walker*. University of Chicago Press, 2016.

Crosby, Andrew, and Jeffrey Monaghan. *Policing Indigenous Movements: Dissent and the Security State*. Fernwood, 2018.

Daggett, Cara. "Petro-Masculinity: Fossil Fuels and Authoritarian Desire." *Millennium* 47, no. 1 (2018): 25–44.

Deloria, Philip J. *Playing Indian*. Yale University Press, 1999.

Democracy at Work. "Ask Prof. Wolff: Finding Common Ground with Canadian Truckers." YouTube video, February 15, 2022. https://www.youtube.com/watch?v=z013VR9C2uI.

Edwards, Jonathan. "A Teenager Allegedly Hit 6 Bicyclists with His Truck, Sending 3 to the Hospital. A Biker Says the Driver Was Harassing Them." *Washington Post*, September 30, 2021. https://www.washingtonpost.com/nation/2021/09/30/texas-teen-coal-rolls-hits-bicyclists/.

Gerson, Jen. "One-Man Oil Sands Advocate, Tired of Smears Against Alberta, Takes On Celebrity Activists in PR War." *National Post*, September 22, 2014. https://nationalpost.com/news/canada/one-man-oil-sands-advocate-tired-of-smears-against-alberta-takes-on-celebrities-in-pr-war.

Huber, Matt. *Lifeblood: Oil, Freedom, and the Forces of Capital*. University of Minnesota Press, 2013.

Indigenous Policing Bureau, Traffic Safety and Operational Support Command. "A Framework for Police Preparedness for Indigenous Critical Incidents." Ontario Provincial Police, revised December 2018. https://commissionsurletatdurgence.ca/files/exhibits/OPP00004566.pdf?t=1668059451.

Johnson, Doug. "What's Needed to Get EVs into Rural Canada?" Weather Network, May 27, 2023. https://www.theweathernetwork.com/en/news/climate/solutions/how-to-get-electric-vehicles-into-rural-and-remote-canada.

Kinder, Jordan B. *Petroturfing: Refining Canadian Oil Through Social Media*. University of Minnesota Press, 2024.

King, Adam D. K. "The Trucker Convoy Is Not a Workers' Revolt." *Passage*, February 4, 2022. https://readpassage.com/p/the-trucker-convoy-is-not-a-workers-revolt/.

LaDuke, Winona, and Deborah Cowen. "Beyond Wiindigo Infrastructure." *South Atlantic Quarterly* 119, no. 2 (2020): 243–68.

Lapierre, Matthew. "Three Indigenous Chiefs Disavow 'Freedom Convoy' and Ceremonies Taking Place in Confederation Park." *Ottawa Citizen*, February 3, 2022. https://ottawacitizen.com/news/local-news/three-indigenous-chiefs-disavow-Freedom-convoy-and-ceremonies-taking-place-in-confederation-park.

Little, Simon, and Kristen Robinson. "No Jail Time for B.C. Man Who Drove through Residential School March, Hitting 4." *Global News*, November 4, 2024. https://globalnews.ca/news/10850386/richard-manuel-sentence-no-jail-time/.

Morrisseau, Miles. "First Nations Speak Out Against Trucker Convoy." ICT News, February 5, 2022. https://ictnews.org/news/first-nations-speak-out-against-trucker-convoy.

Morton, Timothy. *Hyperobjects: Philosophy and Ecology After the End of the World*. University of Minnesota Press, 2013.

Pimentel, Tamara. "Tamara Lich Produces Membership Card for Métis Nation of Alberta." APTN News, March 25, 2022. https://www.aptnnews.ca/national-news/protest-organizer-tamara-lich-produces-membership-card-for-metis-nation-of-alberta/.

POEC (Public Order Emergency Commission). *Report*, vol. 2 (October 14, 2022). https://publicorderemergencycommission.ca/.

POEC (Public Order Emergency Commission). *Report*, vol. 15 (November 2, 2022). https://publicorderemergencycommission.ca/.

POEC (Public Order Emergency Commission). *Report*, vol. 16 (November 3, 2022). https://publicorderemergencycommission.ca/.

POEC (Public Order Emergency Commission). *Report*, vol. 17 (November 4, 2022). https://publicorderemergencycommission.ca/.

POEC (Public Order Emergency Commission). *Report*, vol. 30 (November 24, 2022). https://publicorderemergencycommission.ca/.

Ruiz-Soler, Javier, and Wendy Hui Kyong Chun. "Regionally Alt-Right? #Wexit as a Digital Public Sphere." *Canadian Journal of Communication* 47, no. 1 (2022): 173–96.

Salomons, Tanisha. "Ipperwash Crisis." First Nations and Indigenous Studies, University of British Columbia, 2009. Accessed April 28, 2023. https://indigenousfoundations.arts.ubc.ca/ipperwash_crisis/.

Simpson, Leanne Betasamosake. *A Short History of the Blockade: Giant Beavers, Diplomacy, and Regeneration in Nishnaabewin*. University of Alberta Press, 2021.

Taylor, Casey. "Truck Convoy's Message Muddies the Closer It Gets to Capital." *Bay Today*, January 26, 2022. https://www.baytoday.ca/local-news/truck-convoys-message-muddies-the-closer-it-gets-to-capital-49949.

Tewksbury, Doug. "Petro-Nostalgia and the Politics of Yellow Vests Canada." *Canadian Journal of Communication* 46, no. 4 (2021): 939–59.

Transport Canada, Government of Canada. "Building a Green Economy: Government of Canada to Require 100% of Car and Passenger Truck Sales Be Zero-Emission by 2035 in Canada." News release, June 29, 2021. https://www.canada.ca/en/transport-canada/news/2021/06/building-a-green-economy-government-of-canada-to-require-100-of-car-and-passenger-truck-sales-be-zero-emission-by-2035-in-canada.html.

Watson, Bridgette. "Driver Told Memorial Marchers to 'Get Over' Residential Schools Before Plowing into Crowd, Witness." CBC News, June 9, 2022. https://www.cbc.ca/news/canada/british-columbia/hit-and-run-witnesses-accountability-mission-bc-1.6481532.

Weigel, David. "Rolling Coal." *Slate*, July 3, 2014. https://slate.com/news-and-politics/2014/07/rolling-coal-conservatives-who-show-their-annoyance-with-liberals-obama-and-the-epa-by-blowing-black-smoke-from-their-trucks.html.

Windspeaker Staff. "Saskatoon Police Chief Admits Starlight Cruises Are Not New." *Windspeaker* 21, no. 4 (July 1, 2003): 9. Archived at Aboriginal Multi-Media Society (AAMSA). https://ammsa.com/publications/windspeaker/saskatoon-police-chief-admits-starlight-cruises-are-not-new.

10

Where the Market Dares Not Tread

Mapping Rural Broadband in the United States

CHRISTOPHER ALI

This chapter chronicles the ongoing efforts of the Federal Communications Commission (FCC) to map broadband infrastructure in the United States. The FCC is currently engaged in a massive overhaul of its national broadband map after critiques of its previous broadband map became widely known. The former map grossly overstated the level of connectivity in the country, much to the benefit of incumbent broadband providers who boasted of a digitally connected nation.[1] This discrepancy was disastrous for rural communities, which lost out on federal subsidies for broadband deployment because of their erroneously connected status on the map. The FCC's new map, released in November 2022, is an attempt to correct the previous iteration and offer a more granular and realistic picture of the country.

This chapter's critique of the FCC's mapping efforts, both current and former, offers two interventions in the theorization of media rurality. First, by demonstrating how erroneous maps stifle broadband competition and protect the status quo for incumbent broadband providers, it illustrates the ongoing entrenchment of neoliberal values leading to the compromise of public interest regulation within contemporary American policymaking. Second, it demonstrates how maps themselves are conduits of political economic power that serve the interests of capital and underplay the lived experiences of rural, remote, and Indigenous communities. As the introduction to this edited collection argues, ruralities are dynamic, subjective, and situated. However, the political economy of broadband maps denies rural communities such dynamism, reducing them to singular and objectivized data points—and in some instances not even that.

Broadband is defined by the FCC as a reliable internet connection at or surpassing download speeds of 100 megabits per second (Mbps) and upload speeds of 20 Mbps (commonly depicted as 100/20). The definition of broadband was raised in 2024 from the previous 2015 standard of 25/3, an impoverished standard that failed to incorporate contemporary use, speed, and data needs, such as the Internet of Things and the advent of ultra-high-definition/4K streaming.[2] Using the 25/3 standard, however, the FCC used internet service providers' (ISP) self-reported coverage estimates to determine in 2021 that 14.5 million people in the United States lacked broadband access, including 17 percent (or 11 million) of rural residents.[3] Rural areas have been particularly difficult to connect because of a refusal of private market providers to offer service or upgrade networks because of a lack of return on investment.[4]

Congress's latest initiative to correct this market failure is the USD42.5 billion Broadband Equity Access and Deployment (BEAD) program, which was created in 2021's trillion-dollar Infrastructure Investments and Jobs Act (IIJA).[5] In June 2023, the National Telecommunications and Information Administration (NTIA) used the FCC's revised broadband map to determine how much funding each state gets on the basis of the number of unserved locations, with each state guaranteed at least USD100 million.[6]

The FCC's new broadband map, released in November 2022, was supposed to address prior methodological problems by increasing the precision of the map down to the individual address level, thereby determining who is connected, underconnected, and unconnected in each household in the country.[7] This new map should have revealed where the private market refused to tread, and therefore which areas require public support. Instead, the new map follows the old by depicting a vision of a connected America as self-reported by providers. Put simply, the FCC increased the *precision* of the map without addressing fundamental problems of *accuracy*. The logic behind an exaggerated map is the fear of competition. Incumbent providers have benefited from over a decade of federal subsidy, and by exaggerating their deployment, they eliminate the potential for competition from a publicly subsidized new entrant. In

2020, for instance, Charter successfully petitioned the FCC to remove 2,127 census blocks in New York State from public funding eligibility because the company *intended* to serve these areas by order of New York State for its merger with Time Warner Cable.[8] The maps—both current and former—reinforce this anticompetitive behavior, with reports of service area exaggeration (where a provider claims to offer service where none is actually available) and millions of missing households and buildings. The issue is particularly egregious for rural constituencies, including Tribal communities whose persistent underconnection is compounded by their misrepresentation and, in some cases, erasure by the FCC's broadband map.

The field of critical geography teaches us that maps are imbued with power.[9] We trust maps and make critical decisions based on them. This is no less true in the digital age, as anyone who has followed Google Maps directions has directly experienced; maps powered by geographic information systems (GIS) carry significant authority (even when they fail to account for on-the-ground realities). As Specht and Feigenbaum argue, "As mapping platforms often predetermine places, and their meanings, they shape users' spatial imaginations and limit what is possible to map."[10] Maps are proxies of power.[11] When it comes to broadband mapping, the stakes could not be higher, as the national broadband map instructs the NTIA on how much funding to distribute to each state with inaccuracies in the map that expunge unserved households, resulting in a state receiving less funding to bridge the digital divide.

Drawing on critical political economy and critical geography, I argue that the present iteration of the FCC's broadband map continues to disenfranchise rural, remote, and Tribal communities through a process of what geographer J. B. Harley calls "cartographic silence."[12] As a result, rural communities must expend greater resources—both human and economic—to correct for the map's failure and ensure they are properly counted. Placing the onus of correction on communities themselves, rather than on the map maker (the FCC's private subcontractor, CostQuest), ISPs, or the FCC, serves the interests of incumbent providers, which benefit from tremendous ongoing public subsidies with little accountability. Rural communities lose out by being underserved, undercounted, and silenced; meanwhile, telco incumbents win by paradoxically taking ongoing subsidies for bridging the digital divide while simultaneously keeping out potential competitors by declaring that digital divides do not exist.

Rural Broadband

As noted above, broadband is defined as an always-on, reliable internet connection of no less than 100 Mbps download and 20 Mbps upload.[13] Connectivity can be provided by different technologies, each with its own affordances. Technologies include wired/fixed options (coaxial cable, digital subscriber line [DSL], fiber-optic cable) and wireless options (fixed wireless, satellite, mobile). Rural residents are most familiar with DSL technology.[14] Carried by legacy copper wires deployed by telephone companies, DSL suffers from slow download and upload speeds as well as attenuation (signal loss) the further one is away from a network node or headend. Both cable and DSL networks also suffer from degradation and decay and need regular network maintenance, refreshing, and upgrading. Attention to the importance of upload speeds and symmetric speeds were heightened during the pandemic, when many households found, for example, that while download speeds to watch Netflix were satisfactory, upload speeds, necessary for the massive uptake in live videoconferencing, education, work, and telehealth, often proved problematic.[15] In contrast to other network technologies, fiber-optic networks are labeled future proof, capable of handling an almost unlimited amount of data symmetrically (where upload and download speeds are equal) and delivering speeds of a gigabit per second or more. While a fiber-first strategy is preferred by federal and local officials alike, the downside of fiber is the cost to deploy.[16]

There are three primary actors in the rural broadband policy ecosystem in the country: the FCC, the Department of Agriculture (USDA), and the NTIA. Each administers programs for broadband deployment in rural areas, while the FCC also has responsibility for setting broadband policy, reporting on broadband deployment, and defining broadband. That said, it lacks the authority to force a provider to connect communities when only private capital is used and to regulate broadband rates. The FCC, through its Universal Service Administrative Company, oversees the Universal Service Fund with USD4 to 5 billion in annual funding for rural broadband deployment. The USDA, through its Rural Utilities Service, manages five loan and grant programs with almost USD5 billion at its disposal. The NTIA manages the newest and largest program, with USD42.5 billion in broadband deployment funding as a part of its BEAD program.

The reason extensive public intervention is needed is the high cost of connecting rural, remote, and Tribal communities. Indeed, the programs of the FCC are not called "rural broadband" programs but rather "high-cost" programs, acknowledging the political economic connection between rurality and capital expenditures. The cost is higher in rural areas because distances are vast and the population less dense than in urban areas. Fiber-optic cables, for instance, often cost a minimum of USD27,000 per mile, according to the US Department of Transportation.[17] When population density is lower, the cost per household is significantly higher than in densely populated areas. The tension between return on investment and public service cannot be overstated here. Low levels of rural, remote, and Tribal broadband service are market failures in that private actors have failed to deliver adequate levels of public services.[18] Adding to this failure, alternatives to large providers, notably public options through municipal broadband projects, are prohibited or inhibited in sixteen states, with little in the way of support from the FCC.[19]

The COVID-19 pandemic saw the allocation of hundreds of billions of dollars in relief funds, a portion of which has been made available for broadband projects. The apex of these efforts is the IIJA. The IIJA allocated a total of USD62.5 billion for broadband, including USD42.5 billion for BEAD, USD14 billion for the Affordable Connectivity Program (ACP), and USD2.75 billion for two different digital equity programs. The ACP is administered by the FCC and provides direct subsidies to ISPs of USD30 per month (USD75 per month on Tribal lands) to connect low-income households. The BEAD and digital equity programs are administered by the NTIA and by states themselves who will be the final arbiters of the money. The ACP ended in June 2024 after Congress failed to renew it, and the second Trump administration ended the Digital Equity Act programs a year later.

The BEAD program is a multiphase initiative, including the mandate that each state create a five-year broadband strategic plan. Every state is guaranteed at least USD100 million, with distribution of the remaining funds determined by the number of unserved households within each state.[20] To determine who is un- and underconnected, the NTIA relied on the FCC's national broadband map. The NTIA released its decision on state funding allocations for the BEAD program in June 2023. As the NTIA determines each state's allotment, the states develop their own request for proposals and guidelines for distributing the funds through a competitive award process.

That funding is going to states rather than a federal agency for allocation is a significant departure for federal broadband policy, which heretofore had allocated funding through the FCC and the USDA. This decision by Congress recognizes both the failure of the FCC to adequately and accountably distribute funds and the importance of local decision-making and knowledge in broadband infrastructure. That said, many states are unprepared for the hundreds of millions of dollars that they are going to have to spend because of inadequate broadband office staffing numbers.[21] Some worry that incumbents, with their vast resources, are already garnering the bulk of COVID-19 subsidies for broadband, "hoovering up the lion's share of federal and state funding to fix a problem they helped create."[22] There is also concern about those states that prohibit or inhibit publicly owned and operated municipal broadband and the fact that the NTIA explicitly states that municipalities cannot be discriminated against.[23] Actual service outcomes will depend on how state grant processes will be operationalized, including the establishment of eligibility criteria.

The Power of Maps

Maps have long been the basis on which certain policy decisions are made and policy actions taken.[24] We trust maps because of their supposed ideological neutrality;[25] after all, how could our Google Maps directions to the nearest gas station have a politics? But maps are anything but neutral.[26] "Behind the map-maker," writes J. B. Harley, "lies a set of power relations, creating their own specifications. Whether imposed by an individual patron, by state bureaucracy, or the market, these rules can be reconstructed both from the content of maps and from the mode of cartographic representation."[27] Maps are loci of power, signifiers imbued with meaning by their authors. Maps are both the outcome of powerful relations and the product of them, exuding truth claims that actively guide the viewer in determining what is perceived to be true, factual, and correct. Maps tell us what is there and what is not, and by extension *who* is there and who is not, who counts and who doesn't.

Ideologies embedded within maps became even more blurred with the advent of computer modeling and GIS software. "Computer technology," writes Harley, "has increased [the] concentration of media power."[28] Cartographer Robert Williams adds that *because* maps contain truth claims, the responsibility to get things right is heightened in the digital age: "Since maps are perhaps the most persuasive form of communication, those who make them must accept an unusually high degree of responsibility for their truthfulness. This responsibility is increased when the aura of infallibility of the computer is added to the map."[29] While GIS has been praised for its "democratization" potential, "geospatial data may in fact generate new methods of exclusion and present new technological and societal barriers for community organizations and their members," as Rob McMahon, Trevor James Smith, and Tim Whiteduck explain.[30] One way of intensifying exclusion within the GIS ecosystem, observe McMahon et al., is via the technical expertise and access to technologies required to create and analyze the data.

As instruments of both the state and the market, maps are imbued with significant political economic power. Here, erasure speaks as loudly as presence. The notion of cartographic silence developed by Harley is crucial to this understanding of power: "The notion of 'silences' on maps is central to any argument about the influence of their hidden political messages. . . . It is asserted here that maps—just as much as examples of literature or the spoken word—exert a social influence through their omissions as much as by the features they depict and emphasize."[31] To this Harley adds the term "cartographic censorship," which is the "deliberative misrepresentation of maps designed to mislead potential users, usually those regarded as opponents of the territorial status quo."[32] Harley argues that maps need to be read as socially constructed texts—to be read, as Murton puts it, in "the context of the cartographer; the context of other maps; and the context of society."[33]

The remainder of this chapter takes heed of Harley's methodological provocations, analyzing the national broadband map in the context of its creators, the previous broadband map, and the processes and programs of broadband funding.

Broadband Map 1.0

The FCC has a troubled history when it comes to broadband mapping.[34] The previous iteration of the broadband map was deeply flawed, leading to gross exaggerations of connectivity. Many agree, for instance, that the FCC's 2021 estimates were off by at least 50 percent.[35] BroadbandNow estimates that 42 million homes lack access to a broadband network (compared with the FCC's 2021 estimates that only 14.5 million people lacked connectivity).[36] Alarmingly, Microsoft reported in 2019 that upward of 162.8 million people were underconnected.[37] As I've noted elsewhere, there were three fatal flaws to the FCC's original broadband map.[38] First, data were self-reported by ISPs rather than collected by the FCC or an impartial third party. ISPs have strong incentives to "overstate speed and availability to downplay industry failures."[39] ISPs were required to submit a form (known as Form 477) twice a year noting the areas that received service. This information was then used to populate the national broadband map. Benefits of an exaggerated map include the public relations boon of promoting near-universal coverage and, more egregiously, the ability to stymie new entry competition, as I mentioned above. By exaggerating connectivity, both incumbents and the FCC mitigate the likelihood of regulatory intervention to enforce the policy goals attached to federal subsidies. In addition to self-reportage, ISPs only had to report *advertised* speeds rather than *actual* speeds, leading to a second distortion on the map: connectivity versus underconnectivity. The hypothetical maximum speed of a network often differs substantially from the speeds received at a customer's home. Depending on the technology deployed, there can sometimes be massive differences between the hypothetical top speed of the network and the actual experience of the user. Weather, distance from the node, number of users on the network, and trees blocking the line of sight can all diminish speeds of technologies like fixed wireless, mobile, coaxial cable, and DSL. Fiber-optic systems, on the other hand, are largely immune to these common limitations.

The second flaw is the networks themselves. Not all broadband technologies are equal, but the FCC had such a low threshold for the definition of broadband (25/3) that subpar technologies such as DSL and fixed wireless (both of which are heavily deployed in rural areas) "counted" as broadband.[40] DSL, for instance, delivers median speeds in rural areas only at 10/1—far below the FCC's 2015 baseline of 25/3, and woefully below the new standard of 100/20 set by both the Infrastructure Act and the FCC.[41] The result is a "Swiss cheese" coverage pattern of broadband deployment in rural areas, with many areas receiving broadband through antiquated technologies even as a few others are implementing fiber.[42] Yet all these discrepant systems count as broadband in the FCC's official tally.[43] The result is a plurality of rural communities that continue to rely on DSL and other outdated technologies but are counted as served on the broadband map.[44]

The third flaw of the original map was the granularity of the data. The FCC's map assessed service at the census-block level rather than individual buildings. As long as one building in a census block had service (advertised or otherwise) *or* could be served within ten business days, the entire census block was considered served with broadband. The implications are colossal for rural areas, which have larger census blocks than urban areas. The largest census block in the country, for instance, is found in rural Alaska and is 8,500 square miles—and if a single home within that area *could* be served, then the entire area would be considered served.[45] As Busby and Tanberk explain: "Urban areas, due to dense populations, tend to have census blocks that are narrow in square miles, whereas rural areas, with less population density have wider and less concentrated blocks. This increases the probability of those blocks having outstated coverage, because while one house could reach a wired line, the next house (which could be acres away) might not. Despite this, under the current system, it will still be counted."[46] The consequences of the map's failures were detrimental for rural areas. Being marked "served" on the map rendered areas ineligible for federal subsidies (which at the time amounted to over USD4 billion a year from the FCC alone) and reinforced a deeply digitally divided country. The result of the FCC's pre-2022 maps was a depiction of a connected America with near-ubiquitous high-speed internet. This favors national broadband providers and dismisses the lived, unconnected, and underconnected experiences of rural, remote, and Tribal communities.

The Broadband DATA Act and the Infrastructure Act

Congress took notice of the ongoing misinformation propagated by the FCC's original national broadband map and passed the Broadband Deployment Accuracy and Technological Availability (DATA) Act in 2020. The Broadband DATA Act came on the heels of an FCC initiative known as the Digital Opportunity Data Collection (DODC), "which [was] intended to address many of the issues that currently lead to inaccurate broadband mapping data."[47] According to the Congressional Research Service, one major difference between the FCC's DODC and the Broadband DATA Act is that the FCC wanted its Universal Service Administrative Company, which handles broadband subsidies, to lead the initiative, while the Broadband DATA Act mandated that such changes come from the FCC itself. The Broadband DATA Act (2020) ordered the FCC to "change the way broadband data is collected, verified and reported" and required the creation of a "broadband serviceable location fabric."[48] To fulfill Congress's mandate, the FCC released a request for bids from companies seeking to earn the commission to redraft the national broadband map. After a contentious bidding process (the result of the FCC's lack of transparency), the FCC settled on CostQuest as the winner and awarded it a contract worth USD45 million. CostQuest is best known as an industry consulting firm and has partnered with the telecommunications industry through the industry's trade association, USTelecom.[49] CostQuest beat out another bidder, LightBox, a real-estate data firm, for the FCC's mapping contract.[50]

Congress was so confident in the FCC's abilities to revise its broadband map that it tethered the allocation of the USD42.5 billion BEAD program to the new map; it also required the NTIA to consult the map to determine how many unserved homes each state has, and thus how much funding it will receive.[51] CostQuest retains ownership of the underlying datasets used to create the broadband map.[52] While members of the public can use the graphical interface, they are unable to work with the underlying datasets without procuring a license. Communities are granted a "basic license" to use the map only to review the FCC's findings (but not to utilize the FCC's map for their own planning purposes). According to telecommunications expert Doug Dawson:

Communities or others with the basic license can't use the mapping data for:

- Prepar[ing] for the BEAD program, grant proposals, or other funding initiatives.
- Broadband Mapping.
- Opportunity Analysis.
- Network Planning or Design.
- Marketing purposes, such as sending mailers to addresses or identifying new customers to target marketing efforts.[53]

Even the NTIA, ordered by Congress in the Infrastructure Act to use the FCC/CostQuest map, has had to purchase a license to use the broadband map to make determinations related to the BEAD program. The final cost for the NTIA's license came to USD50 million—which is in addition to the USD45 million in public money spent to create the map in the first place. In the case of the FFC's new broadband map, the federal state surrendered its cartographic authority to a private contractor with ties to the broadband industry, and the result was an instrument optimized for capital accumulation instead of the public interest.

Broadband Map 2.0

The first version of the *new* national broadband map was released in November 2022. The new map has two layers to it. First is the Broadband Serviceable Location Fabric, a dataset that contains the locations of individual structures that can be served with broadband. This includes buildings such as homes and businesses. The hope is that by exchanging the census block determinators for building-level information, a more granular understanding of broadband service will be possible. The second layer is broadband availability, as determined by the providers themselves. The map is interactive, wherein users can input an address, county, street, Tribal area, or census area and receive information on service and availability.

In addition to the licensing issue wherein this publicly funded map is inaccessible to the public without a fee, three major problems are proving detrimental for rural, remote, and Tribal communities. First, data are once again self-reported by broadband service providers and again, advertised speeds are permitted (as is the ten-day service window where a building is considered served so long as the reporting ISP claims it could do so in ten business days). Neither the advertised speed nor the availability claims are verified by the FCC. Second, ISPs have already been shown to claim to cover houses where no such coverage exists or such coverage is overstated.[54] The state of Vermont, for instance, found that over 20,000 addresses had errors in the reported broadband service.[55] This could cause the state to lose out on potentially millions of dollars in broadband funding because of undercounting the unserved. Third, locations were absent from the map—an echo of Harley's cartographic silence.[56] Through this silencing, rural, remote, and Tribal households, streets, and communities have been systematically erased.[57] New York, for instance, reported that some 31,000 locations were missing from the Fabric location.[58] Back in Vermont, a reported 22 percent of locations were absent from the map, representing a shockingly high failure rate for such an expensive endeavor.[59] Some of the most egregious examples of cartographic silencing are found in rural Tribal communities in Alaska. Take, for instance, the town of Koyuk, home of the federally recognized tribe of the Native Village of Koyuk. The town's 2020 population was 243, of whom 98 percent are American Indian and Alaska Native.[60] One does not need to look hard for cartographic information about Koyuk: A Google Maps image finds

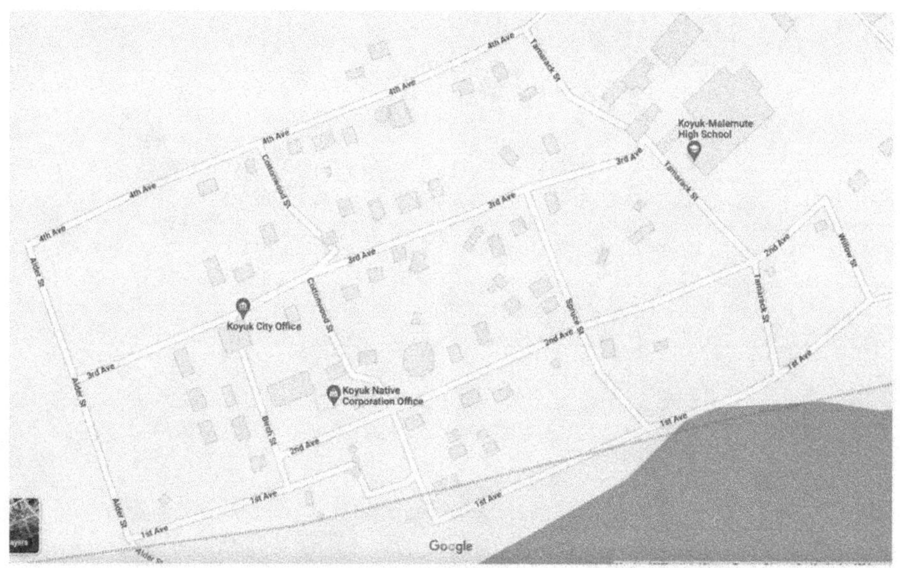

10.1 Detail of Koyuk, Alaska. Source: January 2023 Google Maps screenshot, https://www.google.com/maps.

multiple edifices in a cross section of the town (figure 10.1). An early release of the FCC's map, by contrast, depicted the town of Koyuk without any buildings.

To be sure, this was an egregious example of cartographic silence, but it exemplifies the continuing and deeply problematic erasure of rural, remote, and Tribal communities from the United States' official record of the digital divides facing these same constituencies. At the same time we must give credit to the FCC, which is constantly updating its map according to public challenges. Indeed, the latest iteration of the map finally includes broadband-serviceable locations in Koyuk, Alaska (figure 10.2).

The reason and justification for the maps' early failure are both myriad and circumspect. Critical political economy, with its eye on unequal power dynamics, and theories of regulatory capture suggest that the FCC's broadband maps function in service of the large providers and to stymie competition in their favor.[61] It also circumvents greater regulatory scrutiny or rule making. Heretofore, and with certain exceptions, the commission has been reluctant to rein in broadband providers, as noted when it failed to admonish Frontier and CenturyLink for failing to live up to their commitments.[62]

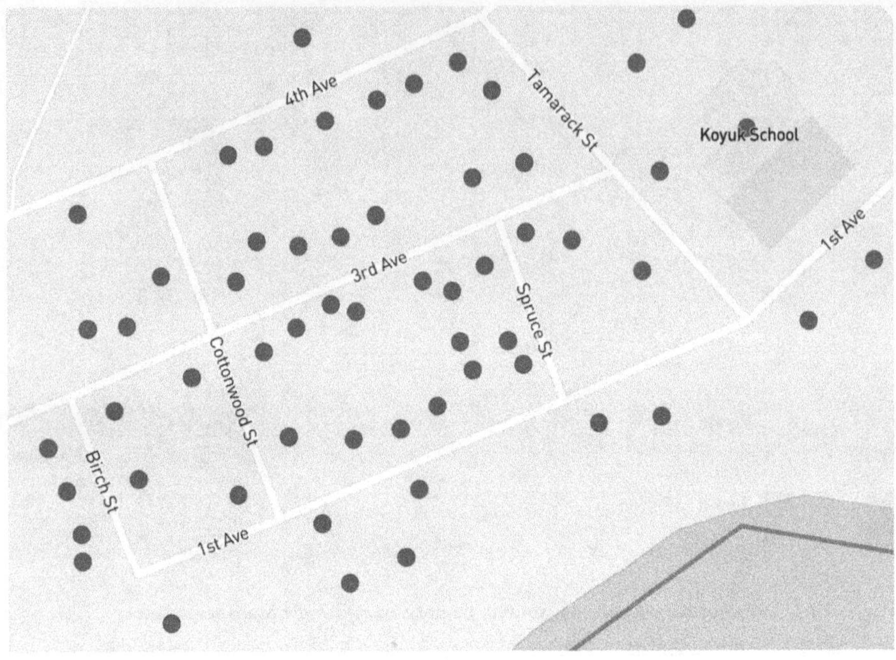

10.2 The FCC's newest broadband map showing broadband-serviceable locations in Koyuk, Alaska. Source: August 2023 screenshot from FCC's broadband map, https://broadbandmap.fcc.gov/home.

Historically, Tribal communities have borne the brunt of cartographic silence and censure.[63] As much as maps are artifacts and instruments of capitalist power, they are also outcomes and propagators of colonialism. "They help create myths which would assist in the maintenance of the territorial *status quo*," especially by forcing the market fundamentalist belief in private property.[64] The problematic political economic intersection between broadband, maps, and colonialism recalls digital technology scholar Marisa Duarte's critique that "policymakers in Washington DC, by not considering the rules of tribal sovereignty develop policies that preclude tribal peoples from creating the programs and services they need for self-governance goals."[65] Broadband, as Duarte brilliantly argues, is a crucial element toward what she calls "network sovereignty." The lack of visibility on the map not only stymies connectivity attempts in rural, remote, and Tribal communities but also directly counteracts Tribal communities' attempts at digital and data sovereignty.

Two challenge processes—availability and location—were made available to those who found themselves incorrectly represented on the FCC's map. The official challenge process available to the public lasted only a few weeks, from late November 2022 to January 13, 2023. The FCC received over 1.11 million "bulk Fabric challenges" from government entities and an undetermined number of availability challenges.[66] States had until March 25, 2023, to submit bulk location challenges.[67] While challenges will be accepted on an ongoing basis, at the time of writing, the map that will be used by the NTIA to determine BEAD allocations has been established, with the NTIA's state funding decisions released in June 2023. Compounding the short duration of the FCC's challenge window was the remarkable problem that the underlying data were not publicly available (and to access the data, states had to give up numerous rights to CostQuest) and that the challenges themselves had to include data that could not possibly exist (for example, at one point requiring the listing of the location ID of a home that does not exist on the map, and thus has no location ID). Furthermore, the adjudication process remains opaque, with no objective criteria being publicly released as to why some challenges are being accepted while the same challenge evidence is rejected for other locations.

If it is a question of broadband service—called an availability challenge—users may challenge the speeds and service claimed by ISPs by clicking an icon above the map and filling out the details. That said, the FCC has repeatedly stated that it will not accept home speed tests as valid evidence for a service challenge, thus depriving local communities of a powerful political tool.[68] Crowdsourced data (such as speed tests) are only permissible for a less formal challenge process where the provider is not required to respond.[69] The efficacy of these challenges remains unknown.

If the concern is about the location itself—a Fabric challenge—more steps are required to lodge a challenge, including pinning the longitude and latitude for the building, and at the very least having postal service address information.[70] While this may be an easy fix in an urban area where one's smartphone can provide the exact location of a missing home, mapping a rural residence is more demanding, given the lack of connectivity or potentially the lack of a postal service delivery address. Mapping an entire community, moreover, takes time, money, and connectivity—something many rural, remote, and Tribal communities do not have.[71] The sad irony is, to be counted on the broadband map, one often needs to have connectivity. It also means that communities must actively monitor the broadband map to ensure correctness—something state broadband offices have been promoting to their constituents. Thus, the labor of accuracy is offloaded from both the FCC and CostQuest (who is paid to be accurate) and placed directly on already underresourced communities (who are expected to provide free labor to CostQuest and the FCC).

This rural neglect reflects a policy architecture of what Thomas et al. call "urbanormativity."[72] For political economic reasons, it is easier to geotag a building in a city than in the countryside. Such ease is built into the system itself, thus ignoring the rural and remote difficulties of (1) gaining access to the map when one already has limited broadband and potentially a lack of community resources (like a library or community center with complementary internet access); (2) the resources necessary to geotag a building, series of buildings, or entire town; and (3) the geographic structure of addresses aligning with the postal service. The issue of addresses in rural Tribal communities was discussed by Chris Teale in an article for the publication *Route Fifty*: "Tribal lands typically do not use standardized home addresses, so they do not fit in the maps' location fabric data, which uses traditional street addresses. Some tribes rely on informal, descriptive addresses, like the blue house 3 miles west of Route 550 just past the Animus River. Others like the Navajo Nation rely on Plus Codes, which use longitude and latitude to produce short digital addresses that are easy to share."[73] Despite these challenges, and to the credit of broadband offices, states, and communities, after the most recent challenge process (March 2023), 1.04 million new locations were added to the Fabric.[74] But some additional challenges are still in process, and many additional incorrect addresses and broadband availability claims have not yet been challenged at all.

Beyond the difficulty of recreating a comprehensive broadband serviceable location map of the United States (something that the census attempts to do every decade), the complications of rural, remote, and Tribal broadband mapping have material, political, and economic consequences for states, communities, and individuals alike. At the state level, the more unserved locations a state reveals, the more funding the state will receive from the NTIA. This also benefits local communities who may see improved service as a direct result of these funding decisions. At the community and individual levels, we encounter issues of representation, power, and dignity. No one should have to prove their existence, but this is often what rural, remote, and Tribal communities must do. The literal and figurative costs of rural broadband mapping fall directly on the communities who need the connectivity the most, with the concomitant problem that if they do not provide this free labor, they may become ineligible for funding to address their digital divide because the official maps claim it doesn't exist.

Conclusion: The Political Economy of Broadband Mapping

The field of political economy of communication positions communication as a resource underscored by power dynamics and inequalities.[75] To study communication resources means to "study . . . the social relations, particularly the power relations, that mutually constitute [their] production, distribution, and consumption."[76] Thinking about broadband as a communication resource centers the map as the key arbiter in resource allocation.

What the past and present of the FCC's mapping efforts demonstrate is an obfuscation of responsibility and a delegation of what should be a public mapping process to the private market and to underresourced communities. Shifting such crucial responsibility from the regulator to industry serves the political economic interests of both CostQuest and private providers. For both, it lessens the number of challenges to process. In the specific case of incumbent providers, they benefit from mistaken claims that an area is served because this prevents competitors from receiving federal grants to serve that area, thus keeping their local monopoly and preventing a competitor from entering the market using public funding. The presence of cartographic silences, cartographic censorship, and urbanormativity contributes to this political economic maneuvering. Rural, remote, and Tribal communities are being erased and thus must have their digital hopes dashed; such processes are a core component of a structure that *maintains* digital divides for these constituencies.

Maps and mapping are powerful, and broadband mapping is no different. States, counties, and communities have taken it on themselves to create their own maps in hopes of correcting the FCC's errors now that NTIA's BEAD funding has been allocated to states. These efforts, however, while localized and participatory, once again require both substantive economic and human resources. These resources may include hiring a consultant to develop a community broadband survey and developing recruitment strategies to reach unconnected community members; it may require the participation of community members to both get the word out and perform individual speed tests to populate the map. It will also be necessary to recruit and hire GIS specialists, as well as the funds necessary to produce and sustain the new map. Many communities are simply unable to meet these political-economic demands. Moreover, while these efforts will make state and community maps more accurate, the truth is that many states have already received less money from the NTIA because of the FCC's mapping methodological shortcomings. The result is the further digital disenfranchisement of rural, remote, and Tribal communities.

Here rural communities are doubly mediated, first by broadband itself and second by their representation on the broadband map. While broadband is indeed a crucial component of everyday life—a utility, necessity, or universal service—its presence also further invites extraction of value from industrialized rural locations under conditions of a capitalist driven market. Tourism, increased subscription to social media companies, and precision agriculture all generate vital data for potential extraction and exploitation. Meanwhile, data centers, notably in rural Texas, consume tremendous amounts of energy and resources while giving little back to the community.[77] While we cannot and must not discount the importance of broadband to contemporary life, its problematic elements must also not be disavowed, especially as they pertain to capital, value extraction, and exploitation.

This tension between the rural and the market—one seen repeatedly in the field of critical rural studies—is now being played out via the broadband map. Such tensions parallel Harvey's articulation of the hopes and dreams attached to maps: "Geographical knowledges have the largely unrealized potentiality to express hopes and aspirations as well as fears, to seek universal understanding based on mutual respect and concern and to articulate firmer bases for human cooperation in a world market by strong geographical differences."[78] In the case of the national broadband map and rural communities, the hope is for a connected countryside; the fear is a private market that refuses to let that happen.

Notes

Acknowledgments: A shorter version of this chapter was published in the Law and Political Economy Project's *LPE Blog* under the title "Putting Rural Communities on the (Broadband) Map." I thank the editors of LPE for permission to repurpose my writing for this chapter.
I thank Sascha Meinrath and Sydney Forde, Bellisario College of Communications at Penn State, for reading, critiquing, and substantially strengthening this chapter.

1 Bode, "How Bad Maps."
2 FCC, "2024 Broadband Deployment Report."
3 FCC, "2021 Broadband Deployment Report."
4 Ali, *Farm Fresh Broadband.*
5 DeFazio, Infrastructure Investment and Jobs Act.
6 NTIA, "Biden–Harris Administration."
7 Wicker, S.1822.
8 Brodkin, "FCC Helps Charter."
9 Specht and Feigenbaum, "From the Cartographic Gaze."
10 Specht and Feigenbaum, "From the Cartographic Gaze," 47.
11 Bargues-Pedreny et al., "Mapping and Politics."
12 Harley, "Maps, Knowledge."
13 FCC, "2024 Broadband Deployment Report."
14 Gallardo and Whitacre, "Look at Broadband Access."
15 Holpuch, "US's Digital Divide."
16 Crawford, *Fiber.*
17 Benson, "USDA's ReConnect Program."
18 Brodkin, "CenturyLink"; Taglang, "CenturyLink and Frontier."
19 Cooper, "Municipal Broadband 2023"; Ali, "Telecommunications Localism."
20 Some argue that this calculation does injustice to the most remote communities because they are the most expensive to reach. In Kansas, for instance, broadband expert Mike Conlow calculated that a remote household may cost over USD16,000 to connect, while a rural household in Maryland may take only USD1,000. By these estimates, Kansas should receive billions for its hard-to-reach places but is earmarked only for USD435 million. Conlow, "Second Look."
21 Hinton et al., "Are States Ready."
22 Bode, "Charter, Comcast."
23 Goovaerts, "States, NTIA Say."
24 McMahon et al., "Reclaiming Geospatial Data."
25 Chrisman, "Full Circle."
26 Wilson, *New Lines*; Harvey, *Spaces.*
27 Harley, "Maps, Knowledge," 135.
28 Harley, "Maps, Knowledge," 142.
29 Quoted in Wilson, *New Lines*, 61.
30 McMahon et al., "Reclaiming Geospatial Data," 431.
31 Harley, "Maps, Knowledge," 136.
32 Harley, "Maps, Knowledge," 135.
33 Murton, "Power," 275.
34 Bode, "How Bad Maps"; Whitacre and Biedny, "Preview."
35 Meinrath, "Broadband Availability"; Busby et al., "BroadbandNow."
36 Busby et al., "BroadbandNow."

37 Kahan, "It's Time." Note that an update from Microsoft in 2021 puts the number at 120 million. Robinson, "Addressing Racial and Digital Inequity."

38 Ali, "Politics of Good Enough"; Ali, *Farm Fresh Broadband.*

39 Bode, "How Bad Maps."

40 Ali, *Farm Fresh Broadband*; Gallardo and Whitacre, "Look at Broadband Access."

41 Gallardo and Whitacre, "Look at Broadband Access."

42 de Wit, "States Risk Leaving Broadband Money."

43 Grubesic and Mack, *Broadband Telecommunications.*

44 Gallardo and Whitacre, "Look at Broadband Access"; Ali, *Farm Fresh Broadband.*

45 Ali, "Politics of Good Enough."

46 Busby et al., "BroadbandNow."

47 Rachfal, "Broadband Data," 6.

48 Rachfal, "Broadband Data," 1.

49 Engebretson, "USTelecom."

50 Goovaerts, "FCC Locks In."

51 DeFazio, Infrastructure Investment and Jobs Act, §60102(c)(1)(A).

52 McGarry, "Panelists"; Engebretson, "CostQuest."

53 Dawson, "Access."

54 Dawson, "Will the FCC Maps Get Better?"

55 Lefrak and Meyer, "State Calls on Vermonters."

56 Harley, "Maps, Knowledge."

57 Demurring, broadband expert Mike Conlow argues in a blog post that "the Fabric actually seems to have *too many locations*, not too few" (emphasis added). In rural areas, for instance, the census counted 24.5 million housing units, while the broadband map finds 25.5 million broadband serviceable locations. Conlow, "Update."

58 Goovaerts, "CostQuest."

59 Barrett, "FCC Has a New Broadband Map."

60 Data USA, "Koyuk, AK."

61 Ali, "Politics of Good Enough"; Horwitz, *Irony*; Popiel, "Let's Talk About Regulation."

62 A notable exception to this rule is the case of the Rural Digital Opportunity Fund, where the FCC revoked the awards of both Starlink and LTD for failing to meet certain criteria. Goovaerts, "FCC Rejects LTD Broadband."

63 Radcliffe, "Geography and Indigeneity."

64 Harley, "Maps, Knowledge," 132; Wainwright and Bryan, "Cartography, Territory."

65 Duarte, *Network Sovereignty*, 104.

66 Rosenworcel, "Letter to Senator Capito."

67 FCC, "Public Notice."

68 Maruri, "FCC Has Released Its Broadband Map."

69 FCC, "Differences."

70 FCC, "Fabric Challenge."

71 Dawson, "Will the FCC Maps Get Better?"

72 Thomas et al., *Critical Rural Theory.*

73 Teale, "Broadband Maps."

74 Engebretson, "CostQuest."

75 Mosco, *Political Economy.*

76 Mosco, *Political Economy*, 2.

77 Dance et al., "Real-World Costs."

78 Harvey, *Spaces*, 232.

Bibliography

Ali, Christopher. *Farm Fresh Broadband: The Politics of Rural Connectivity.* MIT Press, 2021.

Ali, Christopher. "The Politics of Good Enough: Rural Broadband and Policy Failure in the United States." *International Journal of Communication* 14 (2020): 5982–6004. https://ijoc.org/index.php/ijoc/article/view/15203.

Ali, Christopher. "Putting Rural Communities on the (Broadband) Map." Law and Political Economy Project (blog), 2022. https://broadbandnow.com/research/fcc-underestimates-unserved-by-50-percent.

Ali, Christopher. "Telecommunications Localism: The Fight for Control over Local Communication Networks in the United States." Media International Australia, August 30, 2023. https://doi.org/10.1177/1329878X231198318.

Bargues-Pedreny, Pol, David Chandler, and Elena Simon. "Mapping and Politics in the Digital Age: An Introduction." In *Mapping and Politics in the Digital Age*, edited by Pol Bargues-Pedreny, David Chandler, and Elena Simon. Routledge, 2019.

Barrett, Rick. "The FCC Has a New Broadband Map, and You Can Challenge the Results." *Milwaukee Journal Sentinel*, December 13, 2022. https://www.jsonline.com/story/money/business/2022/12/13/new-fcc-broadband-map-allows-you-to-challenge-service-availability/69704357007/.

Benson, Lisa. "USDA's ReConnect Program: Expanding Rural Broadband." Congressional Research Service, updated December 14, 2022. https://www.everycrsreport.com/files/2022-12-14_R47017_a40122c2bc84a3e34f3fa4bbb1fe9d3eab857016.pdf.

Bode, Karl. “Charter, Comcast Continue to Dominate State Grant Awards.” Community Networks, November 3, 2022. https://communitynets.org/content/charter-comcast-continue-dominate-state-grant-awards#.

Bode, Karl. “How Bad Maps Are Ruining American Broadband.” *Verge*, September 24, 2018. https://www.theverge.com/2018/9/24/17882842/us-internet-broadband-map-isp-fcc-wireless-competition.

Brodkin, Jon. “CenturyLink, Frontier Took FCC Cash, Failed to Deploy All Required Broadband.” *Ars Technica*, January 23, 2020. https://arstechnica.com/tech-policy/2020/01/centurylink-frontier-took-fcc-cash-failed-to-deploy-all-required-broadband/.

Brodkin, Jon. “FCC Helps Charter Avoid Broadband Competition.” *Ars Technica*, June 26, 2020. https://arstechnica.com/tech-policy/2020/06/fcc-helps-charter-avoid-broadband-competition/.

Busby, John, Julia Tanberk, and Tyler Cooper. “BroadbandNow Estimates Availability for All 50 States; Confirms that More than 42 Million Americans Do Not Have Access to Broadband.” *Broadband Now*, May 6, 2022. https://broadbandnow.com/research/fcc-broadband-overreporting-by-state.

Chrisman, Nicholas. “Full Circle: More than Just Social Implications of GIS.” *Cartographica* 40, no. 4 (2005): 23–35. https://doi.org/10.3138/8U64-K7M1-5XW3-2677.

Conlow, Mike. “Second Look: New FCC Maps.” Mike’s Newsletter (Substack newsletter/blog), November 27, 2022. https://mikeconlow.substack.com/p/second-look-new-fcc-maps.

Conlow, Mike. “Update: Comparing the New FCC Fabric to the Census.” Mike’s Newsletter (Substack newsletter/blog), January 22, 2023. https://mikeconlow.substack.com/p/update-comparing-the-new-fcc-fabric.

Cooper, Tyler. “Municipal Broadband 2023: 16 States Still Restrict Community Broadband.” Broadband Now (blog), 2023. https://broadbandnow.com/report/municipal-broadband-roadblocks.

Crawford, Susan. *Fiber: The Coming Tech Revolution—And Why America Might Miss It*. Yale University Press, 2019.

Dance, Gabriel J. X., Tim Wallace, and Zach Levitt. “The Real-World Costs of the Digital Race for Bitcoin.” *New York Times*, April 10, 2023. https://www.nytimes.com/2023/04/09/business/bitcoin-mining-electricity-pollution.html.

Data USA. “Koyuk, AK.” 2022. https://datausa.io/profile/geo/koyuk-ak#demographics.

Dawson, Doug. “Access to the FCC Broadband Maps.” POTS and PANS (blog), December 21, 2022. https://potsandpansbyccg.com/2022/12/21/access-to-the-fcc-broadband-maps/.

Dawson, Doug. "Will the FCC Maps Get Better?" POTS and PANS (blog), January 10, 2023. https://potsandpansbyccg.com/2023/01/10/will-the-fcc-maps-get-better/.

DeFazio, Peter. Infrastructure Investment and Jobs Act. Pub. L. No. HR 3684 (2021). https://www.congress.gov/bill/117th-congress/house-bill/3684/text.

Duarte, Marisa. *Network Sovereignty: Building the Internet across Indian Country*. University of Washington Press, 2017.

Engebretson, Joan. "CostQuest Breaks Its Silence on Broadband Map Issues." *Telecompetitor*, March 3, 2023. https://www.telecompetitor.com/exclusive-costquest-breaks-its-silence-on-broadband-map-issues/.

Engebretson, Joan. "USTelecom: Fixing Carrier Data Only Solves Part of the Broadband Mapping Problem; We Know How to Solve the Other Part." *Telecompetitor*, 2019. https://www.telecompetitor.com/ustelecom-fixing-carrier-data-only-solves-part-of-the-broadband-mapping-problem-we-know-how-to-solve-the-other-part/.

FCC (Federal Communications Commission). "2021 Broadband Deployment Report: In the Matter of Inquiry Concerning Deployment of Advanced Telecommunications Capability to All Americans in a Reasonable and Timely Fashion (GN Docket No. 20-269)." Washington, D.C.: Federal Communications Commission, January 19, 2021. https://docs.fcc.gov/public/attachments/FCC-21-18A1.pdf.

FCC. "2024 Broadband Deployment Report: In the Matter of Inquiry Concerning Deployment of Advanced Telecommunications Capability to All Americans in a Reasonable and Timely Fashion (GN Docket No. 22-270)." February 22, 2024. https://docs.fcc.gov/public/attachments/DOC-400675A1.pdf

FCC. "Differences Between Bulk Fixed Availability Challenge Data and Crowdsource Data." BDC Help Center, November 17, 2022. https://help.bdc.fcc.gov/hc/en-us/articles/10390788241307-Differences-between-Bulk-Fixed-Availability-Challenge-Data-and-Crowdsource-Data.

FCC. "Fabric Challenge Category Code Overview." BDC Help Center, October 11, 2022. https://help.bdc.fcc.gov/hc/en-us/articles/9201263798811-Fabric-Challenge-Category-Code-Overview-.

FCC. "Public Notice: Broadband Data Task Force Announces Recommended Best Practices for Challenges to Updated Broadband Serviceable Location Fabric (WC Docket Nos. 11-10 and 19-195)," 2023. https://docs.fcc.gov/public/attachments/DA-23-69A1.pdf.

Gallardo, Roberto, and Brian Whitacre. "A Look at Broadband Access, Providers and Technology." Purdue University: Center for Regional Development, 2019. https://pcrd.purdue.edu/files/media/008-A-Look-at-Broadband-Access-Providers-and-Technology.pdf.

Goovaerts, Diana. "CostQuest: Missing Sites on FCC Broadband Map." *Fierce Network*, November 8, 2022. https://www.fierce-network.com/broadband/costquest-says-ny-locations-missing-fcc-broadband-map-fraction-total-count.

Goovaerts, Diana. "FCC Locks in Key Vendor for Broadband Map Revamp." *Fierce Network*, March 3, 2022. https://www.fierce-network.com/telecom/fcc-locks-key-vendor-broadband-map-revamp.

Goovaerts, Diana. "FCC Rejects LTD Broadband, Starlink RDOF Bids." *Fierce Network*, August 10, 2022. https://www.fierce-network.com/broadband/fcc-rejects-ltd-broadband-starlink-rdof-bids.

Goovaerts, Diana. "States, NTIA Say Municipal Broadband Laws Won't Delay BEAD Funding." *Fierce Network*, April 17, 2023. https://www.fierce-network.com/telecom/municipal-broadband-laws-probably-wont-delay-bead-funding.

Grubesic, Tony H., and Elizabeth A. Mack. *Broadband Telecommunications and Regional Development*. Routledge, 2017.

Harley, J. Brian. "Maps, Knowledge, and Power." In *Geographic Thought: A Praxis Perspective*, edited by George Henderson and Marvin Waterstone. Routledge, 2009.

Harvey, David. *Spaces of Capital: Towards a Critical Geography*. Routledge, 2001.

Hinton, Danielle, Adi Kumar, Blair Levin, et al. "Are States Ready to Close the US Digital Divide?" McKinsey & Company (blog), June 1, 2022. https://www.mckinsey.com/industries/public-sector/our-insights/are-states-ready-to-close-the-us-digital-divide.

Holpuch, Amanda. "US's Digital Divide 'Is Going to Kill People' as Covid-19 Exposes Inequalities." *Guardian*, April 13, 2020. https://www.theguardian.com/world/2020/apr/13/coronavirus-covid-19-exposes-cracks-us-digital-divide.

Horwitz, Robert Britt. *The Irony of Regulatory Reform: The Deregulation of American Telecommunications*. Oxford University Press, 1991.

Kahan, John. "It's Time for a New Approach for Mapping Broadband Data to Better Serve Americans." Microsoft on the Issues (blog), April 8, 2019. https://blogs.microsoft.com/on-the-issues/2019/04/08/its-time-for-a-new-approach-for-mapping-broadband-data-to-better-serve-americans/.

Lefrak, Mikaela, and Tedra Meyer. "State Calls on Vermonters to Help Correct Federal Broadband Map." Vermont Public, 2022. https://www.vermontpublic.org/show/vermont-edition/2022-12-20/state-calls-on-vermonters-to-help-correct-federal-broadband-map.

Maruri, Katya. "The FCC Has Released Its Broadband Map—But Work Is Far from Over." GovTech, November 18, 2022. https://www.govtech.com/network/the-fcc-has-released-its-broadband-map-but-work-is-far-from-over.

McGarry, David B. "Panelists at Broadband Breakfast Event Urge the FCC Mapping Fabric Be Made Public." Broadband Breakfast (blog), September 23, 2022. https://broadbandbreakfast.com/2022/09/panelists-at-broadband-breakfast-event-urge-the-fcc-mapping-fabric-be-made-public/.

McMahon, Rob, Trevor James Smith, and Tim Whiteduck. "Reclaiming Geospatial Data and GIS Design for Indigenous-Led Telecommunications Policy Advocacy: A Process Discussion of Mapping Broadband Availability in Remote and Northern Regions of Canada." *Journal of Information Policy* 7 (2017): 423–49. https://doi.org/10.5325/jinfopoli.7.2017.0423.

Meinrath, Sascha D. "Broadband Availability and Access in Rural Pennsylvania." Center for Rural Pennsylvania, June 2019. https://www.rural.palegislature.us/broadband/Broadband_Availability_and_Access_in_Rural_Pennsylvania_2019_Report.pdf.

Mosco, Vincent. *The Political Economy of Communication*. SAGE, 2009.

Murton, Galen. "Power of Blank Spaces: A Critical Cartography of China's Belt and Road Initiative." *Asia Pacific Viewpoint* 62, no. 3 (2021): 274–80. https://doi.org/10.1111/apv.12318.

NTIA (National Telecommunications and Information Administration). "Biden–Harris Administration Announces State Allocations for $42.45 Billion High-Speed Internet Grant Program as Part of Investing in America Agenda." Press release, June 26, 2023. https://www.ntia.gov/press-release/2023/biden-harris-administration-announces-state-allocations-4245-billion-high-speed.

Popiel, Pawel. "Let's Talk About Regulation: The Influence of the Revolving Door and Partisanship on FCC Regulatory Discourses." *Journal of Broadcasting and Electronic Media* 64, no. 2 (2020): 341–64. https://doi.org/10.1080/08838151.2020.1757367.

Rachfal, Colby Leigh. "Broadband Data and Mapping: Background and Issues for the 117th Congress." Congressional Research Service, updated May 19, 2021. https://sgp.fas.org/crs/misc/R45962.pdf.

Radcliffe, Sarah A. "Geography and Indigeneity I: Indigeneity, Coloniality and Knowledge." *Progress in Human Geography* 41, no. 2 (2017): 220–29. https://doi.org/10.1177/0309132515612952.

Robinson, Vickie. "Addressing Racial and Digital Inequity." Microsoft on the Issues (blog), June 2, 2021. https://blogs.microsoft.com/on-the-issues/2021/06/02/racial-digital-inequity-airband-broadband-access/.

Rosenworcel, Jessica. "Letter to Senator Capito." Federal Communications Commission, 2023. https://docs.fcc.gov/public/attachments/DOC-391046A2.pdf.

Specht, Doug, and Anna Feigenbaum. "From the Cartographic Gaze to Contestatory Cartographies." In *Mapping and Politics in the Digital*

Age, edited by Pol Bargués-Pedreny, David Chandler, and Elena Simon. Routledge, 2019.

Taglang, Kevin. "CenturyLink and Frontier Miss FCC Connect America Fund Broadband Deployment Milestones." Benton Foundation (blog), January 29, 2019. https://www.benton.org/headlines/centurylink-and-frontier-miss-fcc-connect-america-fund-broadband-deployment-milestones.

Teale, Chris. "Broadband Maps for Indian Country Called 'Horrible,' 'Egregious' and 'Negligent.'" *Route Fifty*, January 10, 2023. https://www.route-fifty.com/infrastructure/2023/01/broadband-maps-indian-country-called-horrible-egregious-and-negligent/381675/.

Thomas, Alexander R., Brian Lowe, Gregory Fulkerson, and Polly Smith. *Critical Rural Theory: Structure, Space, Culture*. Lexington, 2013.

Wainwright, Joel, and Joe Bryan. "Cartography, Territory, Property: Postcolonial Reflections on Indigenous Counter-Mapping in Nicaragua and Belize." *Cultural Geographies* 16, no. 2 (2009): 153–78. https://doi.org/10.1177/1474474008101515.

Whitacre, Brian, and Christina Biedny. "A Preview of the Broadband Fabric: Opportunities and Issues for Researchers and Policymakers." *Telecommunications Policy* 46, no. 3 (2022): 102281. https://doi.org/10.1016/j.telpol.2021.102281.

Wicker, Roger. S.1822: Broadband Deployment Accuracy and Technological Availability Act, or The Broadband DATA Act. Pub. L. No. PL No. 116-130. 2020. https://www.congress.gov/bill/116th-congress/senate-bill/1822.

Wilson, Matthew W. *New Lines: Critical GIS and the Trouble of the Map*. University of Minnesota Press, 2017.

Wit, Kathryn de. "States Risk Leaving Broadband Money on the Table." Pew (blog), April 22, 2022. https://pew.org/3v3fXLv.

11

The Virtual Fire

CINDY KAIYING LIN

One late afternoon, I was with government remote sensing scientists and computer engineers in the National Institute of Aeronautics and Space, some thirty minutes outside of downtown Jakarta, the capital city of Indonesia. We were on the last day of a two-day-long workshop on peatland fire detection methods organized by what was then called the National Institute of Aeronautics and Space (hereafter the National Institute), with guests from two well-known US federal science agencies and several other Indonesian ministries, research institutions, and agriculture companies.[1] The workshop was focused on evaluating existing global fire detection instruments and their remotely sensed outputs, called hotspot data.

Parwati Sofan, an Indonesian remote sensing scientist from the National Institute, was showing us what hotspot data could sense—and what they could not. Parwati's first satellite image pictured a scene of fire in Central Kalimantan, a rural province in Indonesia that has experienced massive peatland fires over the past few decades. It showed a large brownish black patch with bright orange edges labeled as peatland fires. She next flashed a second satellite image of a surface wildfire engulfed in thousands of red dots—the conventional representation of hotspot data. These surface wildfires, Parwati noted, are remotely sensed thermal anomalies that existing global fire detection instruments would register as abnormally hot. However, peatland fires in the first image would not be detected to the same extent. These fires can evade detection because they smolder underground and slowly transverse the depths of peatland, rendering their patterns difficult to observe from the sky. The inability to detect such fires has become pressing for the global environmental community because these fires emit an equivalent of 5 percent of anthropogenic-driven greenhouse gas emissions, presenting a climate and health risk that the United Nations has advised requires greater monitoring and control.[2]

With growing demand for more extensive monitoring, Parwati's workshop participants aimed to design new remote sensing technologies for peatland fire detection. For Parwati, it was important to not just reuse existing instruments but also to redesign how they detect peatland fires. She explained that because peatland fires could reignite after being extinguished, firefighters flood burning peat with water for weeks. This placed tremendous stress on already limited water resources and firefighting labor in rural Indonesia, making it necessary to rethink how hotspot data are used. This is when Parwati thought, "That is why we do not need daily hotspot data of peatland fires."

Parwati's proposal to reimagine rural mediation of peatland fires moves away from prior and long-standing uses of remote sensing data infrastructures to either expand a user base throughout Indonesia or regard rural Indonesia as a site for surveillance. Peatlands in so-called rural Outer Island Indonesia have become what the editors of this volume characterize as spaces that are media intensive.[3] Most telling is the installation of more than 729 water loggers in 280 pulpwood and palm oil companies throughout rural Indonesia in December 2019.[4] These loggers were initially designed to monitor water in peatland every ten minutes, sending data to relevant government ministries and agencies in charge of peatland protection and restoration efforts.[5] Collected data would then be showcased in several flashy formats, ranging from dashboards on several LCD screens to bold data visualizations. Geographer Rini Astuti argues that the use of environmental surveillance is part of Indonesia's disciplinary power to regulate plantation permit holders—or what the Ministry of Environment and Forestry has called the "corrective era."[6] This era includes penalizing palm oil companies who illegally clear peatland with fires. As such, rural Indonesia is not distant from the control of Jakarta's central government and remains one of the most policed plantation economies since Dutch colonization.[7]

Parwati's demand for a different remote sensing system, then, speaks to a growing literature in science and technology studies (STS) and the history of science about the role of data and technology in knowing and experiencing the environment. Anthropologist Stefan Helmreich shows how oceanographers equipped with remotely operated vehicles begin to operate as a single body, creating an intimate sensing experience with the ocean's depths.[8] Similarly, the remote sensing redesigns that Parwati and her research team introduced changed the spatiotemporal bounds of remote sensing instruments, producing what historian of science Etienne Benson calls the "virtual field." Like field scientists, who have to gain a keen sense of their terrain up close, developed through tedious field surveys and sometimes forced interactions with residents, Benson argues, the networked scientists of today also access the field through the laboratory or by computational models, be it cleaning and aggregating remote sensing data on a centralized platform or maintaining and repairing water loggers.[9] Networked scientists like Parwati carry their virtual fields with them as laptops and software, with a special attention to the particularities of peatland fires and their mediation (or lack thereof) through fire hotspots and other existing fire detection instruments.

Today, hotspots challenged Parwati and her Indonesian colleagues not only because, as historian of science Paul Edwards argues, "no one lives in a 'global' climate," but also because, like many other tools of projection, hotspots have "degraded" in the face of an increasingly volatile fire landscape in rural Indonesia: tropical peatland fires.[10] Managing the rift between climate destabilizations and local sociopolitical interests, fire hotspots no longer suffice for locating peatland fires. As described to me by one government researcher, 50 percent of these hotspot data in rural Indonesia are false alarms; firefighters sighted no fires in locations of fire hotspot data in various provinces.

Indeed, while the changes in Indonesia's fire regimes cannot be readily observed by global remote sensing instruments, Parwati shows how a careful unsettling of what global visions these instruments prioritize in prior renditions is necessary for designing new forms of rural mediation.[11] Virtual mediations of peatland fires in rural Indonesia, then, are not just a result of US technological expansion into the third world or centralized state surveillance into rural communities but also a virtual field site for Parwati and her colleagues to create political and social possibilities for both rural Indonesia and the scientific professions they belonged to.[12]

Drawing on ethnographic fieldwork conducted in the summers of 2017 and 2018 and 2019–2020, as well as oral histories and archival research, I show how government scientists and engineers (hereafter referred to as government researchers) develop new methodologies of fire detection when hotspots no longer index the location of peat and peatland fires. Then I turn to how the use of US hotspot technology focused on *expanding* its use in remote areas of Indonesia before showing how hotspot usage became *centralized* to specific institutes made responsible for analyzing hotspot data. Last, I describe how government researchers began to build new peatland fire detection algorithms that are focused on combining different visions of remote Indonesia. In my concluding remarks, I reflect on why it is important to foreground practices and histories of rurality and rural places as central sites and successes for global remote and environmental sensing and instrumentation programs.

Expanding Users of Fire Hotspots

Hotspots have been integral to fire management efforts in Indonesia since the 1980s. One of the earliest documented uses happened in 1983, when remote sensing scientist J. P. Malingreau and his colleagues used the National Oceanic and Atmospheric Administration's (NOAA) advanced very high-resolution radiometer (AVHRR) instrument for detecting fire events at scale. Their site of study was East Kalimantan, where fires damaging forests and peatlands went, according to these scientists, "unabated and unmonitored for close to three months in early 1983."[13] While the resolution of this instrument was believed to be too coarse for observing forest fires, these scientists show that the AVHRR thermal channel can "effectively monitor individual heat sources in the tropical forest in East Kalimantan."[14]

This initial foray in East Kalimantan became a test-bed for justifying the *expanded* use of hotspot technology in rural Indonesia. These justifications rest first on how these scientists regard US fire detection instruments as capable of counting fires. When the NOAA AVHRR detected thermal anomalies that reached 322 Kelvin, the pixels in the remote sensing imagery became saturated. They are then mapped as fire points and become hotspots, forming the basis for later versions of fire detection instruments carried by other federal science agencies such as the National Aeronautics and Space Administration (NASA). The discretization of fire as points allowed remote sensing scientists to treat the dynamic phenomena of both naturally occurring and man-made fires as countable. Remote sensing scientists can now chart historical fire trends, making them similar statistical objects as those that were used to measure economic growth and health rates, for instance.

Second, hotspots became early-warning response technologies for firefighting and forest conservation efforts in Indonesia. For instance,

Malingreau and his colleagues used the "movement of the pixel from one day to another" to show how "regions of fire concentration indicate new influxes of settlers along newly opened roads or in new settlements."[15] In other words, hotspots became "good indicators of frontier areas" that, when accompanied with field surveys, can provide assessments of deforestation.[16] In establishing the connection between hotspots and new resource frontiers, Malingreau and his colleagues argue that "fires can be used as alarms for identifying incursions into protected areas."[17] Hotspots became evidence of fires—but they also became central to detecting frontier relations in rural Indonesia.

Because hotspots are deployed as indicators of both fires and resource incursion in remote areas, Indonesia's central government agencies, supported by foreign technical aid, were eager to remake rural Indonesians into hotspot data analysts as early as the 1990s. In 1990, the Ministry of Environment and Forestry in Jakarta expressed interest in using the NOAA AVHRR technology. In 1994, a receiving station was set up at the provincial forestry office in Palangkaraya, Central Kalimantan, Borneo—a city in a mostly forested rural province of Indonesia. Shortly after, three additional aid efforts from Japan, the European Union, and Germany were aimed at decentralizing the collection, processing, and analysis of hotspot data from Jakarta to rural Indonesia before 1997.[18] They reasoned that equipping rural Indonesians directly with firefighting expertise, equipment, and hotspot information meant that those who were most directly affected by fires could act on them quickly.[19] They also cited greater access to geographic information systems software and microcomputers as aiding rural Indonesians to map their own hotspots instead of relying on a bulky mainframe computer in Jakarta's offices.[20]

The use of hotspot technology for fire management and the training of rural Indonesians in the use of this technology suggest that expansion is key to this recent history of rural mediation. Regardless of whether the hotspot technology can fully encompass the complexities of peatland fires, this early deployment of hotspot technology was seen as promissory for both firefighting and conservation efforts in Indonesia. As such, rural Indonesia was media intensive even before peatland water loggers were installed to detect such fires, showing how rurality is unmade and remade when central government agencies expand remote sensing efforts outward from Jakarta to remote areas.

Centralizing Analysis to Jakarta: The Closed World

Such efforts to decentralize the analysis of hotspot data were short-lived. When research institutions and forestry offices in the Outer Islands chose their own criteria for how to process and map fire hotspot data, they also produced conflicting hotspot numbers that concerned foreign experts and researchers in Jakarta. This concern was documented in a bilateral partnership between the European Commission and the regional office of the Ministry of Forestry and Estate Corps in South Sumatra, where foreign experts reported that the adoption of different temperature thresholds to detect fires, the mistaking of hot and bright objects as fires by hotspot sensors, and the type of processing algorithm are some of the reasons why hotspot numbers may differ from fires on the ground.[21] The authors of this report also point out that operators in rural Indonesia are "subjective" in the choice of their algorithms and observation of fires.[22] Despite this subjectivity, they urged that conflicting hotspot numbers are "comparable and all can be accepted as accurate."[23]

Still, US federal agencies continue to promote the use of their global fire-sensing instruments throughout the developing world. As documented by historian Megan Black, the US Department of the Interior promoted the use of satellite data in developing countries while also facilitating foreign mineral extraction for US firms.[24] While fire detection instruments were not promoted by the Department of the Interior, some of these instruments were carried by the same satellites used for mineral prospecting. As such, the US effort to scale up remote sensing in developing countries also meant that global fire detection instruments compromised on the specificity of their fire products. For instance, limited data storage meant that moderate-resolution images and a more conservative fire detection algorithm are used.[25]

The compromised quality of data and desire to reduce operators' subjectivity eventually encouraged Jakarta's government experts to establish the National Geographic Information System forum to standardize mapping practices for the nation from 1991.[26] By 2007, Presidential Decree 85 was issued to establish this system and made the National Institute in Jakarta the primary data recipient and distributor for hotspot data. The centralization of fire hotspot data processing and mapping efforts in the National Institute meant that the rural mediation of fires shifted from expanding hotspot use across regions in Indonesia to contracting its analysis solely to Jakarta. Consequently, the National Institute became responsible

11.1 Environment Law Enforcement control room in Indonesia, 2021. Source: SEADA.

for communicating the necessary information to emergency responders in rural Indonesia today.

This centralized responsibility also meant that remote sensing technologies became ubiquitous in Jakarta's ministerial offices. For instance, several ministerial offices called their control room a war room, or they designed interfaces of their early warning response systems (figure 11.1) according to the cult classic film *War Games* (1983)—a film about an American teenage hacker who takes command of the North American Aerospace Defense Command mainframe system. These cybermilitaristic aesthetics and design choices are no coincidence and can be traced to early proto-AI systems (more specifically, symbolic AI) developed in the 1950s for military think tanks in North America.[27] Carnegie Mellon University computer scientist Allen Newell and political scientist Herbert Simon, both early proponents of bringing together cognitive science and computing, worked together at the military think tank, the RAND Corporation, to simulate the organizational behavior of an air defense control center.[28] Both researchers also developed experiments to test how people responded to different simulated scenarios, undergirded by the assumption that a person is an "information processing system."[29]

Central to these control room experiments is what historian Paul Edwards has called the "closed world." The post–World War II United States witnessed cybernetics, management and organizational theory, and military strategy come together to produce both material and metaphorical senses of closure. The simulated control center of Newell and Simon joins a series of large-scale technological enterprises that embodied the "archetypal closed-world space: enclosed and insulated, containing a world represented abstractly on a screen, rendered manageable, coherent, and rational through digital calculation and control."[30] Such closed and concentrated forms of surveillance and control enable data analysts in the United States and urban Indonesia to monitor fires in rural Indonesia, flattening geographically different and unequal regions and prescribing what the editors of this volume might call "a singular, static, homogeneous, universal, and finished object" of rural fires.

The stakes of detecting peatland fires became prescient as fire regimes in Indonesia changed rapidly. Originally a practice for native swidden farmers to clear and regenerate forests with fire, fires have now increased as a result of the migration of farmers, the widespread expansion of oil palm and rubber plantations, which replace native plants, and extreme weather events. Furthermore, palm oil companies consistently evade responsibility for clearing protected forest and peatlands. Activists, environmental scientists, and scholars of plantations have identified these companies as one of the key drivers of the increased frequency and severity of peatland fires and the displacement of Indigenous groups.[31]

All these factors combine to change the historical patterns of fire in rural Indonesia such that there can be no single point of intervention that can be traced and fixed. Indeed, in 2015, Indonesia's President Joko Widodo created a series of measures and policies to move away from containing fires to preventing them, having noted the difficulties for emergency responders to locate fires based on hotspot data alone. Regardless of the actual drivers of such increased fire frequencies, hotspot inaccuracies have begun to further challenge the legitimacy of the National Institute.

Beyond Expansion or Centralization

To bolster its legitimacy, government researchers at the National Institute such as Parwati have begun to construct alternative forms of rural mediation. This includes exploring designs that go beyond expanding users of hotspot data or concentrating analytic power to state authorities in Jakarta. This was made clear to me at Parwati's co-organized workshop described above. The workshop was centered around a proposal made by a visiting scholar from a prominent US federal science agency whom I will call Bart. Bart introduced a fire hotspot product that can detect smaller-size fires and emphasized the importance of a heavily staffed control room to observe and distribute hotspot data throughout the United States, with a single operator always on call. The operator, in this plan, is tasked with monitoring real-time data of potential emergencies throughout the United States such as tornadoes and fires on a 24/7 basis. Bart described this act of monitoring as similar to tracking "air traffic," where there is "always someone at the fire desk, even during holidays." These operators are tasked with monitoring data coming in every five minutes because "fire doesn't choose." While operators in Jakarta also monitor emergencies daily, their focus is on producing and distributing daily reports instead of providing round-the-clock support to emergency responders. A debate ensued on what was the right extent of monitoring and how such fires could be best managed.

At this moment, Parwati interjected, "Why should we monitor hotspots every day?" Parwati's question was followed by a description of how peatland fires cannot be traced to discrete hotspots. Peatland fires rarely die, and once they start burning on the surface, Parwati continued, they can move and burn underground if enough fuel such as leaf litter is present. Sometimes these fires also reignite on the surface again, making it difficult to figure out who or what has caused these fires, or when they happened. Indeed, the nonlinear nature of peatland fires set off the development of new fire danger rating systems that were aimed at predicting the likelihood of fires based on several drivers of fire: the fuel load, weather, geography, and socioeconomic factors of people who lived in a specific area.[32]

The way peatland fires spread underground and last for some time made monitoring fires from Jakarta on a round-the-clock basis difficult, if not impossible. Parwati described how long peatland fires lasted, noting that some of them burned for weeks. "The longest duration was six months!" Parwati exclaimed before pausing to take in the shocked faces of audience members. Because these fires can subsist for a long time underground, they confuse hotspot sensors when warm ground temperatures trigger sensors but are not visible to the human eye. Firefighters learned to shake and loosen the dead leaf litter with their feet and then check to see if smoke emerged from the ground.

In place of using hotspot for navigation, firefighters now inject their hoses into areas that reveal repeated patterns or clusters of fire hotspots (more than three days). They insert them into smoldering peatland, hoping to cool it down and reduce the spread of fires.[33] Even then, it is difficult to know where these hoses should be injected. These efforts, Parwati explained, are not only resource intensive but also strenuous on firefighters, who already work in life-threatening conditions and must deal with the contingency of peatland fires.

One could interpret the issue with hotspot systems from the United States as a failure to develop an appropriate system for Indonesia. But the issue, as Parwati presents, is that these systems *did not catch up* with the shifts in fire regimes in rural Indonesia—a regime that has increasingly been affected by industrial monocrop expansion and the clearing of peatland. Already the expansion of oil palm plantations has destroyed 24 million hectares of Indonesia's rain forest between 1990 and 2015, leaving only 52 percent of the rainforest intact.[34]

Parwati's choice to move away from monitoring fire hotspots every day breaks the truce formed among scientists and policymakers on the role of fires in Southeast Asia. Daily hotspot reports have been instrumental to the policing of both smallholder farmers and swidden cultivators, wherein environmental law enforcers use hotspots to locate and arrest or fine perpetrators of fires. This policing is not new. Over the colonial and postcolonial era in Indonesia, scientists and decision-makers have largely blamed uncontrolled fires on swidden cultivation, a rotational form of agriculture mostly practiced by ethnic minorities in Outer Island Indonesia.[35] This happens despite growing evidence that swidden cultivation is a more productive use of the forests than commercial logging "in terms of the size of the population supported."[36] Apart from swidden cultivation, smallholder farmers have also been targeted for illegally clearing peatlands to plant cash crops.[37] Environmental groups have pointed out that smallholder farmers bear the brunt of state punishment more than palm oil corporations.[38]

By refusing daily hotspots, Parwati commits to adapt to a fire regime that is increasingly uninhabitable and insuppressible, rather than work against it. Additionally, Parwati's choice of refusing daily hotspots flipped not only the centralized power and surveillance of control rooms but also the treatment of rural Indonesia as a site of criminality by state authorities. Next I turn to how Parwati's redesigned global fire detection algorithm is not simply a localized rendition of a global technology but instead a careful inquiry into how the rural animates new visions in such technologies.

Looking at the Edges

In STS and anthropological studies of scale, scholars discuss how practitioners often use scale comparatively in the tech industry.[39] These comparisons ranged from contrasting the size of organizations to differentiating manual labor from algorithmic prediction. Central to this tension, as anthropologist Nick Seaver has pointed out, is the supposed commonsense division between impersonal scale and careful detail, where scale-up operations are regarded as the opposite of care and specificity. Likewise, the expansion of US global fire detection instruments into Indonesia can be interpreted as a project of scalability. As Anna Tsing notes, "Scalable projects are those that can expand without changing," and as I have shown, they have also "banish[ed] the meaningful diversity" of peatland fires in rural Indonesia.[40] Even more literally, the use of hotspots to surveil rural farmers by central state authorities reflects a "pixelated quality of the expansion-oriented world," which Tsing describes as the ability to zoom in and out of an image without changing the shape and stability of the pixel, and in this context, the fire hotspot.[41]

Yet as Nick Seaver argues, and as Parwati shows, technologists reformulate the oppositions between scale and care by "reimagining the terms of their relationship."[42] Tech enterprises, according to Seaver, have begun to embrace personal contact with customers and manual data work in their business expansion. For many of these enterprises, scaling up can be regarded as separate from doing careful handcrafted work. Similarly, Parwati's redesign of a global remote sensing instrument does not start with a preoccupation that scale is a problem. Parwati challenges the "metronormativity" inherent in the design of global technologies by foregrounding rural Indonesia as a site that is not "small and disengaged" but central to a changing fire regime that has catastrophic effects across the world.[43] With this knowledge, she carefully chooses and combines multiple channels into a single global remote sensing infrastructure to detect peatland fires.

Channels in global remote sensing instruments such as NOAA AVHRR are initially designed to detect distinct wavelengths of the electromagnetic spectrum or light. When NOAA AVHRR first operated in the early 1980s, remote sensing scientists believed that Channel 3 (middle infrared, wavelength 3.5 microns) was best suited for capturing the unique spectral signature of fires globally. Channel 3 is said to have an "atmospheric window" that is extra sensitive to the energy released from flaming wildfires, which is higher than the surrounding area.[44] In this way, remote sensing instruments allow scientists to apply physics and universalize understanding of fires across regions rather than design from and with the specificities of the object.

This resulted in the omission of smaller fires, which later versions of global fire detection instruments focused on detecting, such as NASA's active fire data product called MOD14, which uses confidence values to calculate whether a hotspot is a false alarm, and NASA's Visible Infrared Imaging Radiometer Suite detection instrument, which has higher spatial resolution sensors.[45] While earlier instruments assume that all fires shared a unique spectral property, later renditions attempted to correct such an assumption by using contextual information and higher resolutions to confirm smaller fires. However, the smoldering phase of peatland fires remains one of the most difficult to detect, even with new instruments and data products, because the new products do not distinguish between different wildfire phases.[46]

Bart explained further that because these global fire detection instruments are designed to detect fires in multiple settings, they cannot be specific to any single region. He acknowledged that this risks nondetection of peatland fires in rural Indonesia, which Indonesian researchers wanted to avoid. To them, scaling up is not inherently oppositional to care or detail. Nor was it about using higher-resolution imagery or considering the surrounding pixels of a fire. For Parwati, sensing with a global remote sensing instrument means using multiple visions to detect the changing phases of peatland fires instead of solely focusing on one channel to detect one distinct feature of fires.

Parwati confidently described how she developed a fire detection algorithm to detect the smoldering phases of peatland fires in rural Indonesia. This algorithm, the tropical peatland fire combustion algorithm, or ToPeCAI, is designed to detect nonbright objects with cool temperatures. The algorithm is based on Landsat data, the longest-running US earth observation satellite program, and a free source of data that Parwati regularly tinkered with in her spare time. During one such session, Parwati noticed that the area surrounding the edge of a flaming fire in Central Kalimantan registered as hot, and after further investigation with her colleagues, she realized that particular channels in Landsat could detect both the high-temperature flaming phases and the low-temperature smoldering phases of peatland fires. This was a surprise because Landsat was not the go-to instrument for fire detection because it lacked the middle infrared bands traditionally used to detect fires.[47]

Parwati's chance discovery reveals how early twentieth-century ideas of fire suppression and firefighting have shaped the choice of remote sensing channels for fire detection across the world. These fire detection instruments prioritized detecting high-temperature visible flames over low-temperature smoldering fires, largely because fires in the United States were regarded as a disaster to be fought against and suppressed.[48] This model of fire containment has since been imported from the United States to Indonesia to rate the danger of visible fires, without due consideration of the range of fires Indonesia might experience.[49] Parwati's willingness to pursue unconventional instruments of fire detection meant that she moved away from prior renditions of remote sensing and understandings of fire—to either expand the use of a single channel to detect and control fires, or to conduct more extensive surveillance of rural Indonesia.

Today, Parwati's algorithm is used to validate the accuracy of fire detection instruments developed by NASA and NOAA. While in the eyes of foreign scientists peatland fires were once "useful as a foil for the norm," peatland fires in rural Indonesia have now become the benchmark for deciding how global other fire detection instruments can be.[50] Parwati's algorithm reveals not only that remote sensing instruments have "multiple visions" within a single infrastructure, but also how rurality and development status changes in significance for deciding how global an environment is.[51] That is, the choice of using remote sensing to detect fires in rural Indonesia was once developed with a view that all fires in this region could be readily captured into a single channel and characterization of fire, without due consideration that attending to specific fires in rural Indonesia can also assist in the development and design of a global earth observation infrastructure. In this way, the ToPeCAI algorithm counters the impulse for universalizing a single understanding of fires, instead centering these Indonesian fires in rethinking both metropolitan and imperialistic imaginaries of what the rural constitutes: either a site for resource extraction or a place of criminality.

Conclusion

The use of global remote sensing technology does not simply abstract and quantify social relations held between urban Jakarta and rural Indonesia; it also forces government researchers to reflect on what is possible and imaginable through their own material and ethical interventions. As historians of science and STS scholars have noted, the advent of remote sensing has shaped the ways we constitute global environment—and, I would add, what it prioritizes. Fire hotspots have been historically used by foreign experts to expand a user base across the world or by local officials to surveil rural Indonesia, but this status is currently contested through the failed detection of tropical peatland fires.

Additionally, peat and peatland fires have given the institute and its researchers opportunity to challenge the privileged position North American fire-sensing instruments have. As Stephen Pyne argues, "Fire is what its circumstances make it."—"it synthesizes its surroundings"—and in the case of Indonesia, the technologies designed to observe them.[52] With the advent of monoculture palm oil plantations and the widespread use of fires as a cheap and easy way to clear vegetation, debris, and regrowth, new fire regimes in rural Indonesia are undeniably different from surface wildfires elsewhere. But instead of developing a new fire monitoring system, remote sensing scientists in Indonesia chose to question the truism of global and local, urban and rural, and scale and detail by showing how existing global earth observation programs can account for peatland fires in unexpected ways.[53] As such, remote sensing allows for new modes of mediation and sensing from these seemingly peripheral sites of resource extraction. Far from disconnected and illegible, peat and peatland fires in rural Indonesia have generated new scientific communities and quantification practices.

Notes

1 Researchers are now part of the Ministry of Research and Technology.
2 Koudenoukpo, "As Wildfires Increase."
3 Comprising the islands of Sumatra, Borneo, Bali, Lombok, Celebes, New Guinea, Timor, Ternate, and the Moluccas.
4 Astuti, "Governing the Ungovernable," 381–91.
5 *Peatland restoration* refers to the rewetting of peatlands to decrease its flammability during dry seasons in Indonesia.
6 Astuti, "Governing the Ungovernable," 382.
7 Stoler, *Capitalism and Confrontation.*
8 Helmreich, *Alien Ocean*, 142.
9 Benson, "Virtual Field."
10 Edwards, *Vast Machine*, 4; Petryna, *Horizon Work*, 3.
11 Camprubí and Lehmann, "Scales," 1–5.
12 Haraway, "Promises of Monsters," 64.
13 Malingreau et al., "Remote Sensing," 314.
14 Malingreau et al., "Remote Sensing," 315.
15 Malingreau et al., "AVHRR," 859

16 Malingreau et al., "AVHRR," 859.
17 Malingreau et al., "AVHRR," 858.
18 Dennis, *Review of Fire Projects,* 10, 25–32.
19 Dennis, *Review of Fire Projects,* 26.
20 Dennis, *Review of Fire Projects,* 56.
21 Anderson and Imanda, *Vegetation Fires,* iv–v.
22 Anderson and Imanda, *Vegetation Fires,* iv.
23 Anderson and Imanda, *Vegetation Fires,* 1.
24 Black, *Global Interior,* 148–82.
25 Li et al., "Review," 15.
26 Putra et al., "Toward the Evolution," 263.
27 Katz, *Artificial Whiteness*; Edwards, *Closed World*; Hayles, *How We Became Posthuman,* 464; Bowker, "How to Be Universal"; Haraway, "Manifesto for Cyborgs."
28 Carnegie Mellon University Digital Collections, Allen Newell Collection.
29 Carnegie Mellon University Digital Collections, Allen Newell Collection.
30 Edwards, *Closed World,* 104.
31 For peatland fires, see Ansori, "Fingertips"; Goldstein et al., "Beyond Slash-and-Burn." For displacement, see Chao, *In the Shadow,* 11–15.
32 Sanjaya et al., "Indonesia Fire Danger Rating System," 1–5.
33 Brogan, "Magical Fire Suppressant."
34 UNDP China, "Mapping," 2.
35 Neale et al., "Eternal Flame," 116.
36 Dove, "Theories," 85.
37 Purnomo et al., "Fire Economy," 21–31.
38 Jacques, "Small Farmers Jailed."
39 Avle et al., "Scaling Techno-Optimistic Visions," 237–38; Hong, "Technofutures," 1946.
40 Tsing, "On Nonscalability," 507.
41 Tsing, "On Nonscalability," 507.
42 Seaver, "Care and Scale."
43 Halberstam, *In a Queer Time,* 36; Wang, *Blockchain Chicken Farm,* 7.
44 Riggan et al., "Estimating Fire Properties," 2
45 Wooster et al., "Satellite Remote Sensing"; Tansey et al., "Relationship."
46 Read more in Giglio et al., "Enhanced Contextual Fire Detection Algorithm," 279–80.
47 Cartalis, "Introduction."
48 Petryna, *Horizon Work,* 79–80.
49 Groot et al., "Development," 165–80.
50 Canguilhem and Jaeger, "Monstrosity," 35.

51 Benson, "One Infrastructure," 846.

52 Pyne, "Problems, Paradoxes," 271.

53 In critiques of the rural/urban divide, urban researcher Terry McGee coins the concept of *desakota* ("city-village" in the Indonesian language) to describe places in Asia where rural and urban forms of land use comingle to show how the idea of urban life encroaching on the small-size, personal rural spaces is a division that largely frames Western society. McGee, "The Emergence of Desakota Regions in Asia: Expanding a Hypothesis."

Bibliography

Anderson, Ivan P. and Ifran D. Imanda. "Vegetation Fires in Indonesia: The Interpretation of NOAA Derived Hotspot Data." Pamphlet. Forest Fire Prevention and Control Project, 1999.

Ansori, Sofyan. "The Fingertips of Government: Forest Fires and the Shifting Allegiance of Indonesia's State Officials." *Indonesia* 108 (2019): 41–64.

Astuti, Rini. "Governing the Ungovernable: The Politics of Disciplining Pulpwood and Palm Oil Plantations in Indonesia's Tropical Peatland." *Geoforum* 124 (2021): 381–91. https://doi.org/10.1016/j.geoforum.2021.03.004.

Avle, Seyram, Cindy Lin, Jean Hardy, and Silvia Lindtner. "Scaling Techno-Optimistic Visions." *Engaging Science, Technology, and Society*, vol. 6 (2020). https://doi.org/10.17351/ests2020.283.

Benson, Etienne. "One Infrastructure, Many Global Visions: The Commercialization and Diversification of Argos, a Satellite-Based Environmental Surveillance System." *Social Studies of Science* 42, no. 6 (2012): 843–68. https://doi.org/10.1177/0306312712457851.

Benson, Etienne. "The Virtual Field." *Technosphere Magazine*, November 15, 2016. https://www.anthropocene-curriculum.org/contribution/the-virtual-field.

Black, Megan. *The Global Interior: Mineral Frontiers and American Power.* Harvard University Press, 2018.

Bowker, Geof. "How to Be Universal: Some Cybernetic Strategies, 1943–70." *Social Studies of Science* 23, no. 1 (1993): 107–27. https://doi.org/10.1177/030631293023001004.

Brogan, Caroline. "Magical Fire Suppressant Kills Zombie Fires 40% Faster than Water Alone." Imperial College London, March 12, 2021. https://www.sciencedaily.com/releases/2021/03/210312121306.htm.

Camprubí, Lino, and Philipp Lehmann. "The Scales of Experience: Introduction to the Special Issue *Experiencing the Global Environment.*" *Studies in*

History and Philosophy of Science Part A 70 (2018): 1–5. https://doi.org/10.1016/j.shpsa.2018.05.003.

Canguilhem, Georges, and Therese Jaeger. "Monstrosity and the Monstrous." *Diogenes* 10, no. 40 (1962): 27–42. https://doi.org/10.1177/039219216201004000.

Carnegie Mellon University Digital Collections. Allen Newell Collection. Last modified May 8, 2023. https://digitalcollections.library.cmu.edu/cmu-collection/allen-newell.

Cartalis, Constantinos. "Introduction to Thermal Remote Sensing." Paper presented at the Seventh Advanced Training Course on Land Remote Sensing, 2017. https://eo4society.esa.int/wp-content/uploads/2021/04/2017Land_D2T3- P_Cartalis_Thermal.pdf.

Chao, Sophie. *In the Shadow of the Palms: More-than-Human Becomings in West Papua.* Duke University Press, 2022.

Dennis, Rona. *A Review of Fire Projects in Indonesia (1982–1998).* CIFOR, 1999.

Dove, Michael R. "Theories of Swidden Agriculture, and the Political Economy of Ignorance." *Agroforestry Systems* 1 (1983): 85–99. https://doi.org/10.1007/BF00596353.

Edwards, Paul N. *The Closed World: Computers and the Politics of Discourse in Cold War America.* MIT Press, 1996.

Edwards, Paul N. *A Vast Machine: Computer Models, Climate Data, and the Politics of Global Warming.* MIT Press, 2013.

Giglio, Louis, Jacques Descloitres, Christopher O. Justice, and Yoram J. Kaufman. "An Enhanced Contextual Fire Detection Algorithm for MODIS." *Remote Sensing of Environment* 87, no. 2–3 (2003): 273–82. https://doi.org/10.1016/S0034-4257(03)00184-6.

Goldstein, Jenny E., Laura Graham, Sofyan Ansori, et al. "Beyond Slash-and-Burn: The Roles of Human Activities, Altered Hydrology, and Fuels in Peat Fires in Central Kalimantan, Indonesia." *Singapore Journal of Tropical Geography* 41, no. 2 (2020): 190–208. https://doi.org/10.1111/sjtg.12319.

Groot, William J. de, Robert D. Field, Michael A. Brady, Orbita Roswintiarti, and Maznorizan Mohamad. "Development of the Indonesian and Malaysian Fire Danger Rating Systems." *Mitigation and Adaptation Strategies for Global Change* 12 (2007): 165–80. https://doi.org/10.1007/s11027-006-9043-8.

Halberstam, J. Jack. *In a Queer Time and Place: Transgender Bodies, Subcultural Lives.* NYU Press, 2005.

Haraway, Donna. "A Manifesto for Cyborgs: Science, Technology, and Socialist Feminism in the 1980s." In *Simians, Cyborgs, and Women: The Reinvention of Nature.* Routledge, 1991.

Haraway, Donna. "The Promises of Monsters: A Regenerative Politics for Inappropriate/d Others." In *Cultural Studies,* edited by Lawrence Grossberg, Cary Nelson, and Paula A. Treichler. Routledge, 1992.

Hayles, N. Katherine. *How We Became Posthuman: Virtual Bodies in Cybernetics, Literature, and Informatics.* University of Chicago Press, 1999.

Helmreich, Stefan. *Alien Ocean: Anthropological Voyages in Microbial Seas.* University of California Press, 2009.

Hong, Sun-Ha. "Technofutures in Stasis: Smart Machines, Ubiquitous Computing, and the Future that Keeps Coming Back." *International Journal of Communication* 15 (2021): 1940–60. https://ijoc.org/index.php/ijoc/article/view/15697.

Jacques, Harry. "Small Farmers Jailed for Sumatra Fires as Companies Duck Blame." Reuters, November 24, 2023. https://www.reuters.com/article/us-indonesia-wildfire-crime-farmers-feat/small-farmers-jailed-for-sumatra-fires-as-companies-duck-blame-idUSKBN285070.

Katz, Yarden. *Artificial Whiteness: Politics and Ideology in Artificial Intelligence.* Columbia University Press, 2020.

Koudenoukpo, Juliette Biao. "As Wildfires Increase, Integrated Strategies for Forests, Climate and Sustainability Are Ever More Urgent." *UN Chronicle*, July 31, 2023. https://www.un.org/en/un-chronicle/wildfires-increase-integrated-strategies-forests-climate-and-sustainability-are-ever-0.

Li, Zhan Qing, Yoram J. Kaufman, Charles Ichoku, et al. "A Review of AVHRR-Based Active Fire Detection Algorithms: Principles, Limitations, and Recommendations." In *Global and Regional Vegetation Fire Monitoring from Space: Planning and Coordinated International Effort*, edited by Frank J. Ahern, Johann G. Goldammer, and Christopher O. Justice. SPB Academic, 2001.

Malingreau, J. P., G. Stephens, and L. Fellows. "Remote Sensing of Forest Fires: Kalimantan and North Borneo in 1982–83." *Ambio* 14, no. 6 (1985): 314–21.

Malingreau, J. P., C. J. Tucker, and N. Laporte. "AVHRR for Monitoring Global Tropical Deforestation." *International Journal of Remote Sensing* 10, no. 4–5 (1989): 855–67. https://doi.org/10.1080/01431168908903926.

McGee, T. G. "The Emergence of Desakota Regions in Asia: Expanding a Hypothesis." In *The Extended Metropolis: Settlement Transition in Asia*, edited by Ginsburg, Norton Sydney, Bruce Koppel, and T. G. McGee. University of Hawaii Press, 1991. https://scholar.google.com/citations?view_op=view_citation&hl=en&user=bJ74z9UAAAAJ&citation_for_view=bJ74z9UAAAAJ:u5HHmVD_uO8C.

Neale, Timothy, Alex Zahara, and Will Smith. "An Eternal Flame: The Elemental Governance of Wildfire's Pasts, Presents, and Futures." *Cultural Studies Review* 25, no. 2 (2019): 115–34. https://doi.org/10.5130/csr.v25i2.6886.

Petryna, Adryana. *Horizon Work: At the Edges of Knowledge in an Age of Runaway Climate Change.* Princeton University Press, 2022.

Purnomo, Herry, Bayuni Shantiko, Soaduon Sitorus, et al. "Fire Economy and Actor Network of Forest and Land Fires in Indonesia." *Forest Policy and Economics* 78 (2017): 21–31.

Putra, Tandang Yuliadi Dwi, Yoshihide Sekimoto, and Ryosuke Shibasaki. "Toward the Evolution of National Spatial Data Infrastructure Development in Indonesia." *ISPRS International Journal of Geo-Information* 8, no. 6 (2019): 263. https://doi.org/10.3390/ijgi8060263.

Pyne, Stephen. "The Ecology of Fire." *Nature Education Knowledge* 3, no. 10 (2010): 30. www.nature.com/scitable/knowledge/library/the-ecology-of-fire-13259892/.

Riggan, Philip J., James W. Hoffman, and James A. Brass. "Estimating Fire Properties by Remote Sensing." *IEEE Aerospace and Electronic Systems Magazine* 24, no. 2 (2009): 13–19. https://doi.org/10.1109/MAES.2009.4798987.

Sanjaya, Hartanto, G. Fajar Suryono, Azalea Eugenie, et al. "Indonesia Fire Danger Rating System (INA-FDRS), a New Algorithm for the Fire Prevention in Indonesia." In *2019 IEEE Asia-Pacific Conference on Geoscience, Electronics and Remote Sensing Technology (AGERS)*, 1–5. IEEE, 2019. https://doi.org/10.1109/AGERS48446.2019.9034326.

Seaver, Nick. "Care and Scale: Decorrelative Ethics in Algorithmic Recommendation." *Cultural Anthropology* 36, no. 3 (2021). https://doi.org/10.14506/ca36.3.11.

Stoler, Ann Laura. *Capitalism and Confrontation in Sumatra's Plantation Belt, 1870–1979*. 2nd ed. University of Michigan Press, 1995.

Tansey, K., J. Beston, A. Hoscilo, S. E. Page, and C. U. Paredes Hernández. "Relationship Between MODIS Fire Hot Spot Count and Burned Area in a Degraded Tropical Peat Swamp Forest in Central Kalimantan, Indonesia." *Journal of Geophysical Research: Atmospheres* 113, no. D23 (2008): 1–8. https://doi.org/10.1029/2008JD010717.

Tsing, Anna. "On Nonscalability: The Living World Is Not Amenable to Precision-Nested Scales." *Common Knowledge* 18, no. 3 (2012): 505–24. https://doi.org/10.1215/0961754X-1630424.

UNDP China (United Nations Development Program–China). "Mapping the Palm Oil Value Chain: Opportunities for Sustainable Palm Oil in Indonesia and China." UNDP and Kingdom of the Netherlands, 2020. https://www.undp.org/sites/g/files/zskgke326/files/migration/cn/Palm_oil_report_EN.pdf.

Wang, Xiaowei. *Blockchain Chicken Farm: And Other Stories of Tech in China's Countryside*. FSG Originals, 2020.

Wooster, Martin J., Gareth J. Roberts, Louis Giglio, et al. "Satellite Remote Sensing of Active Fires: History and Current Status, Applications, and Future Requirements." *Remote Sensing of Environment* 267 (2021): 112694. https://doi.org/10.1016/j.rse.2021.112694.

12

Embankment Economies, Soaking Ecologies, and the Conservation Zone of Kaziranga

AYESHA VEMURI

Kaziranga National Park in Assam is one of the oldest wildlife conservation reserves in India. Located in the floodplains of the Brahmaputra River, the park exemplifies what Anuradha Mathur and Dilip da Cunha call a "soaking ecology," a place where water seeps into the land, muddying the lines of separation between river and bank, wet and dry, land and water.[1] Every year, floods deposit silt and nutrients to make these floodplains some of the most fertile agricultural lands in India. However, in recent years, a mixture of increased development, deforestation, climate change, and poorly designed flood-control infrastructure[2] have made the floods increasingly destructive, resulting in widespread erosion in the region. An important symbol of Assamese nationalism and pride, as well as a globally recognized heritage site, Kaziranga National Park and the species it protects—the greater one-horned rhinoceros—emerge as subjects of international concern and care during the floods. Each year, news reports, photographs, cartoons, and other media depict starving, stranded, and drowned rhinos, lamenting their loss and precarity thanks to the floods. The scene in figure 12.1 is typical: It shows an emaciated rhino, accompanied by calves, seeking refuge from the floods on an elevated sandbar. Similar images, accompanied by heartbreaking statistics of rhinos lost to the floods, appear every monsoon in both Indian and international publications, including the BBC, AP News, the *New York Times*, and CNN.[3]

12.1 A greater one-horned rhinoceros and calf wade through floodwaters, 2020. Source: Anupam Nath/AP.

These articles position the rhino as the primary subject of international care, even occasioning British royals to proclaim their "anguish" when they heard news of rhinos drowned in the 2020 floods.[4]

This is why I was surprised when, in 2019, I read about a group of local villagers who were protesting the flood protection infrastructure being constructed near the park.[5] Who could possibly not want to protect Kaziranga, and why? What were the reasons for their opposition to the embankment?

The protestors in question were primarily Indigenous farmers from an organization called Jeepal Krishak Shramik Sangha (Jeepal for short), which was organizing against the Asian Development Bank (ADB)-funded Kaziranga embankment being built near their village, Agoratoli. The embankment is part of a larger initiative undertaken by the Flood and River Erosion Management Authority of Assam (FREMAA) to increase the reliability and effectiveness of flood and riverbank erosion risk management systems in Assam. Financed by the ADB through a USD120 million loan, it includes the construction of embankments and other erosion protection measures in three flood-prone regions in Assam: Dibrugarh, Kaziranga, and Palasbari-Gumi. The project is meant to anticipate and prepare for the risks posed by riverine floods and erosion to some of Assam's financially and culturally significant sites.

Eager to understand why the villagers were against the project, in March and April 2022, I visited Agoratoli to learn more about their concerns. I was fortunate to meet and stay with activists from Jeepal, most of whom belong to the Mising tribe, an Indigenous community that for centuries has lived along the banks of the Brahmaputra, depending on the river and the forest for their sustenance and livelihoods. While FREMAA argued that this embankment, along with other flood-control devices, would protect both the park and the village from destructive floods and erosion, Jeepal contests this claim. As Pranab, Manohar, and Biru, three members of Jeepal, explained to me, their community disagrees with the very idea that embankments are an effective means of managing floods, arguing that research and lived experience over decades have demonstrated that such infrastructures in fact worsen the effects of the floods rather than ease them. For them, the embankments are reflective of a corrupt and ineffective approach to governing the river that ignores the expertise of communities such as theirs, whose long history of living with the floods holds important lessons and technologies for embracing rather than trying to tame the riverine ecology of Kaziranga. As Pranab explained, "Mising means man (*mi*) of the water (*asi*)—we belong to the river and the river is part of us."[6] Despite their expertise and knowledge of the river, the Mising villages that were ostensibly to be protected by the embankment were never consulted before the project was approved, in a process that both encroaches on their legal rights and reveals the Indian state's racist dismissal of Indigenous perspectives. Moreover, although the embankment is built on community-owned land, the farmers have yet to receive compensation for the loss of their fields. Finally, members of Jeepal were worried about the possibility of being forcibly displaced from their homes and losing access to their traditional lifeways and livelihoods because of this infrastructure project.

Underlying all these issues was the fact that the Mising thought that their own lives, livelihoods, and futures were being forfeited to ensure the future of Kaziranga National Park. One elder, Pitoli Doley, exclaimed, "For the government, the animals' lives are more precious than ours! They only care about protecting the rhino because that is what brings them money!"[7] Pitolidi is drawing attention to a history of government interventions in which the needs of her village and community were superseded by the park.[8] Government and development aid had been directed toward the park and not local communities because of the financial and political value of conservation and the tourism it brings to Assam. Pitolidi's words highlight the ways in which flood relief efforts in the Kaziranga area are based on calculations in which certain forms of nature or wilderness are valued over human—specifically Indigenous peoples'—lives. While such calculations cannot be found in any official reports or plans, they are implicit in reports, policies, funding pledges, and other documents that direct resources toward the park by singling it out as a *valuable site* in need of *care*, because it is *at risk* from both human and natural forces.

However, risk is not an uncontested or objective category. It is produced. It emerges from the situated perspectives, experiences, and desires of a particular group. As François Ewald explains, "In insurance, 'risk' designates neither an event nor a type of actual event—'unfortunate' events—but a specific way of treating certain events."[9] For Ewald, this means that events are calculated as being risky or unfortunate when they pose a threat to the value or capital that is being insured. What is considered to be at risk and thus an object whose future is insured is always *situated*, most often reflecting the needs of the powerful. Recognizing the partial perspective represented by the seemingly objective term *risk* behooves us to interrogate the politics scaffolding its construction and management.

In the case of Kaziranga National Park, the dominant discourse of risk construes the site as vulnerable, drawing on a history of mechanisms of protection already at play, while the Mising are construed not only as unimportant but also as *risky to the park*. Examining risk from the perspective of the Mising, in contrast, allows us to notice the ways government-led infrastructure projects in Kaziranga are not merely instances of development funds diverted to the park and not the community. These forms of flood-control infrastructure actively put Mising villages and their futures at risk. The politics of flood-control infrastructure in Kaziranga illuminates contesting definitions of risk as well as contesting ideas of rurality. As the editors of this volume note in the introduction, "*Ruralities* are plural, dynamic, diverse, situated, and emergent." In Kaziranga, ruralities are often also oppositional, with competing interests undergirding the dynamic processes of ruralization that emerge in and around the park. Although wilderness is not typically thought of as rural, looking at where and how sites of wilderness are created, particularly in relation to surrounding agricultural land and practices, suggests otherwise. The story of risk (re)distribution that I trace in this chapter reveals the ways in which the emergence of, and support for, certain ruralities over others is a political process that in this case is supported not only by local and national interests but also by international finance and policy that lend weight to notions of a human-free wilderness over Indigenous coexistence.

In this chapter, I examine how the politics of the ADB-funded embankment favors one kind of rurality (that of a militarized wilderness) over another (the Misings' farmlands and ancestral hunting and foraging rights) and constructs them in opposition to one another. On the basis of my fieldwork and interviews with members of Jeepal, I offer that the flood-resistant infrastructure is designed to protect Kaziranga National Park *at the expense of* the Mising community and their farmlands. In an epistemic context that sees local communities as threats to wilderness, the forms of rurality they each represent are incompatible with one another because a militarized wilderness not only depends on the erasure of Indigenous communities from the park but also requires the ongoing capture of their lands to insure its own future. In this schema of rationality, Indigenous communities are calculated as having lesser value than the park and the rhino, thus justifying a future where the park is saved even though it requires the ongoing expulsion of Indigenous communities and the capture of their lands. Such necropolitical calculations, which render the Misings' lives precarious and unworthy of care compared to Kaziranga's animals, can be understood via the analytical lens of insurance.

Insurance and Risk

François Ewald defines *insurance* as a technology of redistributing risk, a means of offsetting uncertainty and protecting certain futures through financial or other security measures.[10] Insurance is a mechanism of protecting against loss, damage, or an undesirable future by mobilizing forms of compensation in the event of a misfortune. While these forms of compensation are usually financial, particularly in the case of contract-based insurance with a firm, compensation may also be rendered in services or in kind. In every case, it is based on a calculation of the value of the asset that has been insured and the calculation of probabilities of forms of risk and misfortune that may befall that asset. Risk, then, also functions as a technology of designating value, a schema of rationality that ascribes value to certain elements of a given event. As technologies of value creation and power (re)distribution, risk and insurance function as world-making devices. It is instructive about the kinds of world they make that risk and insurance emerged as full-fledged industries during the age of colonialism and transatlantic slavery; insurance was designed to protect traders' profits in the face of uncertain oceanic voyages, often at the cost of the lives of enslaved peoples.[11]

As Ewald indicates, risk and insurance are anticipatory; they function as a means of predicting, planning for, and managing the future. A host of experts—scientists, social scientists, actuaries, statisticians, and numerous others—are engaged in probabilistic analysis to predict future risk. This process of calculating and anticipating high- and low-risk futures affects the present; to some degree, it produces some futures while foreclosing others. Anticipation mediates the relationship between the present and the future; it "names a particular self-evident 'futurism' in which our 'presents' are necessarily understood as contingent on an ever-changing astral future that may or may not be known for certain, but still must be acted on nonetheless."[12] The processes of anticipating risk and making it legible require media and mediation; media technologies and infrastructures do not merely communicate risk but also produce and contain it.[13] In Kaziranga, risk mediation is a two-step process. First, expert reports construct Kaziranga as being at risk, setting up an evidentiary basis for material interventions to contain and redistribute that risk. Second, various policy, law, funding, and infrastructural interventions are introduced that redistribute the risk of floods and erosion to Kaziranga National Park onto the Mising community. Flood-control infrastructure mediates and redistributes risk in such a way that the land, livelihoods, and futures of the Mising are sacrificed in order to insure the future of not only the neighboring park but also the conservation industry as a whole.[14] In order to elaborate this argument further, I turn to Jeepal's account of flood-control infrastructure in Agoratoli.

Agoratoli

Agoratoli, situated along the banks of the Dhansiri River—one of the tributaries of the Brahmaputra—lies opposite Kaziranga National Park. It comprises about forty *chang ghar* (stilt houses), which are characteristic of the Mising community. These homes exemplify the Misings' familiarity with the seasonal shifts of the river, showing how over the centuries they have developed forms of habitation that allow them to live with floods. However, when I spoke with some village elders, I learned that the Misings' ways of living in Brahmaputra's valley were being challenged by the infrastructures that were supposed to protect them from the floods. Manohar's grandfather, Mukheswar dada, told me about how the village had moved several times in his own lifetime, and that the villagers were forced to relocate both because of the changing course of the rivers that make up this landscape and because of the loss of land to erosion.[15] When I asked if the embankment brought some relief and protection from the destructive effects of the floods, he laughed at my naïveté, saying, "What do you think caused all this erosion? It's not just the river. . . . It's the embankments!" He explained that the embankment that I was standing on was the eleventh to be built in in his lifetime. The ten prior embankments had eroded away during the floods, taking with them fertile farmland, whole villages, and parts of Kaziranga.

Despite the ineffectiveness of embankments, Manohar and Biru explained to me, the government persists in building them, both to bolster what Pranab calls the "embankment economy" of Assam and as a means of publicly demonstrating that the government is acting to protect Assam's people from floods and climate change. The provocative term *embankment economy* refers to the unofficial, highly lucrative, and reputedly corrupt economy of annual construction contracts for embankment building, maintenance, and repair in Assam. This economy persists even though local communities, activists, and scholars have shown that the embankments are not merely ineffective but actively harmful. Embankments have been connected with reduced soil fertility in the floodplain, raising the riverbed as a result of deposited silt, and worsening the destructive effects of floods when breached. Moreover, Mukheswarda explained, embankments constrain the river to narrow channels, making the current—and its erosive effects—much stronger. "It was better in my childhood," Mukheswarda told me, "before all the embankments. Then, the river would flood, but the waters would pass underneath our houses and then would recede. Now, the water cannot flow, so it rises and rises, and we must build higher and higher, and even after the rains, the water takes longer to flow away." As Pranab argued, the economic benefits of the embankment economy for elite groups in Assam seem to overshadow the ecological harms they cause.

Mukheswarda's story suggests that embankments constitute a classic example of Beck's "risk society," defined as the "hazards and insecurities induced and introduced by modernization itself."[16] In a risk society, Beck argues, there is often a "boomerang effect" in which technoscientific interventions to reduce uncertainty proliferate and produce the very risks they set out to minimize.[17] In the past, communities that lived along the floodplains of the Brahmaputra did not just adapt to the annual floods; they *depended* on the silt-rich floodwaters for fertilizing their lands and providing them with sustenance. As Gayatri Chakravorty Spivak writes of the Indigenous farmers and fisherfolk who dwelled along the Brahmaputra, they "have been used to living with water, even yearly flooding, forever . . . they have learned to bend with [the floods] and rise again . . . learned to manage [the annual floods], welcome them, and build a lifestyle with respect for them."[18] However, colonial mapping practices, land taxation policies, and postcolonial infrastructure projects altered the relationship to the floods. These ecological interventions transformed a "flood-dependent agrarian regime into a flood-vulnerable one," in the process altering precolonial and Indigenous knowledge and technologies of living with the floods.[19]

In the early postcolonial period, the Indian government was led by Jawaharlal Nehru and other technocrats who believed in the promise of scientific rationality and engineering prowess. Following colonial logics and technologies of categorization, separation, and containment,[20] these leaders hoped to manage floods by building thousands of kilometers of embankments to protect urban settlements and areas with oil and natural gas deposits.[21] However, a soaking landscape such as Assam's ideally represents places where the relationship between land and water continually shifts, best exemplified by "gradients not walls, fluid occupancies not defined land uses, negotiated moments not hard edges."[22] The separation of land and water by representational, legal, and infrastructural means altered Assam's ecology, draining and drying its wetlands, impoverishing its soil, and increasing intensity of floods.[23] In other words, technoscientific interventions transformed the floods from a necessary and welcome occurrence to a risk to be managed through additional technoscientific interventions.[24] Mukheswarda's recollections reveal how the technological solution to a constructed risk in fact *produced* that risk, and that underlying this process was a desire to maximize arable land, increase agricultural yields—and channel funds toward beneficiaries of the embankment economy.

Despite the widely documented negative impacts of flood-control infrastructure, embankments continue to be the primary means of managing floods, and the desire to maximize land use continues to direct flood-control measures, albeit with a shift from agriculture to conservation. The embankment being built in Agoratoli exemplifies this. As

12.2 A sluice gate (*left*) prevents the flow of water, resulting in a stagnant stream (*right*), 2020. Photographs by Ayesha Vemuri.

Manohar explained, the Agoratoli embankment was built in a way that separates two streams of the Dhansiri River. On one side lies the Dhansiri (which is a tributary of the Brahmaputra), and on the other lies a stagnant stream, which the locals call Mora Dhansiri (dead Dhansiri). This death, as Manohar put it, is the result of a plan, executed in 2019, to build three sluice gates along the river (figure 12.2). The ostensible purpose of these sluice gates is to provide a channel for the floodwaters to exit the fields on the village side of the embankment. It was supposed to address the problem that Mukheswarda had named—that embankments restrain the river, preventing floodwaters from spreading out, thus causing more devastating floods. As is often the case, villagers were not consulted by the authorities before the project was commissioned, resulting in what many call the death of the river. Manohar explains: "Sluice gate number 1 [the largest of the three] was built at least a foot or two above the ground level. This means that when floodwaters enter our farms, they cannot exit unless the water levels are higher than a foot. That defeats the whole supposed purpose of the gate." He points out the farmland filled with water, shaking his head, and explained, "This was productive land, good land. And now it is ruined. Good for nothing except for birds to drink from and mosquitoes to lay their eggs in. But maybe that's what they want." While the lack of consultation is a long-standing problem in many rural, "backward" areas slated for development, Manohar's last statement—"But maybe that's what they want"—warrants deeper attention.

What Manohar means is that he and other members of his community think that the sluice gate's design was not an error or oversight but rather is part of a longer-term strategy to convert the village farmlands into wetlands. He explains, "The state wants to expand the Kaziranga National Park, and our presence here is a thorn in their side. But if they convert this land into a wetland and migratory birds come here, then they can declare this area a biodiversity hotspot and make it into a bird sanctuary, and then get the authority to evict us." While forest and Tribal protection laws in India do not allow for the Mising community's lands to be seized by the government, the state does have the authority to declare certain lands to be protected ecological areas, especially when endangered species are present. On this basis, it is Manohar's contention that the infrastructure-enabled flooding of Mising lands allows the state to capture these lands for conservation. As he added, "Our land is like a backup for the park." By this characterization, Manohar signals one way that the village of Agoratoli securitizes the future of Kaziranga National Park. The Latin term *securitas* was used to connote legal guarantee in oaths, pledges, and contracts, essentially functioning as an instance of what has now evolved into the industry of mercantile insurance.[25] In evoking the idea that his community's lands act as a form of security or insurance for the continued future existence of Kaziranga, Manohar is drawing attention to how the Misings' own futures are sacrificed for the sake of the park. The embankment economy, in this case, also functions as instrument of dispossession of Mising lands and as an instrument of insurance for Kaziranga. Over time, villagers will either voluntarily relocate or be forcibly evicted, Manohar tells me. Pranab's experience illustrates this process. "Because of the erosion . . . my family no longer lives by the river. We had to relocate, but by moving we have lost a part of our relationship to the river," he relates. Pranab explains that for many Mising, losing their lands means losing culture, heritage, and livelihood. "I've never seen so many people from my community engaged in casual labor . . . construction, tourism, etc. as today, after losing their homes to erosion and eviction." For Manohar, this process is bound to increase in the coming years, infrastructurally mediated through ill-conceived embankments and sluice gates.

While there is no way to prove Manohar's provocative assertion that this is a deliberate strategy, it does remain true that the government plans to expand the park. As part of a national strategy, India is planning a massive expansion of its network of national parks. In Assam, the government hopes to double the size of Kaziranga National Park and extend it to join other national parks.[26] Eviction orders have already been issued in several villages, and as of this writing, state police have violently evicted two villages, with resisters beaten and fired on by the police.[27] To legitimize these measures, authorities invoke a mainstream, colonial model of conservation based in the belief that human beings are separate from nature and do not belong in spaces designated as conservation zones. This model of conservation, born in the colonial era, continues to dominate wildlife protection models today, and forms of the basis of militarized displacement of Indigenous communities from forests and other wild spaces, in this case disallowing the coexistence of the Mising with the park.[28] The proposed expansion involves relocating 900 villages, which are home to more than 200,000 people who will now be severed from their traditional homes, lands, and ways of living.[29]

While it might seem counterintuitive that a state whose most important industry is agriculture is sacrificing farmland, the expansion of parks like Kaziranga must be seen as part of a larger strategy of balancing development and climate action. Many activists have argued that this push for conservation is a means of deflecting from the government's investments in deforestation, mining, coal extraction, and other environmentally destructive industries.[30] Viewed in this light, investments in conservation act as insurance, offsetting the risks of India's destructive developmentalism by greenwashing its governance of nature. India desires a future in which it can both proceed with its developmental ambitions and present an image of being environmentally progressive. In calculations of climate action, India's investments into conservation act as a means of downplaying the ecological harms of development, especially when it involves the eviction of Indigenous communities from forests and other protected habitats. Furthermore, it is important to situate Jeepal's ongoing actions against flood-control infrastructure in a longer history of national ambitions, international policy, and global conservation industry to understand how they affect the future of the villagers of Agoratoli.

Conservation as Climate Colonialism

The ways in which climate risks are constructed and managed in Kaziranga can be seen as a form of what Farhana Sultana calls "climate coloniality."[31] This names colonialism as the historical and ongoing structure that undergirds the climate crisis, the unequal distribution of the basic conditions that support life, the disproportionate impacts of climate change on formerly and presently colonized and racialized communities around the globe, and the continued impunity and lack of care of people and institutions in the global north when confronted with these injustices.[32] Coloniality is not only the process through which the climate crisis is born; it also persists in mainstream climate action. As Sultana notes, "International climate negotiations falter in addressing climate change without meaningfully reducing fossil fuel dependency, growth models, and hyperconsumption, along with the systems that undergird them across scales. Rather, these spaces become spectacles, one of performance, that erases historical and spatial geopolitics and power relations."[33] This is certainly true in the case of mainstream conservation, an industry that effaces its colonial past while extending its harmful logics in the present.

The paradox of conservation, then, lies in its capacity to simultaneously preserve and erase—to protect certain forms of life while rendering others expendable. This paradox resonates with longer histories of environmental governance and extractive sacrifice. In many ways, the conservation zone evokes the idea of the "sacrifice zone," a concept that is used widely in the environmental justice movement to refer to geographic areas that have been permanently damaged by high levels of ecological pollution and toxicity.[34] If sacrifice zones have been used to denote spaces deemed expendable for the benefit of others, the conservation zone might be thought of as its inverse: spaces marked as sacred or vital to protect—but only at the cost of displacing the very communities that have sustained them. As Indigenous activists worldwide have argued, many of the areas slated for conservation due to their biodiversity and ecological importance are important for conservation precisely because of their Indigenous inhabitants who have tended to the health of these ecologies. Yet the political and ecological spatial formation that is the conservation zone precludes and invisibilizes this history. Attending to the conservation zone as a bordering formation reveals how environmental protection can paradoxically reproduce the very violence it claims to ameliorate, displacing Indigenous life and erasing its ecological histories in the name of saving nature.

Systemic racism pervades both mainstream conservation science and conservation practice and has functioned as a means of dominating Indigenous knowledge and displacing Indigenous and racialized communities.[35] The theory and practice of conservation often disguises the violence through which so-called wilderness is protected, including forced removal, abuse, loss of livelihoods, cultural assimilation, human rights abuses, and death.[36] Despite the opposition to mainstream conservation of scholars, activists, critical conservationists, and environmentalists, it continues to be one of the foremost forms of climate action in contemporary times, supported by funding from governments, large NGOs, philanthropists, and corporations. As Rudd et al. write, "Conservation's colonial underpinnings continue to permit practices that subjugate local people by portraying them as responsible for conservation problems, forcibly removing them from their land in the name of conservation and preventing them from accessing wildlife and protected areas, often by militarized means."[37] The history of Kaziranga National Park exemplifies the coloniality of mainstream conservation as well as the relationship between coloniality and rurality. The assignation of some rural spaces as wilderness, assumed to be pristine and empty of human presence, is a long-standing colonial trope that can be found across previously and presently colonized nations around the world.

This violence of colonial conservation in Assam began with the British-era division and categorization of land into agricultural land, wasteland, and wilderness in an effort to optimize resource extraction as well as recreational opportunities (such as hunting) on colonized land.[38] From 1916 until 1934, Kaziranga was designated as a game preserve, after which hunting was prohibited—the result of the indiscriminate hunting of rhinos to near extinction.[39] In 1950, it was renamed the Kaziranga Wildlife Sanctuary, following which, in 1954, the government of Assam passed a bill imposing heavy penalties for rhinoceros poaching. In 1968, the park was declared a national park; it was given official status as a protected area in 1974. In 1985, Kaziranga was declared a UNESCO World Heritage Site, and in 2006, it was declared a tiger reserve, a designation that indicates the highest level of environmental protection under India's protected areas network. Accompanying the increase in protections for the rhino is the emergence of the rhino as a symbol of Assamese nationalism. In 1948, the one-horned rhino was declared the state animal of Assam, and it then gradually came to represent many different industries and government organizations of the state including the Assam Oil Company, the Assam regiment, the Assam department of tourism, and the face of the Assam government's new investment and "ease of doing business" initiative (figure 12.3). The association of the rhino with Assamese nationalism is so deep-rooted that its precarity has often been mobilized toward political ends. For instance, Narendra Modi, the prime minister of India, invoked its predicament as an argument during the run-up to the 2014 elections. Alleging that the Indian National Congress was colluding with undocumented immigrants from Assam, he said, "Aren't rhinos the pride

12.3 Examples of rhino logos in multiple Assam state initiatives. *Clockwise from top left*: Assam State Oil Corporation, Assam Regiment, Assam Investment Initiative, and Assam Tourism.

of Assam? These days there is a conspiracy to kill it . . . to save Bangladeshis. . . . They are doing this conspiracy to kill rhinos so that the area becomes empty and Bangladeshis can be settled there."[40] Modi's rhetoric suggests that a threat to the rhino is equivalent to a national security threat posed by undocumented immigrants, but also that the marginalized community of Bengali-speaking Muslims (often called Bangladeshis in right-wing political rhetoric) is directly threatening the park, and by extension the very state of Assam.[41] The rhino is so deeply intertwined with the idea of Assamese statehood that risks to its well-being appear to constitute an existential threat to the state as a whole.

Constructing Risk in Kaziranga

The rhino's deep association with Assamese identity is critical to understanding how any perceived risk to Kaziranga activates an extensive protectionist response. In a colonial conservation framework that imagines wilderness as empty of humans in particular, Indigenous communities like the Mising have already been framed as potential threats to the park. Indigenous communities who depend on the forest for their sustenance and livelihoods have been characterized as both impeding the progress of national development and as engaging in ecologically harmful livelihood practices. This builds on bureaucratic and pseudoscientific processes of organizing locals into a "hierarchy of civilization" during the colonial period, in which Indigenous communities such as the Mising were deemed backward and primitive.[42] In postcolonial Kaziranga, local communities have often been cast either as encroachers who prey on habitat rightly belonging to the rhino or as poachers seeking to benefit by illegally trafficking rhinoceros horn. Poaching is indeed a concern in Kaziranga, but the issue has been mobilized in ways that criminalize the life-supporting activities of Indigenous peoples. This characterization of local Indigenous communities as practicing unlawful and harmful ways of life is not limited to Indian media but also appears in highly influential international publications. For instance, a 2001 UNESCO World Heritage Convention report states that "during the winter, the local people enter the Park for community fishing, which is sometimes associated with illegal activities, such as stealing rifles from forest guards and damaging river boats."[43] In response, fishing in the park was banned soon afterward. The same report also claims that antipoaching efforts in the park were impeded by a lack of funds and that more resources were needed to improve the park's protection.

Such reports are enormously influential in both fiscal and policy terms. They are often followed by pledges of funding from international conservation NGOs like the International Union for Conservation of Nature (IUCN) and the World Wide Fund for Nature, as well as policy directives by the Indian and Assam governments. Reports, then, are an important element of the machinery of conservation that draws on global, national, and local laws, funds, discourse, and physical infrastructure as a means of capturing rural and Indigenous lands for the purpose of conservation, often through violent means. The threat of poaching in Kaziranga has been met with a militarized response that exemplifies the policy of militarized governance in Assam and the northeast more broadly.[44] The governance of these "borderlands," as Sanjib Baruah explains, has been characterized by suspicion; the Northeast has always been regarded as not quite Indian, as being overpopulated by tribal communities whose "primitive" belief systems and ways of living set them apart from the "civilized" mainland.[45] Their lack of education and backward beliefs are often characterized as being in need of disciplining, often through the harshest measures.

In Kaziranga, rangers were given the authority—that is, they were ordered—to shoot suspected poachers on sight. At one stage, the park rangers were killing an average of two people every month—more than twenty people a year.[46] This practice was put in place by former park director

M. K. Yadava, who wrote a detailed report about his SMART strategy for tackling poaching: "Kill the unwanted" should be the guiding principle for the guards, he recommends.[47] In the report, he explains his belief that environmental crimes, including poaching, are more serious than murder. "They erode," he writes, "the very root of existence of all civilizations on this earth silently."[48] The report, which rationalizes and venerates conservation in its most colonial form, fails to account for the fact that the area that now makes up the park has historically been inhabited by communities like the Mising, who for centuries have lived alongside the river and have fished, hunted, and foraged in the park. Instead, it figures these villagers as encroaching on natural spaces and gives the forest department license to evict these communities, thus exacerbating conflicts between the park and villages like Agoratoli. Despite several terrible incidents, such as when eight-year-old Akash Orang was shot and left severely disabled after forest rangers mistook him for a poacher, Kaziranga is widely celebrated as a successful conservation story, and many call for the Kaziranga model to be applied to other national parks and sanctuaries across India and in other parts of the world.

In the conservation zone of Kaziranga, humans are seen to have no place apart from a protectionist one, exemplified by the forest ranger, or a consuming one, exemplified by the nature tourist. The Mising farmers and fisherfolk who claim a complex historical and ecological relationship with the park and the river trouble this arrangement, introducing a messy relational rural presence that does not fit into these neat categories and therefore threatening the conservation ideology undergirding the park. Proclaimed a UNESCO World Heritage Site, Kaziranga National Park is described as "one of the last areas in eastern India undisturbed by a human presence."[49] UNESCO World Heritage sites such as Kaziranga are conceived as "belong[ing] to all the peoples of the world, irrespective of the territory on which they are located."[50] Positioning Kaziranga as belonging to all people enables the erasure of the specific historical and ecological relationship that communities such as the Mising have with the park and its nonhuman inhabitants. Indeed, the conceptual emptying of Kaziranga of all human presence is the essence of the colonial model of conservation. This model, moreover, has the political and financial backing of not only the Assamese and Indian governments but also of large environmental NGOs such as the IUCN and the World Wildlife Fund, development banks such as the ADB and the World Bank, international governments such as France, and corporations such as Tata Motors and Reliance Industries.

Over the past several decades, several of these actors have directly and indirectly funded conservation in various ways, including training forest rangers in antipoaching practices. I highlight these investments to demonstrate that Kaziranga is not merely a site of national pride and care but also an object of international care in both discursive and material senses.

In recent years, the care and protectionism dedicated to Kaziranga has been further exacerbated by the increasingly devastating floods that inundate and erode both forest lands and farmlands. Some estimates claim that nearly 150 square kilometers of parkland has been lost to erosion since 1920.[51] While erosion is an issue across the state of Assam, it generates a more urgent response in Kaziranga because of the importance of the park to Assamese identity. The enormous media attention to Kaziranga in the flood season is one indicator of the ways in which it emerges as needing greater care than other spaces in Assam. As unstable monsoons, rising temperatures, and increased melt from Himalayan glaciers cause more intense floods, experts predict heightened risk to the park.[52] A host of international and national studies have identified Assam as one of the most at-risk locations in the world for climate change effects, particularly floods, erosion, migration, and, increasingly, droughts. For instance, a 2021 climate vulnerability report by the Council on Energy, Environment and Water states that Assam has the highest overall vulnerability index in the country.[53] Another report predicts Assam as being at the highest risk for both direct and indirect effects of climate change, including floods, erosion, displacement, migration, and ethnic conflict.[54] In 2023, the Gross Domestic Climate Risk report identified Assam as one of the fifty most vulnerable states globally for climate change.[55] These reports collectively have the effect of *producing* Assam as a site of risk and vulnerability, inviting national and global concern to intervene and redistribute the risk.

Central to my analysis is the idea that climate risk is not simply a given, an objective fact out there to be managed; rather, it is something that is deliberately constructed by interested actors. As such, risk is always *political*, as its calculation has differential impacts on different groups. By singling out of some entities (in this case, the park) as being at risk and in need for protection and expansion, even as others in similar situations (in this case, the Mising villages next to the park) are not, we must ask ourselves to consider what political, financial, moral, and other interests might underlie the construction of risk as well as a means of redistributing that risk. To insure the future of Kaziranga, various infrastructural, legal, policy, and actuarial measures are deployed that transpose the risks of climate change, floods, erosion, development, and conservation onto the villagers. Through the calculations of these experts, the Mising community emerges as a form of security that insures the futurity of both conservation logics and developmental logics. This form of dispossessive expansion of wildlife preserves is a means of extending the logics of colonialism into climate action.[56] The performance and visibility of conservation as a green practice allows the Indian state to claim that it is acting responsibly to meet global climate goals even as it continues to pursue environmentally detrimental development activities. On a global scale, the lands and futures of the Mising community, like other Indigenous groups, are sacrificed to insure the future of the international conservation industry.

Conclusion

The politics of embankments and conservation I learned about in Kaziranga signal the ways that infrastructures of flood management and conservation split rural spaces into wild and cultivated spaces, creating the conditions for a conservation zone that is valued over Indigenous lives and lands. The sluice gate and embankments in Agoratoli tell of the ways that the very machinery of conservation, as well as the legal, physical, and financial infrastructures that support it, eat into agrarian landscapes. These flood management infrastructures mediate the rurality of Agoratoli into a conservation zone in which the Mising have no future. Similar to the sacrifice zone, the conservation zone is often applied to rural sites of extraction. Here, the future of the wildlife of Kaziranga is pitted against the future of the Indigenous Mising community, such that the risks posed by climate change are redirected from the park onto the lands and peoples of Agoratoli. The Mising people's lands—and thus their lives, livelihoods, and futures—are mobilized as security for globalized and nationalist ideologies of growth, development, and conservation in the face of the risks that climate change, floods, and erosion pose to the park. In the current conjuncture of emerging green capitalism and green neocolonialism characterized by a logic of insurance and the redistribution of the risks of climate change, the conservation zone and its related forms of mediation emerge as a kind of signature rural spatiality supported by the embankment economy.

The green calculations that support the conservation zone justify the displacement of thousands of villagers in the name of conserving the rhino—because of the potential profits to be gained from tourism, certainly, but also because the very model of conservation supported by organizations like the IUCN depend on this vision of pristine, human-free wild spaces. Several international conservation agencies have been lobbying world leaders for the 30 × 30 project—a plan to turn 30 percent of the earth into protected areas by 2030.[57] As activists around the world have argued, this plan facilitates a colossal land grab by Western conservation agencies and their corporate and state allies. This is especially true in previously colonized nations around the world, especially on the continent of Africa. As Aby L. Sène writes, "Indigenous and human rights activists are sounding the alarm, comparing the 30 × 30 plan to the second Scramble for Africa, one that would further dispossess, militarize, and privatize the commons in Africa."[58] The expansions and evictions underway in Kaziranga are one instantiation of this larger global effort, signaling the persistent coloniality of climate action.

The model of conservation being promoted in Kaziranga and other parks globally requires the forcible exclusion of Indigenous, rural, and racialized groups while enabling territorial capture by corporate and conservation elites. Its conception of ecological well-being necessitates the removal of people like the Mising from sites like Kaziranga, replicating the process of colonialism under the guise of environmental action and ecological responsibility. An environmentally just model of conservation, in contrast, would necessarily include decolonization as a key strategy. Survival International, an organization that works with Indigenous communities around the world with the goal of decolonizing conservation, notes that around the world, areas designated for protection because of their status as ecological and biodiversity hot spots have always had indigenous inhabitants. They write, "These territories are important conservation zones today precisely because the original inhabitants have looked after their land and wildlife so well." Instead of the "fortress conservation" model, they advocate for a new model of conservation that has tribal rights at its heart.[59]

In Agoratoli, Manohar, Pranab, and their families advocate for this justice-centered model of conservation. In such a model, their own communities would be seen as leaders rather than barriers to the future of the rhino. The technologies, methods, and strategies of living with animals, floods, and erosion long practiced by the Mising could be used to manage the risks of climate change. "The health of Kaziranga," Pranab says, "depends on the health of the Mising community. Our fight is not only for our own lands and futures but is also a fight for the flora and fauna of Kaziranga." Jeepal advocates, then, for a way of seeing the risks to one as necessarily intertwined with the other rather than positioning them as antagonistic. In their vision of the future, rurality is not contained to the narrow and incompatible categories of either wilderness or human inhabitation. Rather, their vision of coexistence and mutual dependence highlights how the conservation zone should instead be transformed into a space of multispecies flourishing. However, insuring the futures of both the Mising and the park would require a radical rearticulation of value, risk, and the structures of expertise through which they are calculated.

Notes

1 Mathur and da Cunha, *Soak*.
2 See Wasson et al., "Flood Mitigation"; Baruah, *Slow Disaster*.
3 BBC News, "Assam Flooding"; Wu, "Rare One-Horned Rhino"; Associated Press, "Flooding in India"; Hussain and Singh, "British Royals Share Anguish."
4 Hussain and Singh, "British Royals Share Anguish."
5 Environmental Justice Atlas, "ADB Embankment."
6 See also Thakur, "Struggles."
7 Interview with Pitoli Doley, April 2022.
8 The term *didi* (or *di* for short) is used as a term of respect and affection for an elder woman.
9 Ewald, *Birth of Solidarity*, 98.
10 Ewald, *Birth of Solidarity*, 96.
11 See Sharpe, *In the Wake*, 29–39; Armstrong, "Slavery, Insurance."
12 Adams et al., "Anticipation," 247.
13 Ghosh and Sarkar, *Routledge Companion*, 4.
14 See Johansen, *Environmental Racism*; Taylor, *Toxic Communities*; Waldron, *There's Something in the Water*.
15 *Dada* (or the shorter *da*) is used a term of respect and affection for an older male.
16 Beck, *Risk Society*, 21.
17 Beck, *Risk Society*, 23.
18 Spivak, "Responsibility," 47–48.
19 For colonial mapping practices, see Da Cunha, *Invention of Rivers*. for land taxation policies, see D'Souza, "Water in British India"; Narain, "New Extreme Reality." For postcolonial infrastructure projects, see Das, "Majuli in Peril"; Baruah, "Suffering for Land"; D'Souza, *Drowned and Dammed*, p 215–16.

20 D'Souza, "Water in British India."
21 Das, "Majuli in Peril."
22 Mathur and da Cunha, *Soak*.
23 See Saikia, *Unquiet River*, chap. 7.
24 Saikia, *Unquiet River*.
25 Hamilton, "Brief Semantic History," 64.
26 Kalita, "Assam."
27 Das, "Assam."
28 Rudd et al., "Overcoming Racism"; Dowie, *Conservation Refugees*.
29 Rowlatt, "Kaziranga."
30 Bathija and Sylvander, "Conservation Regimes."
31 Sultana, "Unbearable Heaviness."
32 For unequal distribution, see Mbembe, "Universal Right." For disproportionate impacts, see Sultana, "Unbearable Heaviness." For the global north's lack of care, see Pulido, "Racism and the Anthropocene"; Pulido and De Lara, "Reimagining 'Justice.'"
33 Sultana, "Unbearable Heaviness," 2.
34 Juskus, "Sacrifice Zone"
35 Rudd et al., "Overcoming Racism."
36 Rudd et al., "Overcoming Racism," 2.
37 Rudd et al., "Overcoming Racism, "3.
38 Sharma, *Empire's Garden*.
39 Saikia, "Kaziranga National Park."
40 Barbora, "Riding the Rhino"; *Business Standard*, "People in Assam."
41 The idea that India, specifically Assam, faces a threat to its security, culture, and language from Bengali-speaking Muslim communities has a long and complex history. For detailed discussions about the history and politics of this accusation, see Murshid, *India's Bangladesh Problem*; Vijayan, *Midnight's Borders*; Sur, *Jungle Passports*; Ghosh, *Thousand Tiny Cuts*.
42 Sharma, *Empire's Garden*, 31.
43 UNESCO, "Kaziranga National Park (India)."
44 See Baruah, *India Against Itself*; Barbora, "Riding the Rhino."
45 Baruah, *India Against Itself*.
46 Rowlatt, "Kaziranga."
47 Yadava, "Detailed Report," 173.
48 Yadava, "Detailed Report," 101.
49 UNESCO World Heritage Centre, "Kaziranga National Park."
50 UNESCO World Heritage Centre, "World Heritage."
51 *Sentinel Assam*, "Is Kaziranga Shrinking?"
52 IPCC 2022, "Summary for Policymakers."
53 Mohanty and Wadhawan, *Mapping India's Climate Vulnerability*.
54 Bhattacharyya and Werz, "Climate Change."

55 XDI, "Gross Domestic Climate Risk."

56 See Sène, "Land Grabs"; Sène, "Western Nonprofits."

57 Smith, "30 × 30."

58 Sène, "Land Grabs."

59 Survival International, "Indigenous People."

Bibliography

Adams, Vincanne, Michelle Murphy, and Adele E. Clarke. "Anticipation: Technoscience, Life, Affect, Temporality." *Subjectivity* 28, no. 1 (2009): 246–65.

Armstrong, Tim. "Slavery, Insurance, and Sacrifice in the Black Atlantic." In *Sea Changes: Historicizing the Ocean*, edited by Bernhard Klein and Gesa Mackenthun. Taylor & Francis, 2012.

Associated Press. "Flooding in India Kills Scores of Animals, Including Endangered Rhinos." *New York Times*, updated July 25, 2020. https://www.nytimes.com/2020/07/25/world/asia/india-floods-rhinos.html.

Barbora, Sanjay. "Riding the Rhino: Conservation, Conflicts, and Militarisation of Kaziranga National Park in Assam." *Antipode* 49, no.5 (2017): 1145–63. https://doi.org/10.1111/anti.12329.

Baruah, Mitul. *Slow Disaster: Political Ecology of Hazards and Everyday Life in the Brahmaputra Valley, Assam*. Routledge, 2022. https://doi.org/10.4324/9781003051565.

Baruah, Mitul. "Suffering for Land: Environmental Hazards and Popular Struggles in the Brahmaputra Valley (Assam), India." PhD diss., Syracuse University, 2016. https://surface.syr.edu/etd/558/.

Baruah, Sanjib. *India Against Itself: Assam and the Politics of Nationality*. Oxford University Press, 2001.

Bathija, Paromita, and Nora Sylvander. "Conservation Regimes of Exclusion: NGOs and the Role of Discourse in Legitimising Dispossession from Protected Areas in India." *Political Geography Open Research* 2 (2023): 100005. https://doi.org/10.1016/j.jpgor.2023.100005.

BBC News. "Assam Flooding: Several Rare Rhinos Die in India's Kaziranga Park." July 19, 2020. https://www.bbc.com/news/world-asia-india-53464643.

Beck, Ulrich. *Risk Society: Towards a New Modernity*. SAGE, 1992.

Bhattacharyya, Arpita, and Michael Werz. "Climate Change, Migration, and Conflict in South Asia." Center for American Progress, December 2012. https://www.americanprogress.org/wp-content/uploads/sites/2/2012/11/ClimateMigrationSubContinentReport_small_execsumm.pdf.

Business Standard. "People in Assam Govt Conspiring to Eliminate Rhinos: Modi." March 31, 2014. https://www.business-standard.com/article/pti-stories/people-in-assam-govt-conspiring-to-eliminate-rhinos-modi-114033100661_1.html.

D'Souza, Rohan. *Drowned and Damned: Colonial Capitalism and Flood Control in Eastern India*. OUP India, 2006.

D'Souza, Rohan. "Water in British India: The Making of a 'Colonial Hydrology.'" *History Compass* 4, no. 4 (2006): 621–28.

da Cunha, Dilip. *The Invention of Rivers: Alexander's Eye and Ganga's Descent*. University of Pennsylvania Press, 2019.

Das, Debojyoti. "'Majuli in Peril': Challenging the Received Wisdom on Flood Control in Brahmaputra River Basin, Assam (1940–2000)." *Water History* 6, no. 2 (2014): 167–85. https://doi.org/10.1007/s12685-014-0098-2.

Das, Mukut. "Assam: Protest over Eviction Drive in Kaziranga Animal Corridors." *Times of India*, February 22, 2022. https://timesofindia.indiatimes.com/city/guwahati/protest-over-eviction-drive-in-kaziranga-animal-corridors/articleshow/89735868.cms.

Dowie, Mark. *Conservation Refuges: The Hundred-Year Conflict Between Global Conservation and Native Peoples*. MIT Press, 2009.

Environmental Justice Atlas. "ADB Embankment Subproject in the Brahmaputra River of Kaziranga, Assam, India." Global Atlas of Environmental Justice, 2019 database. https://ejatlas.org/conflict/adb-embankment-subproject-in-the-brahmaputra-river-of-kaziranga-assam.

Ewald, François. *The Birth of Solidarity: The History of the French Welfare State*. Edited by Melinda Cooper. Translated by Timothy Scott Johnson. Duke University Press, 2020.

Ghosh, Bishnupriya, and Bhaskar Sarkar, eds. *The Routledge Companion to Media and Risk*. Routledge, 2022.

Ghosh, Sahana. *A Thousand Tiny Cuts: Mobility and Security Across the Bangladesh-India Borderlands*. University of California Press, 2023.

Hamilton, John T. "A Brief Semantic History of Securitas." In *Security: Politics, Humanity, and the Philology of Care*, edited by John T. Hamilton. Princeton University Press, 2013. https://doi.org/10.23943/princeton/9780691157528.003.0004.

Hussain, Wasbir, and Indrajit Singh. "British Royals Share Anguish over Indian Rhino Park's Floods." AP News, July 25, 2020. https://apnews.com/article/7dc615ad4ff55cc6f017928b1c1331ea.

IPCC 2022 (Intergovernmental Panel on Climate Change). Pörtner, Hans-Otto., Debra C. Roberts, Elvira Poloczanska, et al. "Summary for Policymakers." In *Climate Change 2022: Impacts, Adaptation, and Vulnerability*, edited by Hans-Otto. Pörtner, Debra C. Roberts,

Melinda M. B. Tignor, et al. Contribution of Working Group II to the Sixth Assessment Report of the Intergovernmental Panel on Climate Change. Cambridge University Press, June 29, 2023. https://doi.org/10.1017/9781009325844.001.

Johansen, Bruce E. *Environmental Racism in the United States and Canada: Seeking Justice and Sustainability*. Praeger, 2020.

Juskus, Ryan. 2023. "Sacrifice Zones: A Genealogy and Analysis of an Environmental Justice Concept." *Environmental Humanities* 15 (1): 3–24. https://doi.org/10.1215/22011919-10216129.

Kalita, Kangkan. "Assam: Government Set to Acquire New Landmass to Extend Kaziranga National Park." *Times of India*, February 17, 2022. https://timesofindia.indiatimes.com/city/guwahati/assam-government-set-to-acquire-new-landmass-to-extend-kaziranga-national-park/articleshow/89648623.cms.

Mathur, Anuradha, and Dilip da Cunha. *Soak: Mumbai in an Estuary*. Rupa, 2009.

Mbembe, Achille. "The Universal Right to Breathe." Critical Inquiry (blog), April 13, 2020. https://critinq.wordpress.com/2020/04/13/the-universal-right-to-breathe/.

Mohanty, Abinash, and Shreya Wadhawan. *Mapping India's Climate Vulnerability: A District-Level Assessment*. Council on Energy, Environment and Water, 2021.

Murshid, Navine. *India's Bangladesh Problem: The Marginalization of Bengali Muslims in Neoliberal Times*. New ed. Cambridge University Press, 2023.

Narain, Sunita. "The New Extreme Reality of Floods." Down to Earth (blog), August 30, 2016. https://www.downtoearth.org.in/blog/natural-disasters/the-new-extreme-reality-of-floods-55430.

Pulido, Laura. "Racism and the Anthropocene." In *Future Remains: A Cabinet of Curiosities for the Anthropocene*, edited by Marco Armiero, Robert Emmett, and Gregg Mitman. University of Chicago Press, 2020.

Pulido, Laura, and Juan De Lara. "Reimagining 'Justice' in Environmental Justice: Radical Ecologies, Decolonial Thought, and the Black Radical Tradition." *Environment and Planning E: Nature and Space* 1, no. 1–2 (2018): 76–98. https://doi.org/10.1177/2514848618770363.

Rowlatt, Justin. "Kaziranga: The Park that Shoots People to Protect Rhinos." BBC News, February 10, 2017. https://www.bbc.com/news/world-south-asia-38909512.

Rudd, Lauren F., Shorna Allred, Julius G. Bright Ross, et al. "Overcoming Racism in the Twin Spheres of Conservation Science and Practice." *Proceedings of the Royal Society B: Biological Sciences* 288, no. 1962 (2021): 20211871. https://doi.org/10.1098/rspb.2021.1871.

Saikia, Arupjyoti. "The Kaziranga National Park: Dynamics of Social and Political History." *Conservation and Society* 7, no. 2 (2009): 113–29.

Saikia, Arupjyoti. *The Unquiet River: A Biography of the Brahmaputra.* Oxford University Press, 2020.

Sène, Aby L. "Land Grabs and Conservation Propaganda." Africa Is a Country, June 15, 2022. https://africasacountry.com/2022/06/the-propaganda-of-biodiversity-conservation.

Sène, Aby L. "Western Nonprofits Are Trampling over Africans' Rights and Land." *Foreign Policy,* July 1, 2022. https://foreignpolicy.com/2022/07/01/western-nonprofits-african-rights-land/.

Sentinel Assam. "Is Kaziranga Shrinking?" July 30, 2019. https://www.sentinelassam.com/more-news/editorial/is-kaziranga-shrinking.

Sharma, Jayeeta. *Empire's Garden: Assam and the Making of India.* Indian ed. Permanent Black, 2012.

Sharpe, Christina E. *In the Wake: On Blackness and Being.* Duke University Press, 2016.

Smith, Julian. "30 × 30: Protect 30% of the Planet's Land and Water by 2030." Nature Conservancy (blog), February 29, 2020. https://www.nature.org/en-us/magazine/magazine-articles/30x30-wyss-foundation-interview/.

Spivak, Gayatri Chakravorty. "Responsibility." *Boundary 2* 21, no. 3 (1994): 19–64.

Sultana, Farhana. "The Unbearable Heaviness of Climate Coloniality." *Political Geography* 99 (2022): 102812. https://doi.org/10.1016/j.polgeo.2022.102638.

Sur, Malini. *Jungle Passports: Fences, Mobility, and Citizenship at the Northeast India–Bangladesh Border.* University of Pennsylvania Press, 2021.

Survival International. "Indigenous People Are the Best Conservationists." N.d. https://www.survivalinternational.org/conservation.

Taylor, Dorceta E. *Toxic Communities: Environmental Racism, Industrial Pollution, and Residential Mobility.* New York University Press, 2014.

Thakur, Nimisha. "The Struggles of a 'River People' in Assam." Sapiens, April 14, 2021. https://www.sapiens.org/culture/mising-river-people-assam-india/.

UNESCO (United Nations Educational, Scientific, and Cultural Organization). "Kaziranga National Park (India)." Decision 25 BUR V.122–125. World Heritage Convention Report. Accessed July 29, 2023. https://whc.unesco.org/en/decisions/5873/.

UNESCO World Heritage Centre. "Kaziranga National Park." Accessed May 27, 2024. https://whc.unesco.org/en/list/337/.

UNESCO World Heritage Centre. "World Heritage." Accessed May 27, 2024. https://whc.unesco.org/en/about/.

Vijayan, Suchitra. *Midnight's Borders: A People's History of Modern India.* Melville House, 2021.

Waldron, Ingrid R. G. *There's Something in the Water: Environmental Racism in Indigenous and Black Communities.* Fernwood, 2018.

Wasson, Robert, Arupjyoti Saikia, Priya Bansal, and Chuah Joon Chong. "Flood Mitigation, Climate Change Adaptation, and Technological Lock-In in Assam." *Ecology, Economy, and Society* 3, no. 2 (2020): 83–104. https://doi.org/10.37773/ees.v3i2.150.

Wu, Huizhong. "Rare One-Horned Rhino at Risk as Indian Flood Waters Rise." CNN, July 12, 2017. https://www.cnn.com/2017/07/12/asia/india-floods-wildlife-rhino/index.html.

XDI (Cross-Dependency Initiative). "Gross Domestic Climate Risk." Search engine for assessing risk worldwide by states and provinces. XDI Benchmark Series, 2023. https://xdi.systems/gross-domestic-risk-dataset/.

Yadava, M. K. "Detailed Report on Issues and Possible Solutions for Long Term Protection of the Greater One Horned Rhinoceros in Kaziranga National Park." Government of Assam, 2014.

Acknowledgments

This project was conceived between two barstools at the Bar Dominion in Montreal, Quebec. This is fitting: *dominion* means the power or right of governing a given territory, or what today is called *sovereignty*. It refers to lands and people who are mastered or ruled, usually without their consent. In 1867, the British crown established the dominion of Canada by confederating four colonial provinces it had established in the northern half of what is now known as North America, thereby legalizing the dispossession of the Indigenous inhabitants of Turtle Island. These included the Kanien'kehá:ka of the Haudenosaunee Confederacy as well as the Wendat, Abenaki, and Anishinaabeg peoples of Tiohtià:ke, as the island of Montreal is known in Kanien'kéha. (It is known as Mooniyang in Anishinaabemowin.) McGill University, where *Media Rurality* took shape, sits on the unceded territory of these people. Montreal was, and remains, a key urban center from which the political and economic power of the dominion of Canada is enforced on far-flung lands, places, and people ruralized by infrastructural mediation—mounted police, courts, railways, roads, seaways, canals, ports, mines, energy grids, and telegraph, broadcast, and data networks, to name only a few. The role of infrastructural mediation in historical, residual, and ongoing practices of displacement and dispossession (and their contestation) in rural places the world over is a central theme of this book. While sitting on those barstools, all of this weighed heavily on our minds. We hope the work we present in *Media Rurality* honors those who lived, are still living, and who remember these histories, and encourages those who continue to refuse unwanted dominion.

Many people contributed to the work of *Media Rurality*, including authors whose names appear in the table of contents. We owe a primary debt of gratitude to Malcolm Sanger, without whom this book would not exist. Malcolm is a brilliant doctoral candidate in communication studies at McGill University, where he studies the infrastructures and mythologies of reforestation in Canada in the context of settler economies and climate change. He has been an indispensable collaborator, interlocutor, fixer, and friend from day one. It doesn't hurt that he has spent considerable time in the woods.

Several of the chapters included in this book were presented at the Media Rurality Symposium, held in June 2022 at McGill's agricultural campus in St. Anne-de-Bellevue, Quebec. In addition to those appearing as chapters in this book, excellent work was presented by Omolade Adunbi, Andrew Curley, Sarah DesRosier, Gaylene Ducharme, Roger Epp, Rahul Mukherjee, and J. T. Roane. The influence of this work runs through the preceding pages first. Several scholars from McGill, Concordia University, Université de Montréal, and the University of Alberta provided outstanding critical responses that sharpened and extended our thinking in ways also reflected in the chapters gathered here. Our sincere thanks go to Hubert Alain, Montiana Ashour, Isabelle Boucher, Laticia Chapman, Janna Franzel, Stacey Haugen, Helen Hayes, Robert Marinov, Laura Pannekoek, Malcolm Sanger, Sanaz Sohrabi, and Hannah Tollefson for their important contributions. Articles developed from several of these responses appear in a series published in 2023 by the journal *Heliotrope*. We are grateful to Mél Hogan at Queen's University for making space for this work, and to Tessa Brown for bringing these pieces into shape. We also received more-than-capable onsite logistical and technical support from Hussain Almahr, Amitsu Huang, Alexandra Jurecko, and Andrew Singer (aka Dorval Drew).

In April 2023, several contributors presented work from the Media Rurality project at a panel titled "Media Rurality in Global Contexts" at the annual conference of the Society for Cinema and Media Studies in Denver, Colorado. Response to the panel not only strengthened the work but also confirmed our sense of the appetite for attention to rural questions within our field. In March 2024, we presented the Media Rurality project at the workshop "Reimagining Ruralities: From Heartlands to Hinterlands," organized by the Rural Imaginations project based at the Amsterdam School for Cultural Analysis and held at the Center for Place, Culture, and Politics at the CUNY Graduate Center. Discussions at the workshop had a formative impact on the introduction and framing of this book. We are grateful to Esther Peeren, Peter Hitchcock, and all participants in the workshop for including us in their vibrant community.

It takes an infrastructure to make a scholarly volume. Funding for this project was generously provided by the Social Science and Humanities

Research Council of Canada, the Global Emergent Media Lab at Concordia University, and by the Office of the Provost at the University of Alberta. At McGill, support was provided by the Faculty of Arts, the Dean of Arts Development Fund, the Centre for Media, Technology, and Democracy at the Max Bell School of Public Policy, the James McGill Professor of Culture and Technology, the Wolfe Chair in Scientific and Technological Literacy, the James McGill Professor in Urban Media Studies, the office of the Associate Vice Principal (Macdonald Campus), and the Post-Graduate Student Society. The Fonds de Recherche du Québec–Société et Culture funded Pat's time at McGill. As we write this, university infrastructures that support critical research in the social sciences and humanities are under unprecedented threat, making it more important than ever to acknowledge their role in making the work we do possible. It is not a coincidence that the political and economic actors leading current attacks on academic freedom, critical inquiry, and scholarly autonomy are simultaneously working to render rural places into zones of extraction and abandonment.

Darin is especially grateful to Pat, whose intelligence, camaraderie, and commitment to engaged inquiry have been inspirational and sustaining. You are a fine scholar, and I look forward to sitting down over a Citywide to hatch our next plan. Thanks also to the Grierson Research Group, whose members are a constant source of motivation, nourishment, and encouragement. Special thanks to Farah Atoui, Helen Hayes, Jordan Kinder, Burç Köstem, Rafico Ruiz, Malcolm Sanger, Hannah Tollefson, and Ayesha Vemuri for buying in and sticking around. Katherine Strand's exceptional work on the history of dryland agriculture in Palliser's Triangle sets a standard for "dirt research" to which I aspire in my own work. Thanks also to Hanneke Stuit for one of those fleeting scholarly conversations whose influence long outlasts the encounter. "Three guys, one good thing"—you know who you are—has been a constant reminder that scholarship and friendship belong together. Mary Stone, Eva Stone-Barney, and William Stone-Barney give me the gift of intelligent conversation at home and remind me what really matters every day. The debt I owe to Ken Eshpeter and Roger Epp is too big to repay. Whatever I know about the countryside—including that I don't know much at all—I learned from them.

Pat would like to thank Darin for his heroic support and critical voice throughout this project and time at McGill. This was a truly harebrained scheme hatched on barstools (and later over many Covid-era Zoom calls), formed alongside our shared recognition that rurality needed our sustained and focused attention—even more than it was already getting. Like Darin, I'm a wayward rural guy, as much as it took me well into my twenties to admit that where I grew up in the suburbs of Philly was more farm country than it was Eagles country (though it is proudly both). The pungent and pervasive smell of manure and mushroom agriculture wafting through the surrounding countryside should've been an earlier tip-off. The mistreated migrant labor apparatuses of these growing operations, the interstate gas pipeline cutting through my parents' neighborhood, the devastating opioid crisis, and the high-water marks of the precrash building boom—all signatures of so-called rural pathologies in the United States—would shape my life and politics more than I could have imagined. I want to reiterate my thanks to friends and comrades in the Grierson Research Group for their support and participation throughout this project, especially Malcolm, a great friend and scholar, and Jordan, whose robust, generous challenging of the rural as our defining category of analysis helped sharpen the edges of our analysis. Appreciation is owed also to Kay Dickinson, who always pushed me to keep space (and political economy) at the center of my research. To my friend and colleague Patrick Bresnihan, thanks for always maintaining our focus grounded in politics and the construction of a better world—my approach to rurality developed in step with our work together. Thanks to friends at Concordia, McGill, and Montréal more widely during this project, who were there with me during the COVID-19 pandemic and often joined in desperate retreats to the countryside. Pat, Lea, Ben, Cath, Gui, Audrey, Sima, Nik, Fadi, Viviane, Sadie, I'm enormously grateful for you all. The latter stages of my work on this project were completed at University College Dublin in the School of Information and Communication Studies. Thanks to my colleagues and the research supports of the Ad Astra Fellows program for providing the space and research time to finish this project. Finally, thanks to my family—Mom (Anne), Riss, Katie, Liam, Grandma Jean, and all my proliferating nieces and nephews. Thank you, Eimear, for your love, support, and critical voice. It's a blessing to have a partner who will tell me so clearly when I'm talking shite and need to rein it in. Finally, I want to thank my dad, Dave, who passed away in 2022. Thanks for everything, Dad, and I'm glad you got to see (and make fun of) the early stages of my prodigal return to rurality. I miss you every day.

Working with Duke University Press has been a dream. We are grateful to our colleagues for their early reviews and comments. Their work has made *Media Rurality* better. The same goes for Laura Jaramillo, whose expertise and professionalism have made the work of making this book a genuine pleasure. Finally, what can be said about Courtney Berger that has not already been said in the acknowledgments of countless books she has made possible and shaped? The debt of gratitude we owe to her for her support and stewardship of this project is huge, but it is only a small fraction of what all of us owe to her collectively for the work she does to ensure critical ideas, grounded in good research and expressed with clarity, find their way into the world. *Merçi beaucoup*, Courtney.

A closing word about the title. We can't remember who came up with it, but we do remember the moment it stuck. We were looking for a place to share a meal with symposium participants when we wandered into a local establishment in St. Anne-de-Bellevue that we thought might fit the bill. After we described our plans, the bartender asked what the event was called. "Media Rurality," we said. She paused, then said, "Media rurality—so, like, *media reality*, only rural."

Exactly.

Contributors

CHRISTOPHER ALI is the Pioneers Chair in Telecommunications and professor of telecommunications in the Donald P. Bellisario College of Communications at Penn State University. His research and writing focus on broadband policy, planning, and deployment, especially in rural and remote areas. He is author of *Farm Fresh Broadband: The Politics of Rural Connectivity* (2021).

DARIN BARNEY is professor and Grierson Chair in Communication Studies at McGill University. His current research reflects materialist approaches to media and communication, infrastructure, environment, and politics and focuses on the Canadian resource economy, including emerging resource formats in the Canadian oil patch and new energy storage technologies. He is author and editor of several books and journal editions, most recently *Solarities: Seeking Energy Justice* (2022, with Ayesha Vemuri) and "Solarity," a special issue of *South Atlantic Quarterly* (2021, with Imre Szeman). He is a member of the Petrocultures Research Group and the McGill Energy Centre, and a founding member of the After Oil Collective.

PATRICK BRESNIHAN is assistant professor in the Department of Geography at Maynooth University. He works across the interdisciplinary fields of political ecology, science and technology studies, and environmental humanities. His research looks at different but related concerns around water, land, and energy in Ireland and how these speak to broader questions of colonial and postcolonial development, environmental politics, and the green transition. With Naomi Millner, he is coauthor of *All We Want Is the Earth: Land, Labour and Movements Beyond Environmentalism* (2023).

PATRICK BRODIE is assistant professor and Ad Astra Fellow in the School of Information and Communication Studies at University College Dublin. His research focuses on the environmental politics of digital media infrastructures, with a particular focus on dynamics of energy and extractivism. He has published articles and chapters widely in media, communication, STS, and geography books and journals. He is author of *Wild Tides: Media Infrastructure and Financial Crisis in Ireland* (Duke University Press, 2026) and coauthor (with Patrick Bresnihan) of *From the Bog to the Cloud: Dependency and Eco-Modernity in Ireland* (2025). He is the current editor of *Journal of Environmental Media*.

JENNA BURRELL is an affiliate at Data & Society and previously served as the organization's director of research. Before that, she was a professor at the School of Information at UC Berkeley. Her research focuses on how marginalized communities adapt digital technologies to meet their needs and to pursue their goals and ideals. She earned a PhD in sociology from the London School of Economics and a BA in computer science from Cornell University.

JORDAN B. KINDER is assistant professor in the Department of Media, Culture, and Communication at New York University. He previously held postdoctoral fellowships at McGill University (2020–22) and Harvard University (2022–23) and is a citizen of the Métis Nation of Alberta. His first solo-authored book, *Petroturfing: Refining Canadian Oil through Social Media* (2024), studies the emergence of the pro-oil movement in Canada as it took shape on social media during the 2010s.

BURÇ KÖSTEM is assistant professor in Media and Technology Studies at the University of North Carolina, Chapel Hill. He is interested in infrastructure studies, media studies, political ecology, theories of value, affect theory, and Marxist thought. His work has recently appeared in journals such as *Cultural Politics, Rethinking Marxism,* and *Cultural Studies.*

CINDY KAIYING LIN is the Stephen Fleming Early Career Assistant Professor at Georgia Institute of Technology. Her first single-authored book project explores statecraft and computing practices in the environmental and mapping sciences in Indonesia and the professional identities and government institutions that emerged from these efforts.

EMILY NG is assistant professor in the Department of Anthropology at the University of Pennsylvania. She is the author of *A Time of Lost Gods: Mediumship, Madness, and the Ghost after Mao* (2020). As part of the Rural Imaginations project at the Amsterdam School for Cultural Analysis, University of Amsterdam, she has been exploring the politics and aesthetics of rurality in contemporary China.

LISA PARKS is distinguished professor of Film and Media Studies at the University of California at Santa Barbara and was formerly professor of Comparative Media Studies and Science and Technology Studies at MIT. She is author of *Rethinking Media Coverage: Vertical Mediation and the War on Terror*, *Cultures in Orbit: Satellites and the Televisual*, and coeditor of *Media Backends: Digital Infrastructures and Sociotechnical Relations* and *Signal Traffic: Critical Studies of Media Infrastructures*, among other books. Parks directs the Global Media Technologies and Cultures Lab at UCSB and is a 2018 MacArthur fellow.

ANNE PASEK is associate professor and Canada Research Chair (Tier II) in Media, Culture and the Environment at Trent University's Department of Cultural Studies and the Trent School of the Environment. She researches the cultural politics of climate change and the political ecology of the information and communications technology sector.

ESTHER PEEREN is professor of cultural analysis at the University of Amsterdam. She is the author of *Intersubjectivity and Popular Culture: Bakhtin and Beyond* (2008) and *The Spectral Metaphor: Living Ghosts and the Agency of Invisibility* (2014), and coeditor of *Planetary Hinterlands: Extraction, Abandonment and Care* (2024). As PI of the European Research Council–funded Rural Imaginations project, she has been exploring what aspects of globalized rural life become visible in popular cultural imaginations and how this affects the rural's political mobilization.

NICOLE STAROSIELSKI, professor of film and media at UC Berkeley, is author or coeditor of over thirty articles and five books on media, infrastructure, and environments, including *Media Hot and Cold* (2021), *Signal Traffic: Critical Studies of Media Infrastructure* (2015), and *Assembly Codes: The Logistics of Media* (2021). She is coeditor of the Elements book series at Duke University Press.

ISHITA TIWARY is assistant professor and Canada Research Chair at the Mel Hoppenheim School of Cinema, Concordia University. She directs the research lab Raah. Her recent monograph, *Video Culture in India: The Analog Era*, came out in 2024.

HUNTER VAUGHAN is assistant professor of visual and media arts at Emerson College. An environmental media scholar and cultural historian focusing on the relationship between media technologies, social justice, and the environment, Vaughan is author of *Hollywood's Dirtiest Secret: The Hidden Environmental Costs of the Movies* (2019), codirector of the Global Green Media Network, co-PI of the Sustainable Subsea Networks project, and a founding editor of the *Journal of Environmental Media*.

AYESHA VEMURI earned a PhD in Communication Studies at McGill University in 2025. Her research focuses on the construction of risk in the context of climate change induced flooding and migration in Assam, India. Her work has appeared in *Communication, Culture and Critique, Feminist Media Theory,* and *Gender, Place, and Culture.* She is co-editor of *Solarities: Seeking Energy Justice* (2022).

MEGAN WIESSNER is a critical environmental media studies scholar and postdoctoral fellow at the Digital Technology for Democracy Lab at the University of Virginia. She researches digital media systems and the politics of sustainability, with a special focus on the role of computation in ecosystem management and environmental design.

ASSATU WISSEH approaches her work in critical media studies including media histories, media criticism, archival practice, and digital humanities through the lenses of Black feminisms and critical race theory. Examining the historical interactions between coloniality and media infrastructures, apparatuses, and geographies is her focus. Her scholarly research centers on the displacement and marginalization of Black and/or Indigenous bodies, spaces, and cultures. She is a Society of American Archivists–certified digital archives specialist and professional educator.

Index

Note: Page numbers in *italics* refer to figures.

www.ingramcontent.com/pod-product-compliance
Lightning Source LLC
Chambersburg PA
CBHW020454270326
41926CB00008B/603
* 9 7 8 1 4 7 8 0 3 3 2 5 7 *

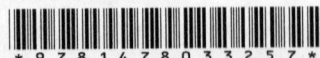